LONDON MATHEMATICAL SOCIETY LECTURE NOTE SERIES

Managing Editor:

Professor M. Reid, Mathematics Institute, University of Warwick, Coventry CV4 7AL, United Kingdom

The titles below are available from booksellers, or from Cambridge University Press at
http://www.cambridge.org/mathematics

London Mathematical Society Lecture Note Series: 408

Complexity Science
The Warwick Master's Course

Edited by
ROBIN BALL
University of Warwick

VASSILI KOLOKOLTSOV
University of Warwick

ROBERT S. MACKAY
University of Warwick

CAMBRIDGE
UNIVERSITY PRESS

CAMBRIDGE
UNIVERSITY PRESS

University Printing House, Cambridge CB2 8BS, United Kingdom

One Liberty Plaza, 20th Floor, New York, NY 10006, USA

477 Williamstown Road, Port Melbourne, VIC 3207, Australia

314-321, 3rd Floor, Plot 3, Splendor Forum, Jasola District Centre, New Delhi - 110025, India

103 Penang Road, #05-06/07, Visioncrest Commercial, Singapore 238467

Cambridge University Press is part of the University of Cambridge.

It furthers the University's mission by disseminating knowledge in the pursuit of education, learning and research at the highest international levels of excellence.

www.cambridge.org
Information on this title: www.cambridge.org/9781107640566

First published 2013

A catalogue record for this publication is available from the British Library

Library of Congress Cataloging in Publication data
Complexity science : the Warwick master's course / edited by Robin Ball, University of Warwick, Vassili Kolokoltsov, University of Warwick, Robert S. Mackay, University of Warwick.
pages   cm. – (London Mathematical Society lecture note series ; 408)
Based on lectures at the University of Warwick.
Includes bibliographical references and index.
ISBN 978-1-107-64056-6 (Paperback)
1. Complexity (Philosophy) 2. Science–Philosophy. I. Ball, Robin, editor of compilation. II. Kolokol'tsov, V. N. (Vasilii Nikitich) editor of compilation.
III. MacKay, Robert S., 1956- editor of compilation.
Q175.32.C65C656 2013
501–dc23
2013028846

ISBN 978-1-107-64056-6 Paperback

Contents

Preface

Complexity Science is the study of systems with many interdependent components.

There is an urgent global need from industry, commerce, research institutions, academia, government and public services for a new generation trained to understand how complex systems behave, how to live with them, to control them and to design them well. We see this in public service management, transport, public opinion, epidemics, riots, terrorism, weather and climate. Relevant technological developments include distributed computing, data management, process control, personalised medicine, disease management, environmental sensor swarms, complex materials and nanobiotechnology.

Stimulated by problems from such a wide range of scientific disciplines, it presents great challenges and opportunities for Mathematics. Mathematics is essential for a deep understanding of complex systems and how to quantify their behaviour, for conclusions of genuine value to end-users, because of its powers for description, abstraction, deduction and prediction.

A range of Complexity Science concepts unify the field across disciplines: dynamics and diffusion, interacting agents and networks, coherent structures, emergence and self-organisation, upscaling and model reduction, quantification of complexity, scaling and extreme events, probabilistic modelling and statistical inference, feedback and control, diversity, optimisation and evolution.

This volume presents coherent introductions to the mathematical treatment of some areas of Complexity Science. It is based on some of the lecture modules of the Warwick EPSRC Doctoral Training Centre in Complexity Science.

Chapter 1 by Mario Nicodemi, Yu-Xi Chau, Christopher Oates, Anas

Rana and Leigh Robinson introduces the key themes of Self-Organisation and Emergence. It presents some of the basic examples and tools to illustrate and analyse these phenomena.

Chapter 2 by Yulia Timofeeva treats Complexity in Deterministic Dynamical Systems. Dynamical systems are represented by mathematical models describing phenomena whose instantaneous state changes over time. Examples are mechanics in physics, population dynamics in biology and chemical kinetics in chemistry. One basic goal of the mathematical theory of dynamical systems is to determine or characterise the long-term behaviour of the system using methods for analysing differential equations and iterated mappings. This chapter introduces some of the techniques used in the modern theory of dynamical systems and the concepts of chaos and strange attractors, and illustrates a range of applications to problems in the physical, biological and engineering sciences.

Chapter 3 by Stefan Grosskinsky treats Stochastic Dynamics of Interacting Particle Systems. These are lattice-based stochastic models of complex systems, describing the time evolution of a large number of interacting components or agents, which are simply called particles. The notes provide an introduction to their mathematical description using Markov semigroups and generators, and to basic probabilistic tools for their analysis. The techniques are used to understand collective phenomena and phase transitions as a result of local motion and interaction of the particles for several classes of models. This discussion is mainly example-based. It involves the role of symmetries and conservation laws and provides a connection to concepts from equilibrium statistical mechanics discussed in Chapter 4.

Chapter 4 by Ellák Somfai treats Statistical Mechanics of Complex Systems. This chapter starts by introducing equilibrium statistical mechanics via the maximum entropy principle. This is followed by a phenomenological description of phase transitions and various applications where dynamics plays a critical role, including interface growth and collective biological motion.

Chapter 5 by Colm Connaughton treats Numerical Simulation of Continuous Systems. This chapter provides a foundation in practical methods of obtaining numerical solutions of partial differential equations that arise in complexity science applications. The focus is on understanding the advantages and limitations of numerical methods generally and on selecting and validating an appropriate numerical algorithm when faced with a particular problem. It starts with a basic outline of timestepping methods for ordinary differential equations, then proceeds to cover finite difference methods for hyperbolic and parabolic equations, explicit versus implicit timestepping, issues related to stability, stiffness and singularities, fast Fourier transform and pseudo-spectral methods. It is example-based.

Chapter 6 by Vassili Kolokoltsov is on Stochastic methods in Economics and Finance. It presents theory for utility, risk, optimisation, portfolios, derivatives, fat tails, option pricing and credit risk.

Chapter 7 by Robert MacKay is on Space-Time Phases. The objective is to put the concept of "emergence" onto a firm foundation in the context of dynamics on large networks. The key notion is space-time phases: probability distributions for state as a function of space and time that can arise in systems that have been running for a long time. The chapter has two sections, the first treating the stochastic case of probabilistic cellular automata and the second the deterministic case of coupled map lattices.

Chapter 8, also by Robert MacKay is on Selfish Routing. The chapter is a summary of the very interesting theory of the gap between free market and centrally controlled solutions for many agent systems in an idealised case of traffic flow, following the excellent book by Roughgarden.

We are most grateful to Dayal Strub for typing up the notes of RSM, preparing the figures for RSM and VNK, putting all the files together into the required style and sorting out many issues with typesetting.

This work was supported by the Engineering and Physical Sciences Research Council [grant numbers EP/I01358X/1 and EP/E501311/1]

R.C. Ball
V.N. Kolokoltsov
R.S. MacKay

Contributors

Mario Nicodemi[*], Yu-Xi Chau[†], Christopher Oates[†], Anas Rana[†] and Leigh Robinson[†]
[*]*Istituto Nazionale Fisica Nucleare, Napoli, 80126, Italy*
[†]*Centre for Complexity Science, University of Warwick, Coventry, CV4 7AL, UK*

Yulia Timofeeva *Department of Computer Science and Centre for Complexity Science, University of Warwick, Coventry, CV4 7AL, UK*

Stefan Grosskinsky *Mathematics Institute and Centre for Complexity Science, University of Warwick, Coventry, CV4 7AL, UK*

Ellák Somfai *Wigner RCP SZFI, H-1121 Budapest, Konkoly-Thege M. u. 29-33, Hungary*

Colm Connaughton *Mathematics Institute and Centre for Complexity Science, University of Warwick, Coventry, CV4 7AL, UK*

Vassili N. Kolokoltsov *Department of Statistics, University of Warwick, Coventry, CV4 7AL, UK*

Robert S. MacKay *Mathematics Institute and Centre for Complexity Science, University of Warwick, Coventry, CV4 7AL, UK*

1

Self-organisation and emergence

Mario Nicodemi, Yu-Xi Chau, Christopher Oates,
Anas Rana and Leigh Robinson

Abstract

Many examples exist of systems made of a large number of comparatively simple elementary constituents which exhibit interesting and surprising collective emergent behaviours. They are encountered in a variety of disciplines ranging from physics to biology and, of course, economics and social sciences. We all experience, for instance, the variety of complex behaviours emerging in social groups. In a similar sense, in biology, the whole spectrum of activities of higher organisms results from the interactions of their cells and, at a different scale, the behaviour of cells from the interactions of their genes and molecular components. Those, in turn, are formed, as all the incredible variety of natural systems, from the spontaneous assembling, in large numbers, of just a few kinds of elementary particles (e.g., protons, electrons).

To stress the contrast between the comparative simplicity of constituents and the complexity of their spontaneous collective behaviour, these systems are sometimes referred to as *"complex systems"*. They involve a number of interacting elements, often exposed to the effects of chance, so the hypothesis has emerged that their behaviour might be understood, and predicted, in a statistical sense. Such a perspective has been exploited in statistical physics, as much as the later idea of "universality". That is the discovery that general mathematical laws might govern the collective behaviour of seemingly different systems, irrespective of the minute details of their components, as we look at them at different scales, like in Chinese boxes. While the single component must be studied on its own, these discoveries offer the hope that we might understand different classes of complex systems from their simpler examples.

A univocal definition of "complexity" can be elusive, but the above criteria hopefully draw a line to distinguish, in a more technical sense, "complex" from the much broader category of "complicated" systems. Here we introduce some of the basic mathematical tools employed to describe their emergent behaviours. We discuss some basic concepts and several applications (e.g., Brownian motion in physics, asset pricing in finance) of the theory of stochastic processes, which is presented more generally in Chapter 3. We also consider some more advanced topics such as statistical mechanics and its applications to define the emergent properties in interacting systems. The foundations of statistical mechanics are discussed in more detail in Chapter 4. Finally, we introduce more recent topics such as self-organised criticality and network theory.

The course was taught by Mario Nicodemi. The notes that form this chapter were written by Yu-Xi Chau, Christopher Oates, Anas Rana and Leigh Robinson, four students of the Complexity Science Doctoral Training Centre of the University of Warwick who attended the lectures in 2009.

1.1 Random walks

1.1.1 Introduction

Put simply, a **random walk** is a mathematical formalisation of a path a "particle" traces out after taking a sequence of random steps. The idea of a random walk is central to the modelling of a wide range of phenomena, including financial modelling, the diffusion of gases, genetic drift, conformation of polymers, and a large number of other applications where the phenomenon in question evolves by a random process in time.

Various different types of random walk exist but can be grouped into broad categories depending on what the random walker is said to "walk on", and how the time evolution is defined. For example, a random walker may be defined on a graph that evolves in discrete time, moving from one node to another in one discrete time step, or just as well defined would be a random walker that moved in continuous time along the whole real line, $\mathbb{R}$. We shall give no further thought to these kinds of random walks and restrict our discussion to ones that occur along the integers, $\mathbb{Z}$ in discrete time steps.

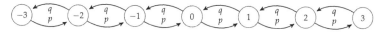

Figure 1.1 A discrete time walk on the integers, $\mathbb{Z}$. Probabilities p and q show how likely that transition is from state to state.

To understand perhaps the simplest example of a random walk we imagine a particle that can inhabit one of the integer points on the number line. At time 0 the particle starts from a specific point and moves in one time step to its next position in the following way: we flip a coin with the result governing how the particle moves. If the coin comes up heads then the particle moves *one* position to the right while if it comes up tails then the particle moves *one* place to the left. If we make n such coin tosses then what will be final position of the particle? Obviously being a random process we can't predict exactly where it will end but we can say a good deal about the distribution of possible outcomes.

1.1.2 One-dimensional discrete random walk

To try and answer such questions we need to introduce some formalism. We define independent random variables, X_i, that can take the values -1 and 1, with $P(X_i = 1) = p$ and $P(X_i = -1) = 1 - p = q$. The X_i represent the direction of the ith step of our random walk. Pictorially we can represent this arrangement as shown in Fig. 1.1. To see how such a system behaves statistically we calculate the first and second moments and variance of X_i as follows:

$$\langle X_i \rangle = \sum k P(X_i = k)_k = 1 \times P(X_i = 1) - 1 \times P(X_i = -1) \quad (1.1)$$
$$= p - (1 - p) = 2p - 1. \quad (1.2)$$

With the second moment and variance given by

$$\langle X_i^2 \rangle = p + (1 - p) = 1, \quad (1.3)$$

$$Var[X_i] = \langle X_i^2 \rangle - (\langle X_i \rangle)^2 = 1 - (2p - 1)^2 \quad (1.4)$$
$$= 1 - 4p^2 + 4p - 1 = 4p(1 - p). \quad (1.5)$$

We now define a new random variable, Z_n, as the sum of n such X_i variables and this defines the distribution of the value of the random walk after n steps:

$$Z_n = \sum_{i=1}^{n} X_i, n > 0. \tag{1.6}$$

We can now look at some of the statistics of Z_n, in particular the average position, $\langle Z_n \rangle$, and the variance: $Var[Z_n]$ of this position:

$$\langle Z_n \rangle = \langle \sum_{i=1}^{n} X_i \rangle = \sum_{i=1}^{n} \langle X_i \rangle = n(2p-1), n > 0. \tag{1.7}$$

Since by definition each of the X_i's are independent the second moment can be easily calculated,

$$\langle Z_n^2 \rangle = \langle (\sum_{i=1}^{n} X_i)^2 \rangle = \sum_{i=1}^{n} \langle X_i^2 \rangle + \sum_{i=1}^{n} \sum_{j=1 i \neq j}^{n} \langle X_i X_j \rangle \tag{1.8}$$

$$= n + n(n-1)(2p-1)^2. \tag{1.9}$$

Hence,

$$Var[Z_n] = \langle Z_n^2 \rangle - (\langle Z_n \rangle)^2 = 4np(1-p). \tag{1.10}$$

In particular, notice that

$$Var[Z_n] \propto n, \tag{1.11}$$

which gives us the result that the variance increases as we walk for more steps. This has important consequences for finance as we shall see later.

Unbiased random walk

So far we have been considering a random walk with general transition probabilities, p and q. The special case where $p = q = 1/2$ is called *unbiased* – since each decision is equiprobable. For these random walks the statistical properties collapse to

$$\langle Z_n \rangle = 0, \tag{1.12}$$

$$Var[Z_n] = \langle Z_n^2 \rangle = n. \tag{1.13}$$

From (1.13) we note that the root-mean-square of Z_n is simply $\sqrt{n}$, which hints that the average absolute distance moved after n steps, $E[|Z_n|] = O(\sqrt{n})$. This is indeed the case, but will not be proven here. Trajectories for a collection of unbiased random walkers are shown in Fig. 1.2.

A further interesting property of unbiased random walkers is the notion of recurrence. Imagine we choose any point, $i \in \mathbb{Z}$. How many times would you expect the random walker to cross this point if the walker could travel forever? Perhaps surprisingly the answer is that the walker will cross any selected point an infinite number or times. From a financial perspective this property has been termed *gambler's ruin*, since if you are a gambler betting in a casino on a *fair* game then your current wealth will evolve as a unbiased random walk and as such must eventually cross 0. At this time you have lost all your money and the "walk" cannot continue. All the casino has to do to force this win is to have substantially more money than you (to absorb the periods where you are winning) and entice you to keep on playing! It seems there is wisdom in the adage "quit while you are ahead".

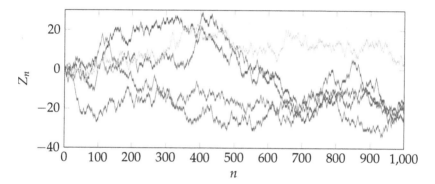

Figure 1.2 Recording the sequence of pairs (n, Z_n) we can plot them on the plane to generate the path of the random walker in time steps. Here we see five unbiased walkers all starting from 0. Note that they remain somewhat centred around 0 – which is what we expect from the results derived in Section 1.1.2

1.1.3 Applications to finance

The simple one-dimensional random walk encountered so far has found great application to finance. Financial forecasting has been labouring under the so-called *random walk hypothesis*, which simply states that stock market prices evolve over time as a random walk.

To see how this works, consider the log price, $S = \log(\text{price})$ of an asset or group of assets, along with ΔS_0, some average price change after a time period of Δt. Economic data suggests that for $\Delta t > 30$ seconds the random walk hypothesis seems valid, as we observe that the correlation between successive prices (the **autocorrelation**) sampled above this threshold is zero.

d	$p(d)$
1	1
2	1
3	0.34
4	0.19
5	0.13
6	0.1

Table 1.1 *Probabilities, $p(d)$ of returning to the origin of a $\mathbb{Z}^d$ random walker initialised at the origin*

Using historical price data you can estimate the value of p and ΔS_0 and so obtain a random walker that behaves *statistically* like the historical data. If you carry out this analysis you will quickly discover that your estimate for p will be approximately 0.5.

1.1.4 Random walks on $\mathbb{Z}^d$ lattices

So far we have restricted our discussion to random walks that occur in one dimension – that is there is only a choice to move left or right along the integers. Formally this is a random walk on the *lattice* defined by $\mathbb{Z}$. An obvious way to extend the concept to higher dimensions is to allow each dimension, d to have its own independent integer random walker, such a random walker is said to walk on the lattice $\mathbb{Z}^d$. An example of a multidimensional random walker is illustrated to two dimensions in Fig. 1.3.

Interestingly not all members of this family of random walkers are recurrent, which may be a little surprising given that they are built from d independent random walks that are recurrent. In fact, for $d > 2$ no such random walks are recurrent. More formally, if we define $p(d)$ as the probability that the random walk on $\mathbb{Z}^d$ starting at 0 will return to 0 some time in the future then we can numerically calculate the probabilities in Table 1.1.

1.1.5 Conformation of polymers

A *polymer* is a large macromolecule composed of repeating structural units typically connected by covalent chemical bonds. Informally we can consider a polymer of length L to be made up of n freely joined con-

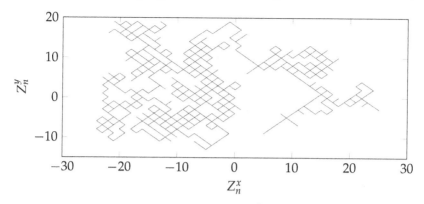

Figure 1.3 An unbiased random walk on $\mathbb{Z}^2$ for $n = 1000$, initialised as the origin.

nected elastic rods of length $l = L/n$, where l is the *persistence length* of the polymer. With this $= L/n$, where l is simplification in mind a naive model of how free floating polymers arrange themselves in space is suggested by a three-dimensional random walker of length n. This model predicts that such a polymer would take up a region of space bounded by a radius, $R \propto \sqrt{n}$. Unfortunately, experimentalists disagreed with the value predicted by the model. Instead of seeing $R \propto n^v$ with $v = 0.5$ they were consistently measuring $v \approx 0.6$. The discrepancy between the experimental value and the model prediction was resolved by the realisation that polymers obviously cannot self-intersect in the way a random walker can return to previously visited states. A simple modification to create the so-called *self-avoiding* random walker yields a value of $v \approx 0.6$ – which agrees much better with the experimental value.

The assumption that the polymer is free to float in space is not always valid. DNA for example can self-interact by forming bonds with sites downstream of the molecule forming loop structures (Fig. 1.4). Introducing the possibility of such interactions into the model gives us our first system that exhibits emergent behaviour and phase transitions. Let E_0 be the energy required to break a self-interacting bond, and let T be the temperature of the environment. At higher temperatures, molecules have more energy and it is easier for self-interacting bonds to be broken. We see that there is a "folding point" temperature above which we have our usual free self-avoiding polymer, but below which we see a polymer that is tightly packed together into a looped structure and needs self-interactions to explain its behaviour (Fig. 1.5).

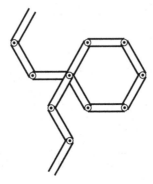

Figure 1.4 Self-interaction of large molecules.

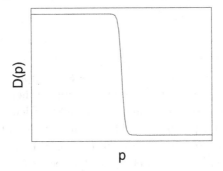

Figure 1.5 Folding point.

This gives us our first important lesson: once we introduce interactions into a model we must be prepared for emergent behaviour to take hold; behaviour that would not be predicted by considering the individual parts of the system alone without considering the interactions. *Interactions are the key.*

1.2 Markov processes

Complex systems often exhibit behaviour which is extremely difficult to quantitatively predict. Whilst initially this might seem problematic, we can turn things to our advantage by constructing models *based* on the apparent randomness. In the previous section we saw how random walks may be used to model simple random processes. In this section we discuss a more general class of models known as stochastic processes and proceed to explore some of the ways in which these models have been successfully used to describe scientific phenomena. In particular, we will consider Markov processes within cellular biology, where these stochastic models will provide us with a framework from which to derive the dynamics of some prototypical biological systems.

1.2.1 Definitions

To begin with, we extend the random walk model introduced in the previous section to encompass continuous time. Define a **stochastic process** with state space χ to be a collection

$$\{Y(t) : t \in T\}$$

of χ-valued random variables indexed by $t \in T$. Here t is the generalisation of the number of steps n taken by the random walk and may be considered to represent some measure of time. Notice that the random walks of the previous section correspond to the special case where $T = \{0, 1, 2, \dots\}$. In this chapter we will explore the natural extension to continuous time $T = \{t \in \mathbb{R} : t \geq 0\}$.

A stochastic process can be expressed in terms of its joint density function

$$p((y_1, t_1), (y_2, t_2), \dots, (y_n, t_n)). \tag{1.14}$$

Conversely, given a function $p((y_1, t_1), \dots, (y_n, t_n))$ how can we decide if p defines a stochastic process? It can be shown that the following constraints represent sufficient criteria:

(i) $p((y_1, t_1), \dots, (y_n, t_n)) \geqslant 0 \;\; \forall \; n, \, y_j, \, t_j$

(ii) $p(\dots, (y_i, t_i), \dots, (y_j, t_j), \dots) \;=\; p(\dots, (y_j, t_j), \dots, (y_i, t_i), \dots)$
$\forall \, i \neq j$

(iii) $p((y_1, t_1), \dots, (y_{n-1}, t_{n-1})) = \iint p((y_1, t_1), \dots, (y_n, t_n)) dy_n dt_n$

(iv) $\iint p(y_1, t_1) dy_1 dt_1 = 1.$

Many of the concepts which were introduced in the context of the random walk can be naturally extended to continuous-time stochastic processes. For example, we can define **moments**

$$\langle Y(t_1) \cdots Y(t_n) \rangle = \iint y_1 \cdots y_n p((y_1, t_1), \dots, (y_n, t_n)) dy_1 \dots dy_n \tag{1.15}$$

and **covariances**

$$K(t_1, t_2) = \langle (Y(t_1) - \langle Y(t_1) \rangle)(Y(t_2) - \langle Y(t_2) \rangle) \rangle. \tag{1.16}$$

An important concept in stochastic processes is stationarity. Specifically, we say that p^S is a **stationary distribution** if for all $\tau > 0$ we have

$$p^S((y_1, t_1 + \tau), \ldots, (y_n, t_n + \tau)) = p^S((y_1, t_1), \ldots, (y_n, t_n)). \tag{1.17}$$

We can think of a stochastic process given by a stationary distribution p^S as being *in a time-independent steady state* – the probability of an event does not depend on when the event is scheduled to happen.

1.2.2 Markov property

We say that a stochastic process $\{Y(t) : t \in T\}$ is a **Markov process** if $Y(t)$ has the property that

$$p((y_{n+1}, t_{n+1})|(y_n, t_n), \ldots, (y_1, t_1)) = p((y_{n+1}, t_{n+1})|(y_n, t_n)), \tag{1.18}$$

for all n and all pairs $(y_1, t_1), \ldots, (y_{n+1}, t_{n+1})$, where the times $t_1 \leq t_2 \leq \cdots \leq t_n \leq t_{n+1}$ form an increasing sequence. This is commonly known as the **Markov property**. Intuitively it means that the stochastic process is *memoryless* in the sense that the future behaviour (y_{n+1}, t_{n+1}) of the process depends only on the present (y_n, t_n) and not on the past $(y_{n-1}, t_{n-1}), \ldots, (y_1, t_1)$. This is a nice property because it allows us to factorise $p((y_{n+1}, t_{n+1}), \ldots, (y_0, t_0))$ as

$$p((y_{n+1}, t_{n+1})|(y_n, t_n)) \times \cdots \times p((y_1, t_1)|(y_0, t_0)) \times p((y_0, t_0)), \tag{1.19}$$

meaning that the entire process can be completely characterised by simply stating the function $p((y_{n+1}, t_{n+1})|(y_n, t_n))$ and the **initial distribution** $p((y_0, t_0))$. Notice that the random walks of the previous section all obey the Markov property – the distribution of the next state Z_{n+1} depends only on the state Z_n in which you happen to lie.

For the special case where $n = 2$ we have

$$p((y_2, t_2), (y_1, t_1), (y_0, t_0)) = p((y_2, t_2)|(y_1, t_1))p((y_1, t_1)|(y_0, t_0))p((y_0, t_0)), \tag{1.20}$$

and integrating the equation over (y_1, t_1) produces

$$p((y_2, t_2), (y_0, t_0))$$
$$= p((y_0, t_0)) \iint p((y_2, t_2)|(y_1, t_1))p((y_1, t_1)|(y_0, t_0))dy_1 dt_1. \tag{1.21}$$

Dividing through by $p((y_0, t_0))$ we have shown that all Markov processes obey the **Chapman–Kolmogorov** equation

$$p((y_2, t_2)|(y_0, t_0)) = \int\!\!\int p((y_2, t_2)|(y_1, t_1))p((y_1, t_1)|(y_0, t_0))dy_1 dt_1.$$

(1.22)

Conversely, it can be shown that any solution of (1.22) defines a Markov process. This is an important observation because it allows us to derive a **master equation** for the dynamics of a given Markov process.

For simplicity we firstly consider the case where the state space χ is discrete and define the **transition rates** $W_{n,m}$ to be the rate of movement of probability flux from state m to state n, where $m \neq n$, obeying

$$W_{n,m} \geq 0 \quad \text{and} \quad \sum_n W_{n,m} = 1.$$

(1.23)

The transition rates are the continuous-time analogue of the transition probabilities and, as with the random walk model, Markov process are often naturally *defined* in terms of their transition rates. In this case the master equation has the form

$$\frac{\partial p(n, t)}{\partial t} = \underbrace{\sum_m W_{n,m} p(m, t)}_{\text{gain term}} - \underbrace{\sum_m W_{m,n} p(n, t)}_{\text{loss term}},$$

(1.24)

where the *gain term* represents the net probability flux into state n and the *loss term* represents the net probability flux out of state n. The master equation has earned this nomenclature by containing all the information we need to completely determine the time-evolution of the probabilities for the system to occupy a particular state, given some initial distribution.

From (1.24) we can read off a sufficient (but not necessary) condition for stationarity, namely the **detailed balance** equations

$$W_{n,m} p(m, t) = W_{m,n} p(n, t),$$

(1.25)

which if satisfied for all n, m, t clearly results in the stationarity condition

$$\frac{\partial p(n, t)}{\partial t} = 0.$$

(1.26)

If we now allow the state space χ to be continuous then the master

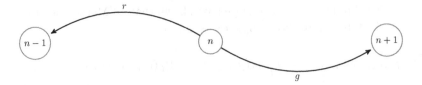

Figure 1.6 Continuous-time random walk.

equation (1.24) becomes

$$\frac{\partial p(y,t)}{\partial t} = \underbrace{\int W(y,x)p(x,t)dx}_{\text{gain term}} - \underbrace{\int W(x,y)p(y,t)dx}_{\text{loss term}}, \qquad (1.27)$$

where by analogy $W(y,x)$ is the rate of movement from state x to state y. For the interested reader, a derivation of the master equation Eqn. 1.27 may be found in [21].

1.2.3 Continuous-time random walks

In this section we will explore a simple class of Markov process known as **one-step processes**. These are Markov processes on a structured state space such as a network, where the transition rate between any pair of non-neighbouring states is zero – in other words we can only move to neighbouring states.

We are ready to generalise our one-dimensional random walk model to the case of continuous-time $T = \{t : t > 0\}$. As with the discrete-time random walks we work on the state space with $Y(t) \in \mathbb{Z}$. Let $r > 0$ represent the rate of moving to a lower state ("reduction") and g be the rate of moving to a higher state ("gain") (Fig. 1.6). This might represent an extension to our stock price model where the changes in price need not occur at regular time steps.

The transition rates in this example are

$$W_{n,m} = \begin{cases} 0 & \text{if } |n - m| > 1, \\ r & \text{if } n = m - 1, \\ g & \text{if } n = m + 1, \end{cases}$$

and hence the master equation (1.24) for the dynamics of the system is simply

$$\frac{\partial p(n,t)}{\partial t} = \underbrace{rp(n+1,t) + gp(n-1,t)}_{\text{gain term}} - \underbrace{(r+g)p(n,t)}_{\text{loss term}}.$$

We can use the master equation to explore the dynamics of this system:

$$\frac{d\langle Y(t)\rangle}{dt} = \sum_n n\frac{\partial p(n,t)}{\partial t}$$

$$= \sum_n n\left[rp(n+1,t) + gp(n-1,t) - (r+g)p(n,t)\right]$$

$$= r\sum_n (n+1)p(n+1,t) - r\sum_n p(n+1,t)$$

$$\quad + g\sum_n (n-1)p(n-1,t) + g\sum_n p(n-1,t)$$

$$\quad - (r+g)\sum_n np(n,t)$$

$$= r\langle Y(t)\rangle - r + g\langle Y(t)\rangle + g - (r+g)\langle Y(t)\rangle = g - r,$$

and therefore

$$\langle Y(t)\rangle - \langle Y(0)\rangle = (g-r)t.$$

So if $g > r$ then our probability flux tends to move to the right over time and if $g < r$ then our probability flux tends to move to the left over time. A similar calculation shows that

$$\text{Var}[Y(t)] = \left\langle (Y(t) - Y(0))^2\right\rangle - \langle Y(t) - Y(0)\rangle^2 = (g+r)t,$$

so that our flux diffuses over time for all choices of g and r, implying that there cannot exist a stationary distribution for this system. This result is the continuous-time analogue of the $O(\sqrt{N})$ diffusion rate for the discrete-time random walk (1.11).

There is a similar approach involving the probability generating function $F(z,t) = \langle z^{Y(t)}\rangle$, which allows us to completely solve for the dynamics of the random walk:

$$\frac{\partial F}{\partial t} = \frac{\partial}{\partial t} \sum_n z^n p(n,t) = \sum_n z^n \frac{\partial p(n,t)}{\partial t}$$

$$= \sum_n z^n \left[rp(n+1,t) + gp(n-1,t) - (r+g)p(n,t) \right]$$

$$= \frac{r}{z} \sum_n z^{n+1} p(n+1,t) + gz \sum_n z^{n-1} p(n-1,t) - (r+g) \sum_n z^n p(n,t)$$

$$= \left[\frac{r}{z} + gz - (r+g) \right] F(z,t) \equiv A(z)F(z,t),$$

so that if we further assume that all the probability begins at the origin $(p(n,0) = \delta_{n,0})$ then

$$F(z,t) = e^{A(z)t}.$$

In particular, we can recover our earlier results, e.g.

$$\langle Y \rangle = \langle Y z^{Y-1} \rangle_{z=1} = (\partial_z F)_{z=1} = A(1)F(1,t)t(\partial_z A)_{z=1} = (g-r)\,t.$$

The random walks of the previous example are often referred to as *simple* random walks owing to the fact that the transition rates are independent of state. In general, we will have transition rates of one of the following forms:

1. $r_m = r$, $g_m = g$ $\forall\, m \in \chi$ (constant)
2. $r_m = am + b$, $g_m = cm + d$ $\forall\, m \in \chi$ (linear)
3. $r_m = e(m)$, $g_m = f(m)$ $\forall\, m \in \chi$ (non-linear)

and discrete state space of one of the following forms:

1. $\chi = \mathbb{Z}$ (no boundaries)
2. $\chi = \{0, 1, 2, \dots\}$ (one boundary)
3. $\chi = \{0, 1, 2, \dots, N\}$ (two boundaries).

We have just seen how the dynamics of a one-step Markov process with constant transition rates and unbounded state space may be fully recovered. In general, we are also able to completely solve for the dynamics of processes with transition rates of type 2; however, transition rates of type 3 may pose much greater challenges with exact solutions being difficult to obtain.

The general form of the master equation for one-step Markov processes is

$$\frac{\partial p(n,t)}{\partial t} = \underbrace{r_{n+1}p(n+1,t) + g_{n-1}p(n-1,t)}_{\text{gain term}} - \underbrace{(r_n + g_n)p(n,t)}_{\text{loss term}}, \quad (1.28)$$

and a derivation similar to that which we saw for the continuous-time random walk gives that

$$\frac{d\langle Y(t)\rangle}{dt} = \langle g_{Y(t)}\rangle - \langle r_{Y(t)}\rangle, \quad (1.29)$$

where

$$\langle g_{Y(t)}\rangle = \sum_n g_n p(n,t),$$

$$\langle r_{Y(t)}\rangle = \sum_n r_n p(n,t).$$

Indeed it is possible to compute each of the moments of $Y(t)$ in the case where r_m and g_m are either constant or linear in m; for example,

$$\frac{d\langle Y(t)^2\rangle}{dt} = \langle Y(t)(g_{Y(t)} - r_{Y(t)})\rangle + \langle g_{Y(t)} + r_{Y(t)}\rangle. \quad (1.30)$$

The difficulty comes when the coefficients r_m and g_m are non-linear since, for the second moment of $Y(t)$, (1.30) would require us to already know *at least* the second moment to begin with – the system of differential equations is no longer **closed**.

It would be nice to know the conditions under which a stationary distribution derived from the detailed balance equation (1.25) was unique. If we define the step operator $\mathbb{S}$ by

$$\mathbb{S}f(n) = f(n+1),$$

then the master equation (1.28) may be rewritten as

$$\frac{\partial p(n,t)}{\partial t} = (\mathbb{S} - 1)(r_n p(n,t)) + (\mathbb{S}^{-1} - 1)(g_n p(n,t)),$$

and so the stationarity condition (1.26) becomes

$$(\mathbb{S} - 1)\left[r_n p(n,t) - \mathbb{S}^{-1}(g_n p(n,t))\right] = 0.$$

Since the kernel of $(\mathbb{S}-1)$ contains only the constant functions we deduce that for all n

$$r_n p(n,t) - \mathbb{S}^{-1}(g_n p(n,t)) = -J,$$

where the constant J is equal to the net flux of probability through an arbitrary state. In other words, we can re-interpret stationarity as the *equilibrium of fluxes*, i.e., when each state experiences the same net flux passing through per unit time.

Consider now the special case where the state space has a boundary (for example, our gene activation model below has a state space which is bounded at the origin). Observe that since the net flux through a boundary is zero we must have $J = 0$ and hence

$$r_n p(n, t) = g_{n-1} p(n - 1, t),$$

so we recover the detailed balance equations! In other words, whenever we have a boundary in the state space we automatically know that any distribution which is a solution of the detailed balance equations *must* be the unique stationary distribution for the system.

1.2.4 Applications of Markov processes

We illustrate the generality of the one-step Markov process in the remainder of the section with two examples taken from cellular biology.

Gene activation

Let us proceed to develop a stochastic model for the biological phenomenon of gene activation.

Human DNA consists of 23 pairs of chromosomes, each chromosome containing many hundreds of genes. We say that a gene is **active** if it is able to produce mRNA (messenger RNA) via a chemical processes known as **transcription**, which we shall discuss later. In a simple model of gene activation, at any particular time a given gene is either active or inactive, but in general a gene may switch between active and inactive states. This behaviour defines a complex system since the activation of one particular gene may lead to the activation or deactivation of several other genes – in other words, the system contains **feedback**. We will use Markov processes to try to understand how many genes may be active at any given moment.

Let α_+ denote the rate at which a given inactive gene becomes active and let α_- denote the rate at which a given active gene becomes inactive. We will model $X(t) \in \{0, 1, \ldots, N\}$, the number of active genes at time t using a one-step Markov process with the transition rates shown in Fig. 1.7. Notice that this process is qualitatively different from the simple random walk example because our transition rates depend linearly on the state.

Assuming that there are no active genes at time $t = 0$, (1.29) gives

$$\frac{d \langle X(t) \rangle}{dt} = N\alpha_+ - (\alpha_+ + \alpha_-) \langle X(t) \rangle,$$

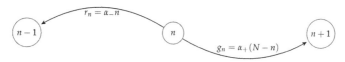

Figure 1.7 Gene activation model.

and hence

$$\langle X(t) \rangle = N\alpha_+\tau \left(1 - e^{-t/\tau}\right),$$

where $\tau = (\alpha_+ + \alpha_-)^{-1}$ is known as the **characteristic time** of the process. So as $t \to \infty$ we have $\langle X(t) \rangle \to \langle X(\infty) \rangle = N\alpha_+\tau$. For prokaryotes (cells with no nucleus) experiments have shown that $\tau \approx 1$ second.

We may also calculate

$$\mathrm{Var}[X(\infty)] = N\alpha_+\tau(1 - \alpha_+\tau).$$

It should be intuitively clear that the system is settling down to a stationary distribution as we let $t \to \infty$. We can solve for this distribution $p^S(n)$ using the detailed balance equation (1.25) to obtain

$$p^S(n) = \frac{g_{n-1}}{r_n} p^S(n-1) = \frac{g_{n-1}}{r_n} \cdot \frac{g_{n-2}}{r_{n-1}} p^S(n-2) = \cdots$$

$$\cdots = p^S(0) \prod_{k=1}^{n} \left(\frac{g_{k-1}}{r_k}\right)$$

$$= p^S(0) \binom{N}{n} \left(\frac{\alpha_+}{\alpha_-}\right)^n$$

$$= \binom{N}{n} (\alpha_+\tau)^n (1 - \alpha_+\tau)^{N-n},$$

which is a binomial distribution (Fig. 1.8) after we normalise by taking

$$p^S(0) = (\alpha_-\tau)^N.$$

In particular, p^S has mean $N\alpha_+\tau$ and variance $N\alpha_+\tau(1 - \alpha_+\tau)$, just as we would expect from our earlier calculation. Moreover, since the state space has a boundary at zero (we cannot have a negative number of active genes), our earlier reasoning tells us that we have discovered the *unique* stationary distribution. By reducing the innumerable chemical reactions which govern gene activation to a tractable Markov process, we are able to generate testable predictions for the statistical behaviour of this system.

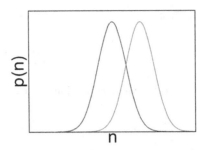

Figure 1.8 Stationary distribution for gene activation is binomial. Left curve: $\alpha_+\tau$ small. Right curve: $\alpha_+\tau$ large.

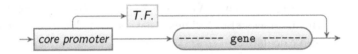

Figure 1.9 mRNA is transcribed from an active gene by a transcription factor beginning from the core promoter.

mRNA transcription

The complex mechanism for mRNA production may be crudely summarised as follows. A protein known as a **transcription factor** attaches itself onto an area of DNA upstream of the gene known as the **core promoter**. The presence of a transcription factor is able to change the rate at which the DNA is read and copied as mRNA. In this model only active genes are able to host a transcription factor and thereby produce mRNA (Fig. 1.9). This process is known as **transcription**.

In this model, unlike the previous one, we will be treating the number of active genes $X(t) \equiv X$ as a constant in time. Let β_+ denote the rate at which a given active gene produces an mRNA polymer and let β_- denote the rate at which a given mRNA polymer degrades. If $Y(t) \in \{0, 1, 2, \dots\}$ denotes the number of mRNA polymers at time t then we have a Markov process with transition rates shown in Fig. 1.10.

Proceeding with the usual analysis under the assumption that there are no mRNA at time $t = 0$ we obtain

$$\langle Y(t) \rangle = \beta_+\eta X \left(1 - e^{t/\tau}\right),$$

where $\eta = (\beta_-)^{-1}$ is another **characteristic time**. For prokaryotes experiments have shown $\eta \approx 10$ seconds. In the limit $t \to \infty$ we have

$$\langle Y(\infty) \rangle = \beta_+\eta X,$$

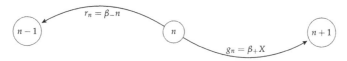

Figure 1.10 mRNA production model.

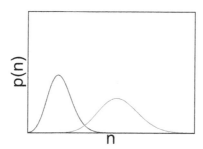

Figure 1.11 Stationary distribution for mRNA production is Poisson. Left curve: $\beta_+\eta X$ small. Right curve: $\beta_+\eta X$ large.

and similarly we are able to derive

$$\mathrm{Var}(Y(\infty)) = \beta_+\eta X.$$

From the detailed balance equation (1.25) we obtain

$$p^S(n) = p^S(0)\prod_{k=1}^{n}\left(\frac{g_{k-1}}{r_k}\right)$$
$$= p^S(0)\,(\beta_+\eta X)^n\,\frac{1}{n!}$$
$$= (\beta_+\eta X)^n\,\frac{1}{n!}e^{-\beta_+\eta X},$$

which we recognise to be a Poisson distribution (Fig. 1.11) with parameter $\beta_+\eta X$ after making the normalisation

$$p^S(0) = e^{-\beta_+\eta X}.$$

Moreover, since the state space is bounded at the origin we deduce that this is the unique stationary distribution for the model.

In the gene-activation example we treated the number $X(t)$ of active genes at time t as a variable, but in the mRNA production example we treated $X(t)$ as a constant. Such an assumption reduces the complexity of the model, allowing us to easily perform a complete analysis and make predictions. However, this approximation is valid only when the characteristic time η for mRNA production is much smaller than the characteristic time τ for gene activation and therefore the approximation would *not* be valid for modeling prokaryote cells, for example, since here $\eta \approx 10$ and $\tau \approx 1$. Indeed for prokaryotes this represents an over-simplification to the extent that the complex behaviour in which we are interested is suppressed by the model. To make further progress we need a more sophisticated approach to dealing with the mRNA master equation in the case where $X(t)$ is itself a random variable.

The central dogma of gene expression

Gene expression is classically described in three stages. The first and second stages – gene activation and mRNA transcription – we have already studied. The third is known as **translation** and concerns the usage of an mRNA sequence as a template with which to guide the synthesis of a chain of amino acids in order to form a protein. Let γ_+ denote the rate at which a given inactive gene becomes active and let γ_- denote the rate at which a given active gene becomes inactive. If $Z(t) \in \{0, 1, 2, \dots\}$ denotes the number of proteins present at time t then we can construct a Markov process with transition rates $g_n = \gamma_+ Y$ and $r_n = \gamma_- n$, since proteins may become denatured or be broken down by enzymes. If we assume that the number $Y(t) \equiv Y$ of mRNA remains constant then the analysis of this Markov process is identical to that of mRNA production and we will not repeat it here – both are Poisson processes under our assumptions.

We have examined these three stages individually, but in doing so we sacrificed much of the complexity within the model. In full generality, the system that we wish to solve is

$$X(t) - 1 \xleftarrow{\alpha_- X(t)} X(t) \xrightarrow{\alpha_+ (N - X(t))} X(t) + 1,$$

$$Y(t) - 1 \xleftarrow{\beta_- Y(t)} Y(t) \xrightarrow{\beta_+ X(t)} Y(t) + 1,$$

$$Z(t) - 1 \xleftarrow{\gamma_- Z(t)} Z(t) \xrightarrow{\gamma_+ Y(t)} Z(t) + 1.$$

This constitutes a representation via a stochastic model of the *central dogma* of gene expression – that the protein production dynamics may be summarised within these coupled Markov processes. This assumption means that any other cellular process which may influence the system's dynamics has been absorbed into the transition rates $\alpha_\pm, \beta_\pm, \gamma_\pm$. At this point we leave the realm of one-dimensional processes since our state space has become $\{0, 1, \ldots, N\} \times \mathbb{N} \times \mathbb{N}$. Since this is countable we will have a master equation of the form (1.24), where the right-hand side is linear in $X(t), Y(t), Z(t)$ and may be solved using the generating function method discussed earlier. It turns out that for reasonable choices of the parameters $\alpha_\pm, \beta_\pm, \gamma_\pm$ there exists a stationary distribution $p^S(n_1, n_2, n_3, t)$ whose moments we can easily calculate.

As an alternative approach we may simply write down a system of ordinary differential equations for the average values

$$\langle X(t) \rangle = \sum_{n \in \{0,1,\ldots,N\}} n p_X(n, t),$$

$$\langle Y(t) \rangle = \sum_{n \in \mathbb{N}} n p_Y(n, t),$$

$$\langle Z(t) \rangle = \sum_{n \in \mathbb{N}} n p_Z(n, t),$$

which in addition to a set of initial conditions specifies the behaviour of the averages for all time:

$$\frac{d \langle X \rangle}{dt} = \alpha_+ (N - \langle X \rangle) - \alpha_- \langle X \rangle,$$

$$\frac{d \langle Y \rangle}{dt} = \beta_+ \langle X \rangle - \beta_- \langle Y \rangle,$$

$$\frac{d \langle Z \rangle}{dt} = \gamma_+ \langle Y \rangle - \gamma_- \langle Z \rangle.$$

Techniques from the theory of dynamical systems may be applied to this system in order to investigate the behaviour for various values of the parameters $\alpha_\pm, \beta_\pm, \gamma_\pm$.

These result are interesting and serve to highlight the trade-off between model fit and model simplicity. Experimental measurements indicate that protein production actually occurs in short, sharp bursts – that is, we do not reach an equilibrium in the usual sense. In this respect the model developed above is too simple. For a critical discussion of these models and the central dogma of gene expression see [15].

1.3 Lattice models

A large number of physical systems can be represented by interacting entities occupying a regular lattice. These lattice models were originally an attempt to reproduce the microscopic fluctuations which occur on the atomic level in physical systems. Since there will be a great many of these interactions for any physically meaningful system, the model also needs to be sufficiently tractable to admit either numerical or analytical solutions.

One of the most important lattice models still capturing the interest of the scientific community is the Ising model. Proposed by Ising and Lenz in the 1920s, the model was initially studied as a simplistic description of ferromagnetism. However, after the exact solution of the critical exponents in two dimensions and the subsequent extension to higher dimensions, it has been widely used in a large number of application outside of physics, including biology and the social sciences.

1.3.1 Lattice gas model

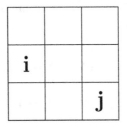

Figure 1.12 Square lattice representation of the torus, with $L = 3$. There are two independent random walkers labelled i and j.

Before diving into the Ising model we give a basic example of the use of a lattice. Consider a lattice on the torus, equipped with two random walkers (Fig. 1.12). In a simple model which doesn't involve interactions (i.e. the random walkers are independent) the long-run probability of occupying state s is $p(s) = 1/N$ where N is the number of possible states. This follows immediately from the symmetry of the walk and independence of the walkers. In the specific case of a 3×3 square lattice

$p(s) = 1$ for all states s. The interesting behaviour comes when we introduce interactions, as we shall soon see.

For a lattice of interacting particles we define the **Hamiltonian** of a state s to be

$$H(s) = -E_0 \sum_{(i,j)} s_i s_j - \mu \sum_i s_i, \qquad (1.31)$$

where the sum ranges over adjacent site pairs (i, j). We may think of $E_0 \sum_{(i,j)} M_i M_j$ as the energy lost when a bond forms between adjacent particles and $-\mu \sum_i M_i$ as the chemical potential energy (each particle contributes $-\mu$). For example, in the previous case of two random walkers we would need to modify (1.31) to remove the interacting term, which corresponds to setting $E_0 = 0$.

1.3.2 Ising model of ferromagnetism

Following in the footsteps of Ising and Lenz, we will use the Ising model to study ferromagnetic properties of metals such as iron. We consider a finite square lattice with the property that each site may adopt either an "up" magnetisation or a "down" magnetisation. We will use the notation $s_i = \pm 1$ with $i \in \{1, \ldots, N\}$ for the spin of site i, adopting the convention that $+1$ refers to up and -1 to down. The Hamiltonian of this model is then

$$H(s) = -J \sum_{(i,j)} s_i s_j - h \sum_i s_i, \qquad (1.32)$$

where $-h \sum_i s_i$ is intended to represent the effect of an uniform external magnetic field and $-J \sum_{(i,j)} s_i s_j$ measures the energy of the magnetic interactions, which is lowest when all sites are polarised in the same direction.

The methodology of statistical mechanics provides us with an expression for the stationary probability distribution of a system [4]:

$$p(s) = \frac{\exp\left(-\frac{H(s)}{k_B T}\right)}{Z(k_B T)}, \qquad (1.33)$$

where k_B is the Boltzman constant and Z is the **partition function**

$$Z(k_B T) = \sum_s \exp\left[-\frac{H(s)}{k_B T}\right]. \qquad (1.34)$$

Z is analogous to the generating function introduced in Section 1.2 and it captures several important properties of the system as we shall see.

To make subsequent calculations cleaner, we will absorb constants into our variables in the following way:

$$J \mapsto \frac{J}{k_B T} \quad , \quad h \mapsto \frac{h}{k_B T} \tag{1.35}$$

$$\Rightarrow Z = \sum_s \exp\left[J \sum_{<i,j>} s_i\, s_j + h \sum_i s_i \right]. \tag{1.36}$$

Infinite range Ising model

In what follows we will consider the **infinite range** Ising model, which assumes interaction of *all* pairs (i,j). This is advantageous because it allows us to rewrite the Hamiltonian as

$$H(s) = -\frac{J}{2} \sum_{i \neq j} s_i s_j - h \sum_i s_i, \tag{1.37}$$

where we have removed the awkward summation over neighbours. Despite this somewhat drastic change to the model the overall properties of the system remain similar. Notice that

$$\frac{1}{2} \sum_{i \neq j} s_i\, s_j = \frac{1}{2}\left(\sum_{i,j} s_i\, s_j - \sum_i s_i^2 \right) = \frac{1}{2}\left(\left(\sum_i s_i\right)^2 - N \right),$$

where N is the number of sites, allowing us to rewrite the partition function (1.37) as

$$Z = \sum_s \exp\left[\frac{J}{N}\left(\sum_i s_i\right)^2 - J + h \sum_i s_i \right]$$

$$= \sum_s \exp\left[JN \left(\frac{\sum_i s_i}{N}\right)^2 - J + h \sum_i s_i \right].$$

The nonlinear term $\left(\sum_i s_i/N\right)^2$ in the exponential is difficult to evaluate, motivating the use of a **Hubbard–Stratanovich** transformation in order to linearise the square. In its general form the transformation can be written as

$$e^{\lambda y^2/2} = \sqrt{\frac{\lambda}{2\pi}} \int dm\, e^{-\frac{\lambda m^2}{2} + \lambda m y}. \tag{1.38}$$

where m will turn out to be, in our case, the overall magnetisation of the system. The cunning choice of

$$y = \frac{\sum_i s_i}{N} \quad , \quad \lambda = 2NJ,$$

leads to

$$e^{NJ\left(\frac{\sum_i s_i}{N}\right)} = \sqrt{\frac{NJ}{\pi}} \int dm \, \exp\left[-NJm^2 + J\sum_i s_i m\right],$$

and consequently to the new form for the partition function

$$Z = \sqrt{\frac{NJ}{\pi}} \sum_s \int dm \, \exp\left[-NJm^2 + J\sum_i s_i m\right] \times \exp\left[-J + h\sum_i s_i\right]$$

$$(1.39)$$

$$\Rightarrow Z = \int dm \, \{\exp[-J]\} \times \sum_s \exp\left[-NJm^2 + (Jm + h)\sum_i s_i\right],$$

$$(1.40)$$

where the term $((Jm + h)\sum_i s_i)$ is known as the **effective** Hamiltonian. We are nearly there, but we still have one simplification trick up our sleeve. Since

$$\sum_s e^{A\sum_i s_i} = \sum_s \prod_i e^{As_i}$$

$$= \prod_i 2\cosh(A) = (2\cosh(A))^N$$

$$= \left(\sum_{s_i=\pm 1} e^{As_i}\right)^N = \exp\left[N\log\left(\sum_{s_i=\pm 1} e^{As_i}\right)\right],$$

the partition function may be reduced further to leave

$$Z = \Xi \int dm \, \exp\left[-NJm^2\right] \exp\left[N\log\left(\sum_{s_i=\pm 1} e^{(Jm+h)s_i}\right)\right]$$

$$= \zeta \int dm \, e^{-Nf(m)},$$

$$(1.41)$$

where Ξ and ζ are used to hide terms which will not prove significant in the $N \to \infty$ limit. In particular, we have defined

$$f(m) = Jm^2 - \log\left(\sum_{s_i=\pm 1} e^{(Jm+h)s_i}\right).$$

$$(1.42)$$

The main contribution to the integral occurs at the point m_0 where $f(m)$ achieves its minimum. A Taylor expansion about m_0 gives

$$f(m) = f(m_0) + mf'(m_0) + \mathcal{O}(m^2) = f(m_0) + \mathcal{O}(m^2) \quad (1.43)$$
$$\Rightarrow Z \approx \zeta e^{-Nf(m_0)}. \quad (1.44)$$

In order to obtain m_0 we proceed in the usual manner:

$$\frac{df}{dm} = 0 \quad \Rightarrow \quad m_0 = \frac{\displaystyle\sum_{s_i = \pm 1} s_i e^{(Jm_0 + h)s_i}}{\displaystyle\sum_{s_i = \pm 1} e^{(Jm_0 + h)s_i}},$$
$$= \langle s_i \rangle, \quad (1.45)$$

where $\langle s_i \rangle$ denotes the expected magnetism of a single site taken with respect to the probability distribution (1.33). In particular

$$m_0 = \frac{e^{(Jm_0 + h)} - e^{-(Jm_0 + h)}}{e^{(Jm_0 + h)} + e^{-(Jm_0 + h)}} = \tanh\left(Jm_0 + h\right)$$
$$\approx (Jm_0 + h) - \frac{(Jm_0 + h)^3}{3}. \quad (1.46)$$

Suppose $h = 0$ so that there is no external field. Then if $J < 1$ the only solution is $m_0 = 0$, corresponding to negligible overall magnetisation. However, for $J > 1$ we have three solutions $m_0 = 0, \pm\sqrt{\frac{3(J-1)}{J^3}}$. In other words, for $J < 1$ the system is not magnetised but as we increase J to the critical point $J = 1$, small clusters of aligned spins start to appear. Depending on the particular random fluctuations at the critical point, as we further increase J the system either tends towards a "mostly up" or "mostly down" state, an emergent behaviour known as **symmetry breaking**. Note that (1.46) is a Taylor expansion so it only holds in the neighbourhood of the critical point $J \approx 1$. It can be shown that the full solution m_0 behaves as in Fig. 1.13.

For lattice models, behaviour in the neighbourhood of a critical point is typically characterised by power laws. For our example we might consider the magnetisation m and the magnetic susceptibility χ, which behave according to

$$m \sim \pm\sqrt{\frac{J}{T} - 1} \sim \pm(T_C - T)^\beta, \quad \beta = \frac{1}{2},$$
$$\chi \equiv \frac{d \geq s_i >}{dh}\bigg|_{h=0} \sim |T_C - T|^{-\gamma}, \quad \gamma = 1, \quad (1.47)$$

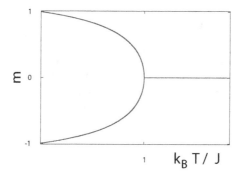

Figure 1.13 Spontaneous magnetisation m as a function of temperature T.

where T_C denotes the critical temperature. Indeed we do observe power law behaviour for the magnetisation and magnetic susceptibility around T_C. The indices β, γ are referred to as **critical exponents**.

Interesting applications of such **phase transitions** in biological systems can be found for instance in [13], with extensions to non-thermal systems in [6, 17].

1.3.3 Percolation

In this section we consider a mathematical model for percolation [19]. Consider an $L \times L$ square lattice whose sites are either occupied or unoccupied. We define a **cluster** to be a group of adjacent occupied squares. A cluster is said to be **percolating** if it connects opposite ends of the lattice, but this is more of an intuitive statement which we will later extend to encompass the limit $L \to \infty$ [5]. Percolation occurs when the liquid has permeated through the porous material.

Suppose a proportion p of the squares on the lattice are occupied. The quantity of interest here is the probability p_∞ that there exists a path from one side of the lattice to the other; the probability that there exists a percolating cluster. It can be shown numerically (and in certain cases analytically) that p_∞ exhibits threshold behaviour about some critical value $p = p_C$, and that as $L \to \infty$ the percolation threshold becomes more and more pronounced (Figs. 1.14 and 1.15).

At $p = p_C$ we actually observe one giant fractal cluster with **fractal dimension** $D < 2$. For such a cluster define a **red bond** to be a square of the lattice whose removal causes the disconnection of the cluster. It

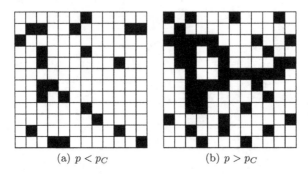

(a) $p < p_C$ (b) $p > p_C$

Figure 1.14 Percolation on a 12×12 lattice.

can be shown that the red bonds themselves form a fractal with fractal dimension $D_R \leq D$.

A second quantity of interest is $n(s)$, the number of clusters of size s per unit area. For large lattices one finds that $n(s)$ scales as

$$n(s) \sim \begin{cases} e^{-\frac{s}{s_0(p)}} & \text{for } p < p_C, \\ s^{-\tau} & \text{for } p = p_C, \\ e^{-\frac{s}{s_0(p)}+\delta(s-s_L)} & \text{for } p > p_C, \end{cases} \qquad (1.48)$$

where $s_0(p)$ is the average cluster size, $s_L = \mathcal{O}(L^2)$ is the size of the system and δ is simply the Kronecker δ-function. Notice that this

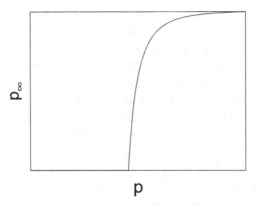

Figure 1.15 Percolation threshold around p_C follows the power law $p_\infty \sim |p - p_C|^{\beta}$

quantity also scales according to a **power law** in the neighbourhood of the critical point.

Constrain a random walker to move only on the occupied squares of the lattice and let $R(n)$ be the average displacement after n steps. If we look in the regime of probabilities smaller then the critical probability $p < p_C$, the random walk is confined to a single isolated cluster. Initially $R(n) \sim \sqrt{n}$, but the walker soon reaches the cluster boundary and so $R(N) \leq B$ for all n, where B is the diameter of the cluster. Conversely when $p > p_C$ most of the lattice is occupied and we regain something resembling the behaviour of a random walk on $\mathbb{Z} \times \mathbb{Z}$.

Percolation in one dimension

In order to illuminate the discussion with an example, let us examine percolation on a one-dimensional lattice of size L. On this lattice we can define p to be the probability that a given square is occupied and p_∞ to be the probability that a spanning cluster exists. Clearly in one dimension we have

$$p_\infty = p^L. \tag{1.49}$$

We can also derive an expression for $n(s)$, the number of clusters of size s:

$$
\begin{aligned}
n(s) &= (1-p)p \cdots p(1-p) \quad \text{(as exactly size } s\text{)} \\
&= p^s(1-p)^2 \\
&= (1-p)^2 e^{\frac{-s}{s_0}},
\end{aligned}
\tag{1.50}
$$

where $s_0 = -1/\log(p)$ is the average cluster size. Expanding about p_C we obtain

$$s_0 = -\frac{1}{\log p} = -\frac{1}{\log\left(p_c + (p - p_C)\right)} \approx \frac{1}{p_C - p} = (p_C - p)^{-1}, \tag{1.51}$$

so that, as usual, s_0 follows a power law around $p = p_C$.

We could define the correlation function $g(r)$ as the probability that a square a distance r from the origin is connected to the origin by occupied squares. Then

$$g(r) = p^r = e^{-r/\zeta}, \tag{1.52}$$

where $\zeta = -(\log p)^{-1}$. Correlation functions are are a useful tool, for the

Ising model, for example:

$$g(r) = \langle s(r)s(0)\rangle - \langle s(r)\rangle \langle s(0)\rangle. \qquad (1.53)$$

Let us also look at the average number of clusters per unit length n_T:

$$n_T = \sum_{s=1}^{\infty} n(s) = \sum_{s=1}^{\infty} (1-p)^2 p^s = p(1-p). \qquad (1.54)$$

Finally, we might consider the probability $W(s)$ that an occupied square belongs to a cluster of size s:

$$W(s) = \frac{sn(s)}{\sum_{s=1}^{\infty} sn(s)}, \qquad (1.55)$$

which has an expectation

$$\langle W(s)\rangle = \frac{1+p}{1-p}.$$

1.3.4 Universality

There exist different systems which display the same behaviour during phase transitions. Physicists would say that these systems belong to the same **universality class**. For example, experimental results suggest that the magnetisation of ferromagnetic materials belongs to the same universality class as the Ising model with a uniform external field. Such an observation justifies our use of the Ising model to simulate the magnetisation of a ferromagnetic material, but more broadly the notion of universality ensures that the Ising model is applicable to all systems which lie in the same universality class [18].

As a second example, consider percolation on a hexagonal lattice. In this case one obtains values for p_C which are different from those of a square lattice, but the *critical exponents are unchanged* (1.48) – the choice of lattice does not effect the overall behaviour. These critical exponents are features of a robust universality class to which all percolation models and similar systems belong. Typically it is possible to make several changes in the details of the model without leaving the universality class. The extension of percolation concepts to spin-systems is discussed in [12]; other applications of the universality concepts are covered in [7].

1.4 Self-organised criticality

1.4.1 Overview

The term **self-organised criticality** (SOC) was introduced to describe the dynamics of many-body systems that appear to reach a critical state without the need to fine-tune their parameters [10]. Several models claim to exhibit SOC, including sand piles, forest fires, magnetic flux line dynamics in superconductors, water droplets on a surface and the dynamics of magnetic domains.

We have already discussed criticality in the context of the Ising model. Here the parameter which decides whether or nor we are at criticality is the temperature T. Under strict laboratory conditions we can observe criticality by **tuning** the temperature to the critical value T_C. This is not an example of SOC, which requires the system to reach the criticality independently and without tuning. In other words, the critical state should be **attracting**.

The study of SOC is mostly based on cellular automata which reach statistically stationary states. Such states are typically characterised by spatial and temporal correlation functions which demonstrate power-law behaviour.

1.4.2 The OFC model

It is claimed that the Olami–Feder–Christensen (OFC) earthquake model which we will proceed to describe demonstrates SOC [14]. Consider a regular array of wooden blocks on a rough surface, such that each block is connected with springs to its four nearest neighbours (Fig. 1.16). A single rigid driving plate connects the blocks to each other via an arrangement of springs. By moving the driving plate the blocks may be slowly driven until one block overcomes its static friction with the rough surface and "slips". The slipping of one block may redefine the position of neighbouring blocks, depending on the distribution of force to the surrounding blocks and the sensitivity of the other blocks to the additional force.

To map this process into the language of cellular automata we define an $L \times L$ square lattice of blocks (i, j), where $i, j \in \{1, 2, \ldots, L\}$. The displacement of each block from the current stationary position on the lattice is written as $dx_{i,j}$. Then the total force exerted on a given block $F_{i,j}$ may be calculated as

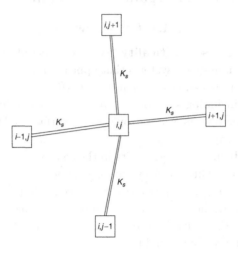

Figure 1.16 A schematic view of the spring model.

$$F_{i,j} = K_s[4dx_{i,j} - dx_{i-1,j} - dx_{i+1,j} - dx_{i,j-1} - dx_{i,j+1}] + K_p dx_{i,j}. \quad (1.56)$$

The physical visualisation of the spring-block model is represented in Fig. 1.16. The spring constant between blocks and the spring constant between the block and the homogeneous driving plate are represented here by K_s and K_p respectively. The rule that the automaton must follow – the **relaxation** – is then

$$F_{i\pm1,j\pm1} \rightarrow F_{i\pm1,j\pm1} + \delta F_{i\pm1,j\pm1},$$

$$\delta F_{i\pm1,j\pm1} = (\frac{K_s}{4K_s + K_p})F_{i,j} = \alpha F_{i,j},$$

$$F_{i,j} \rightarrow 0. \quad (1.57)$$

We have introduced α to denote the proportion of force that is distributed to the surrounding. One result of this simple derivation is that force is not conserved in the system. In a conserved system $\alpha = 0.25$, but for this model $\alpha \leqslant 0.25$ depending on the value of K_p.

In the automaton the position transition $dx_{i,j}$ is substituted by $F_{i,j}$, which can be thought of stress, energy or any physical parameter.

To initialise the simulation each component on the array is randomly

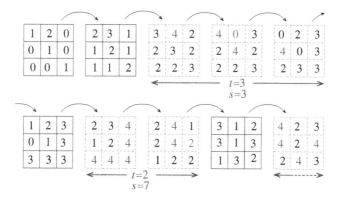

Figure 1.17 A schematic view of the OFC model.

assigned a value $F_{i,j} \in [0, F_{crit})$. The system then follows the relaxation rule

$$F_{i\pm1,j\pm1} \to F_{i\pm1,j\pm1} + \alpha F_{i,j},$$
$$F_{i,j} \to 0, \tag{1.58}$$

if for any block (i,j) we have $F_{i,j} \geq F_{crit}$. The relaxation ceases when all (i,j) satisfy $F_{i,j} < F_{crit}$ and instead the automaton follows the driving rule

$$F_{i,j} \to F_{i,j} + \delta F \qquad \text{for } i,j = 1, \ldots, N, \tag{1.59}$$

until some $F_{i,j}$ exceeds F_{crit} again.

The boundary of the OFC model is said to be **open**, which means that all energy transferred across the boundary will be forever lost. For a schematic view of the process (Fig. 1.17) consider the OFC model with $L = 3$, $F_{i,j} \in \{0, 1, 2, 3\}$, a homogeneous drive $\delta F = 1$, $\alpha = 0.25$ (conserved) and $F_{crit} = 4$. The sizes of avalanches can be measured by the difference in energy before and after the relaxation process.

In the second iteration of (Fig. 1.17), one cell reaches $F_{crit} = 4$ and the driving terminates. The critical cell then follows the relaxation rule and discharges equally to each of its four neighbours whilst decreasing to zero. This then triggers two surrounding cells to themselves discharge. The process stops and driving continues on to the fifth iteration. In this example the discharge time is $t = 3$ and the size of the discharge is $s = 4$.

The implementation of the algorithm may be made more efficient by finding the $\max(F_{i,j})$ and homogeneously increase each cell by F_{crit}

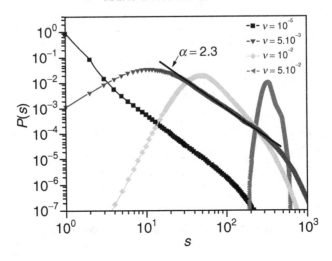

Figure 1.18 The power-law distribution disappears as the driving $\delta F = \nu$ increases.

— $\max(F_{i,j})$, rather than driving in small steps of δF until the next earthquake.

Power-law distributions are observed as $\delta F \to 0$, the physical significance of which is discussed in the next section. The power-law distributions disappear as the driving increases (Fig. 1.18). It should be noted that the OFC model intentionally sets the driving δF to be small in order to faithfully mimic a slowly driven many-body system.

1.4.3 Empirical data and the OFC model

The OFC model shows that a power-law distribution is observed in the energy of earthquakes (see references in [7]). The results are compared to empirical data in Fig. 1.19.

In our example $F_{i,j}$ represents the total force acted on a single component, but it could potentially represent another physical quantities. For example, $F_{i,j}$ can be related to "factor of safety" in a physical landslide [16] (1.60). The Mohr–Coulomb criterion then states that a landslide occurs when the shear stress is higher than a maximum threshold, analogous to F_{crit} in the OFC model:

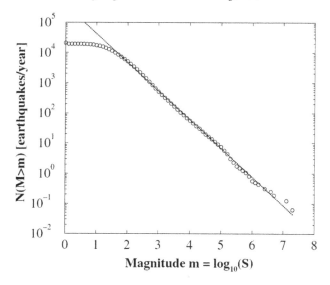

Figure 1.19 This displays the logarithmic-logarithmic plot of the magnitude of earthquakes in a region of California spanning $20°N - 45°N$ and $100°W - 125°W$ in the period 1984–2000. The total number of recorded earthquakes is 335076. The number of earthquakes $N(M > m)$ with a magnitude larger than m per year (open circles). The dashed line is the Gutenberg-Richter law $log_{10}N(M > m) \propto -bm$ with $b = 0.95$. [2]

$$FS = \frac{\tau_{max}}{\tau},$$
$$FS > 1 \rightarrow \text{stable},$$
$$FS < 1 \rightarrow \text{unstable}. \tag{1.60}$$

Similarly, a power distribution is observed on the empirical data for landslide energy (Fig. 1.20).

1.4.4 Does SOC exist?

Ever since the term SOC was coined, the associated concept has been subject to a great controversy. At its most ambitious the theory claims to have solved the underlying mechanisms of many physical processes, including earthquakes and landslides.

The classification of a state of SOC must be inferred from the phenomenological observation that a system has some power-law distribution "without tuning". One might expect this to come from systems of

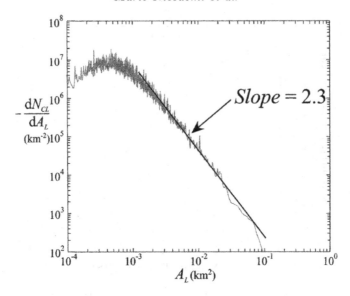

Figure 1.20 The empirical distribution of 11 000 landslides trig-
gered by the 17 January 1994, Northridge, California earthquake.
A logarithmic–logarithmic plot is displayed with the non-cumulative
frequency of landslides $-dN_{CL}/dA_L$ against area A_L. A power-law
slope $-dN_{CL}/dA_L = C' A_L^{-\beta}$ is fitted with $\beta = 2.3$ and $C' = 1.0$. [8]

interacting entities that are slowly driven, but the lack of a rigorous
mathematical formulation for the phrase "without tuning" makes the
identification of SOC difficult.

Conventionally SOC is regarded as a subfield of non-equilibrium statistical mechanics. Despite having a more phenomenological emphasis
than mainstream statistical mechanics, the ambitious claims of the theory have forced us to realise that thresholds and fluctuations play a
key role in the spatiotemporal behaviour of a general class of many-
body systems. This new insight is important to justify and inspire more
theoretical, observational and experimental research into self-organised
criticality.

1.5 Networks

Consider a collection of entities which interact to form a self-contained
whole. If this whole exhibits (in some sense) unexpected, emergent behaviour then we are dealing with a complex system. In this section we

focus on the word *interact* and discuss how the topology of the entity level interaction can influence the emergent behaviour that we observe at large scales.

1.5.1 Definitions

A (countable) **graph** or **network** is a collection (V, E) of vertices

$$v \in V = \{v_1, \ldots, v_n\}, n \in \{0, 1, \ldots\} \cup \{\infty\},$$

and edges

$$e \in E \subset \{1, \ldots, n\} \times \{1, \ldots, n\}.$$

This is a very general definition which encapsulates all the possible ways that n entities might be interconnected by edges. In this section we will only consider **undirected** graphs where $(i, j) \in E \iff (j, i) \in E$ and henceforth we will write (i, j) to mean both edges (i, j) and (j, i). One example of a graph which we have already encountered is the one-dimensional infinite lattice. Here $n = \infty$, $V = \mathbb{Z}$ and $E = \{(i, i+1) : i \in \mathbb{Z}\}$.

We define the **degree** $d(v_i)$ of a vertex v_i to be the number of vertices v_j $(j \neq i)$ which are connected to v_i via a single edge $e \in E$. Equivalently, $d(v_i)$ is equal to the number of edges of the form $e = (i, x)$ for $x \in \{1, \ldots, n\}$. We can also define a measure of distance $D(v_i, v_j) \in \{0, 1, \ldots\} \cup \{\infty\}$ between any pair v_i, v_j of vertices to be the number of edges in a shortest path (if one exists) between v_i and v_j. (If no path between the vertices exists we say that v_i and v_j are **disconnected** and define $D(v_i, v_j) = \infty$.) The equivalence relation $v_i \equiv v_j \iff D(v_i, v_j) < \infty$ defines **connected components** of the graph.

We illustrate some of these definitions using the simple example of the Franklin graph in Fig. 1.21. This connected graph has 12 vertices of degree 3 and 18 edges. The maximum distance $D(v_i, v_j)$ between any two vertices (the **diameter** of the graph) is 3 and there is only one connected component.

1.5.2 Random graphs

In this section we will give a very brief introduction to the theory of **random graphs**, with emphasis on how the local structure of the vertices can affect the global properties of the network. Why are we

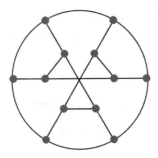

Figure 1.21 The Franklin graph.

interested in randomly generated graphs? For exactly the same reason that we introduced stochasticity into our models in Section 1.2; we can embrace the unknown intricate network structure of a complex system by modeling it as a single realisation of a random graph process.

There are many approaches which could be taken to produce a random graph and we outline the two main techniques here:

- **The Erdös (pronounced "air-dish") random graph with parameter p.** Fix the vertex set V. For each pair $v_i, v_j \in V$, $v_i \neq v_j$, include the edge (i, j) in the edge set E with probability p.

- **Preferential attachment random graphs.** Begin from a single vertex v_1 and add additional vertices one at a time according to the following rule: if there are n vertices $v_1, \ldots, v_n$ then add v_{n+1} and assign an edge from v_{n+1} to one of the original vertices with probability of picking v_j equal to $d(v_j) / (2(n-1))$, (if $n = 1$ we must choose v_1). In other words, the arriving vertex is biased towards joining an existing vertex of high degree.

These random constructions typically produce qualitatively different networks. For instance, the preferential attachment tends to produce a few vertices of large degree (or **hubs**) and many vertices of small degree, whereas the Erdös random graph typically has a much smaller range of vertex degrees. See Fig. 1.22 where the hubs have been shaded in grey. We will make this precise in the following section. Notice how an observation of the local interaction between vertices (specifically the mechanism for generating the networks) allows us to make this global characterisation of the whole network. It is the central goal of complexity science to understand the connection between local and global behaviour in many-body systems.

Random graphs in themselves do not constitute complex systems. However, every complex system can be described as a network of entities which lie at the vertices of a graph, with inter-entity interactions occurring precisely along the edges of the graph. (This is a tautology – take E to be the set of pairs of entities which interact within the system.) We shall discover that consideration of the network structure is fundamental to the understanding of any complex system, beginning with the following example.

1.5.3 Degree distribution

When we go on holiday to Japan from Coventry, UK, we do not fly from Coventry airport. This is not because of the passenger service or the quality of the food, it is because Coventry airport is a vertex of small degree: from Coventry there are only so many destinations to which you can travel, none of which are Japan. Instead we book our flight from one of the major terminals such as London Heathrow or Birmingham International, initially extending our journey to the airport by car or train. Meanwhile, business analysts in Coventry have an idea – why not charter a regular flight to Amsterdam Schiphol (a vertex of large degree, from which you *can* fly to Japan) in order to capitalise on the local demand for flights to Japan? If other airports also followed Coventry's example, Schiphol would become an even greater international transport hub. This discussion illustrates how a preferential attachment network might evolve in the real world, where the airports are vertices, flight paths are edges and the major terminals such as Heathrow, Birmingham and Schiphol are hubs.

Choose one of the methods for generating a random graph and define $p(n)$ to be the probability that a given vertex has degree n. The type of random graph makes a big difference to the form of p:

- For the Erdös method $p(n)$ is binomial and for large n we have, $p(n) \sim \exp(-n)$ (Fig. 1.23).
- However, for the preferential attachment method we have $p(n) \sim k^{-\gamma}$ (Fig. 1.24). This result gives rise to the nomenclature **scale free**, since if we generate a huge preferential attachment graph ($n \to \infty$) and examine only a certain proportion of it then the degree distribution

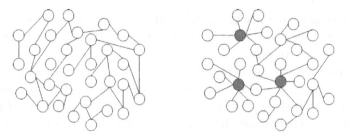

Figure 1.22 Erdös (left) and preferential attachment (right) networks. The preferential attachment network displays more structure than the Erdös graph with the emergence of hubs (grey) [3].

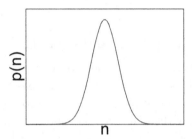

Figure 1.23 Degree distribution of the Erdös random graph is binomial.

of that proportion will have the same form as the degree distribution of the entire graph.

Of course there are some complex systems for which we *do* know one or two things about the network structure. In Section 1.3 we examined

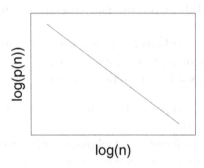

Figure 1.24 Degree distribution of a scale free graph obeys a power law.

in some detail the behaviour of the two-dimensional Ising model which exists on the lattice $\mathbb{Z} \times \mathbb{Z}$. This lattice was very simple to work with because of its uniformity but in principle one could imagine a much more structured graph – a **hierarchy**. One example, displayed in Fig. 1.25, is given by the local topology of the World Wide Web. Given any website we can easily list the sites which are one hyperlink away, two hyperlinks away, etc. from our chosen site. If we look at the sites up to four hyperlinks from the original site, for example, then we would expect to see a tree structure as in the figure. (This will not apply to web *pages* because websites usually contain many internal hyperlinks between its pages, so that the graph of linked web pages will contain many triangles.) We proceed to describe an alternative way in which a hierarchical network might arise within society.

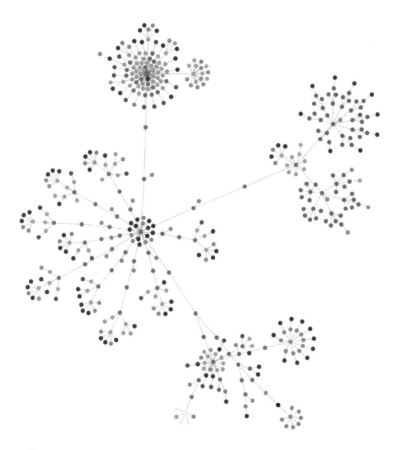

Figure 1.25 Local topology of the website graphs is hierarchical.

Voter models

On the first day of his premiership the new Prime Minister announces that there will be a referendum on the controversial issue of fox hunting. Each vertex (British citizen) must choose to answer "Yes" or "No" after an agreed period of campaigning on the part of several lobby groups, which naturally include the two leading political parties on opposite benches. Since the average citizen has very little experience in hunting foxes, their instantaneous opinion will be based purely on the balance of views that they have been subjected to up to that point. That is, if a person notices many "No to fox hunting" remarks and few "Yes to fox hunting" comments then they will be inclined to vote "No". This is a complex system since people will discuss the issue with friends and colleagues who all attempt to influence their opinion on the matter, so that a person's opinion might change with time up until the vote.

This simple **spin model** from physics ignores some basic facts about everyday life – people get information not just from their friends but from the media. So a better model would have several special vertices called "BBC", "The Sun", "Roadside Adverts", etc., which are joined to those people who expose themselves to the social engineering of the editors. Moreover, several of these special vertices might be controlled by "media moguls" who ensure that their brands support the same campaign, giving an even higher layer of structure to the network.

The point to appreciate is not that the graph is simply "big", but that the interaction network for this system displays structure. Of course this is a mathematical model – people are not normally this impressionable and the quality of the two arguments would be an important factor – but it illustrates how we might come across complicated hierarchical structures in real life.

1.5.4 Clustering coefficient

Typically hierarchical graphs are formed by clusters combined in an iterative manner, resulting in a scale free degree distribution as in Fig. 1.24. We can quantify this observation by defining the **clustering coefficient** $c(v)$ to be the number of triangles in the graph which have v as a vertex, so that $c(v)$ is some measure of "the size of the cluster containing vertex v". We then define $C(n)$ to be the average clustering coefficient, where the average is taken over all vertices v of degree n. It is possible to show that $C(n) \sim n^{-1}$ for most hierarchical graphs, whereas for Erdös random graphs and preferential attachment random graphs $C(n)$ is independent of n. This relation provides a practical test for hierarchical structure, relating the local property of the clustering coefficient to the global property of network structure.

The internet

The internet is a network where the vertices are computers and edges are electrical or fibre optic cables. Nobody knows how the whole internet is wired together – in any case the structure is continually changing as redundant machines are taken offline and new technologies are plugged in. Of course we know where the major underwater connections are located and therefore we have the rough outline of the global structure, but when we demand more detail we run into problems. It is possible to write a computer program to discover local connectivity, but extending this to a global search is time-consuming and the results are difficult to validate.

Fortunately we are now in a position to overcome this difficulty. Armed with the observations that local degree distributions seem to obey a power law we can postulate that the *global* structure of the internet is that of a scale free network (potentially with hierarchical structure) and in particular not an Erdös random network.

Indeed the internet is well known to exhibit hierarchial structure: in addition to the underwater cables generating global level structure, individual countries will host several internet service providers (ISPs) who in turn will transfer the connections to smaller stations. Finally the smaller stations provide the internet connections to individual homes and offices.

Scale free networks have a habit of occurring in many situations at the human level. For example, the collaboration graph of Hollywood actors, the citation graph of the scientific literature and the network of sexual partners (data taken from a small Sweedish town in 1996, see [11]) are all purported to display the power-law degree distribution characteristic of a scale free network.

1.5.5 Network vulnerability

The degree distribution has important consequences in determining the vulnerability of a network. To continue with our airport analogy, suppose that in the year 2050 there will be a terrorist attack at precisely one of the world's many airports. If the terrorists wish to cause maximum damage they will target the busiest airport (which in 2050, due to excellent business management, happens to be Coventry). Since Coventry is now a central hub for worldwide transport the effect of the attack will be to severely disrupt a large proportion of global travel. It is undesirable for the transport authorities to allow these hubs of activity to form, since they compromise the robustness of the global network to attack.

If the hubs in Fig. 1.22 were removed we would be left with 16 connected components. Conversely there is no choice of three vertices which when removed from the Erdös random graph will create as many connected components. An Erdös random graph is likely to be more robust against a targeted attack than the real-life scale free network.

The Erdös random graph is very similar to the percolation lattice. This can be seen by taking the lattice points to be vertices and letting each face of a lattice square correspond to four edges in the graph. The percolation theory then predicts that if we remove vertices at random to the point that a certain critical fraction p_C of the vertices are left then we will see only tiny clusters of vertices and no large component. Interestingly this fraction is greater than the corresponding fraction for scale free networks – that is, if vertices fail *at random* then the scale free networks stay connected for longer. We conclude that the most suitable network for an application depends on whether the failures will occur because nodes are targeted (for their connectivity) or whether the failures occur at random through out the network.

1.5.6 The small-world effect

The most famous example of a global property of networks has to be the **six degrees of separation**, which refers to the idea that, if a person is one step away from each person they know and two steps away from each person who is known by one of the people they know, then everyone is at most six steps away from any other person on Earth. There have been many variants of this idea, most recently the student pastime of "six degrees of Hitler" in which a contestant selects at random a page from the online encyclopedia *Wikipedia* and attempts to navigate to Adolf Hitler's entry using no more than six clicks of the mouse. In the more precise language of Section 1.5.1 we would say that it is possible to have n very large whilst keeping $D(v_i, v_j)$ very small for all pairs v_i, v_j. Indeed for the Erdös random graph the average distance $\langle D \rangle$ between

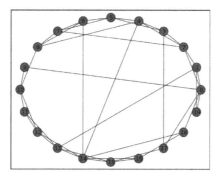

Figure 1.26 Realisation of a Watts–Strogatz graph ($n = 20$, $k = 4$) [9].

two vertices is

$$\langle D \rangle = \frac{\log (n)}{\log \langle d \rangle},$$

where $\langle d \rangle$ is the average vertex degree. For example, at the time of printing *Wikipedia* hosts about $3,000,000$ articles in English and each article contains hyperlinks to about 20 other pages of the encyclopedia. If we assume that the encyclopedia is the result of an Erdös random graph process then we obtain that the average separation between pages of the encyclopedia is roughly 5, meaning that, for a skilled player, the "six degrees of Hitler" game may be more frequently won than lost.

In their frequently cited paper of 1998, Watts and Strogatz [22] discuss their own random graph process:

We start with a ring of n vertices, each connected to its k nearest neighbours by undirected edges. [...] We choose a vertex and the edge that connects it to its nearest neighbour in a clockwise sense. With probability p, we reconnect this edge to a vertex chosen uniformly at random over the entire ring, with duplicate edges forbidden; otherwise we leave the edge in place. We repeat this process by moving clockwise around the ring, considering each vertex in turn until one lap is completed. Next, we consider the edges that connect vertices to their second-nearest neighbours clockwise. As before, we randomly rewire each of these edges with probability p, and continue this process, circulating around the ring and proceeding outward to more distant neighbours after each lap, until each edge in the original lattice has been considered once.

One particular realisation of a **Watts–Strogatz** random graph is shown in Fig. 1.26.

The result that interests us is shown in Fig. 1.26 which compares the average path length $\langle D(p) \rangle$ between two vertices against the average

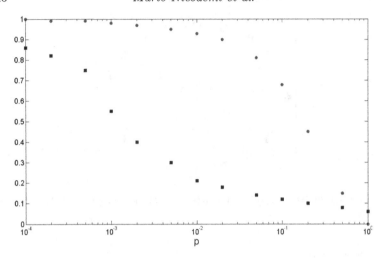

Figure 1.27 The small world effect. $\langle D(p)\rangle/\langle D(0)\rangle$ in shown in squares, $\langle c(p)\rangle/\langle c(0)\rangle$ in circles. There is an interval $[p_1, p_2]$ on which $\langle D(p)\rangle/\langle D(0)\rangle$ is small yet $\langle c(p)\rangle/\langle c(0)\rangle$ is large.

clustering coefficient $\langle c(p)\rangle$ (where this time, unlike $C(n)$, the average is taken over all vertices). As p increases $\langle D(p)\rangle$ rapidly drops, corresponding to the onset of the small-world phenomenon. Meanwhile $\langle c(p)\rangle$ is almost constant at its value for the regular lattice, indicating that the transition to "small world" is almost undetectable at the local level.

The inquisitive reader could do far worse than to consult the original papers of Barabasi and Bonabeau [1], Watts and Strogatz [22] and Strogatz [20].

References

[1] Barabasi, A. L. and Bonabeau, E. 2003. Scale-free networks. *Scientific American*, 288(5), 60–70.

[2] Bak, P., Christensen, K., Danon, L. and Scanlon, T. 2002. Unified scaling law for earthquakes. *Physical Review Letters*, 88(17), 178501.

[3] Carlos, C. 2004. Effective web crawling. http://en.wikipedia.org/wiki/File:Scale-free_network_sample.png.

[4] Chandler, D. 1987. *Introduction to Modern Statistical Mechanics*. Oxford University Press.

[5] Christensen, K. and Maoloney, N. R. 2005. *Complexity and Criticality*. Imperial College Press, London.

[6] Coniglio, A., Fierro, A., Herrmann, H.J. and Nicodemi, M. (eds). 2004. *Unifying Concepts in Granular Media and Glasses: From the Statistical*

Mechanics of Granular Media to the Theory of Jamming. Elsevier, Amsterdam.

[7] de Arcangelis, L., Godano, C., Lippiello, E. and Nicodemi, M. 2006. Universality in solar flare and earthquake occurrence. *Physical Review Letters*, 96(5), 051102.

[8] Guzzetti, F., Malamud, B., Turcotte, D. and Reichenbach, P. 2002. Power-law correlations of landslide areas in central Italy. *Earth and Planetary Science Letters*, 195, 169–183.

[9] Horvath, A. 2008 (May). Watts–Strogatz graph. http://en.wikipedia.org/wiki/File:Watts_strogatz.svg.

[10] Jensen, H. J. 1998. *Self-Organized Criticality*. Cambridge University Press.

[11] Liljeros, F., Edling, C. R., Amaral, L. A. N., Stanley, H. E. and Aberg, Y. 2001. The web of human sexual contacts. *Nature*, 411, 907–908.

[12] Nicodemi, M. 1997. Percolation and cluster formalism in continuous spin systems. *Physica A*, 238, 9.

[13] Nicodemi, M. and Prisco, A. 2007. A symmetry breaking model for X chromosome inactivation. *Phys. Rev. Lett.*, 98(10), 108104.

[14] Olami, Z., Feder, H. J. S., and Christensen, K. 1992. Self-organized criticality in a continuous, nonconservative cellular automaton modeling earthquakes. *Physical Review Letters*, 68(8), 1244–1247.

[15] Paulsson, J. 2005. Stochastic gene expression. *Physics of Life*, 2, 157–175.

[16] Piegari, E., Cataudella, V., Di Maio, R., Milano, L. and Nicodemi, M. 2006. A cellular automaton for the factor of safety field in landslides modeling. *Geophysics Research Letters*, 33, L01403.

[17] Richard, P., Nicodemi, M., Delannay, R., Ribière, P. and Bideau, D. 2005. Slow relaxation and compaction of granular systems. *Nature Materials*, 4, 121.

[18] Sethna, J. P. 2006. *Statistical Mechanics: Entropy, Order Parameters and Complexity*. Oxford University Press.

[19] Stauffer, D. and Aharony, A. 1994. *Introduction to Percolation Theory*. Taylor and Francis, London.

[20] Strogatz, S. H. 2001. Exploring complex networks. *Nature*, 410, 268–276.

[21] van Kampen, N. G. 1992. *Stochastic Processes in Physics and Chemistry*. Elsevier, Amsterdam.

[22] Watts, D.J. and Strogatz, S.H. 1998. Collective dynamics of 'small-world' networks. *Nature*, 393, 409–410.

2

Complexity and chaos in dynamical systems

Yulia Timofeeva

Abstract

Dynamical systems are represented by mathematical models that describe different phenomena whose state (or instantaneous description) changes over time. Examples are mechanics in physics, population dynamics in biology and chemical kinetics in chemistry. One basic goal of the mathematical theory of dynamical systems is to determine or characterise the long-term behaviour of the system using methods of bifurcation theory for analysing differential equations and iterated mappings. Interestingly, some simple deterministic nonlinear dynamical systems and even piecewise linear systems can exhibit completely unpredictable behaviour, which might seem to be random. This behaviour of systems is known as deterministic *chaos*.

This chapter aims to introduce some of the techniques used in the modern theory of dynamical systems and the concepts of chaos and strange attractors, and to illustrate a range of applications to problems in the physical, biological and engineering sciences. The material covered includes differential (continuous-time) and difference (discrete-time) equations, first- and higher order linear and nonlinear systems, bifurcation analysis, nonlinear oscillations, perturbation methods, chaotic dynamics, fractal dimensions, and local and global bifurcation.

Readers are expected to know calculus and linear algebra and be familiar with the general concept of differential equations.

2.1 Dynamical models

Difference equations

A difference equation (or an iterated map) can be written as

$$x_{n+1} = f(x_n),$$

where x_n is the value of a variable x at some time n, and f is a function that describes the relationship between the value of x at time n and $n+1$. This equation can be analysed using a simple geometric interpretation, called the *cobweb* construction below. Given an initial condition x_0, draw a vertical line until it intersects the graph of f: that height is given by x_1. Then return to the horizontal axis and repeat the procedure to get x_2 from x_1.

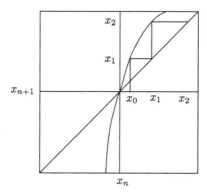

Differential equations

Differential equations describe the evolution of systems in continuous time, whereas iterated maps describe systems evolved in discrete time. A system of ordinary differential equations (ODEs):

$$\dot{x}_1 = f_1(x_1, \ldots, x_n),$$

$$\vdots$$

$$\dot{x}_n = f_n(x_1, \ldots, x_n).$$

Here $\dot{x}_i \equiv \mathrm{d}x_i/\mathrm{d}t$.

A single ODE is a mathematical equation for an unknown function of one independent variable that relates the values of the function itself and of its derivative. In general, ODEs define the rates of change of the variables in terms of the current state. To solve a system of ODEs means to find continuous functions $x_1(t), \ldots, x_n(t)$ of the independent variable t that, along with their derivatives, satisfies the system of equations.

Autonomous systems: $f_i(x_1, \ldots, x_n)$.
Nonautonomous systems: $f_i(x_1, \ldots, x_n, t)$.

2.2 First-order systems

$$\dot{x} = f(x), \quad x \in \mathbb{R}$$

Reminder: Linear ODE

$$\dot{x} = -\frac{x}{\tau}, \qquad x(0) = x_0.$$

Solution: $x(t) = x_0 e^{-t/\tau}$

One of the most basic techniques of dynamics is to interpret a differential equation as a vector field.

Example 2.2.1 $\quad \dot{x} = \sin x$

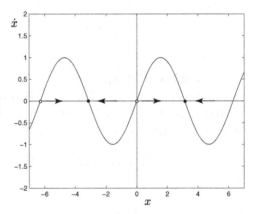

2.2.1 Fixed points and stability

A solution $x(t)$ of differential equation starting from the initial condition x_0 is also called a *trajectory*. The fixed points (equilibrium solutions, steady-states) are defined by $f(x^*) = 0$.

- A fixed point is defined to be *stable* if all sufficiently small disturbances away from it damp out in time.

  - If all solutions of the dynamical system that start out near an equilibrium point x^* stay near x^* forever, then x^* is *Lyapunov stable*.
  - More strongly, if all solutions that start out near x^* converge to x^*, then x^* is *asymptotically stable*.
  - The notion of *exponential stability* guarantees a minimal rate of convergence.

- If disturbances grow in time a fixed point is defined to be *unstable*.

Example 2.2.2 Find all fixed points of the following equation and classify their stability

$$\dot{x} = x^2 - 1.$$

Example 2.2.3 **Population growth**. Classify the fixed points of the logistic equation

$$\dot{N} = rN\left(1 - \frac{N}{K}\right).$$

2.2.2 Linear stability analysis

Let x^* be a fixed point and $\eta(t) = x(t) - x^*$ be a small perturbation, then

$$\dot{\eta} = f(x^* + \eta)$$

Reminder: Taylor series of f about the point x^*

$$f(x) = \sum_{n=0}^{\infty} f^{(n)}(x^*) \frac{(x - x^*)^n}{n!},$$

where $f^k(x^*)$ is the kth derivative of the function f evaluated at x^*.

Then $f(\eta + x^*) = f(x^*) + f'(x^*)\eta + O(\eta^2)$ $\qquad \Rightarrow \qquad \dot{\eta} = f'(x^*)\eta$

- $\eta(t)$ grows exponentially if $f'(x^*) > 0$ $\qquad \Rightarrow \qquad x^*$ is unstable
- $\eta(t)$ decays exponentially if $f'(x^*) < 0$ $\qquad \Rightarrow \qquad x^*$ is stable

2.2.3 Local existence and uniqueness (in $\mathbb{R}$)

Example 2.2.4

$$\dot{x} = x^{1/3}, \qquad x(0) = 0.$$

The point $x = 0$ is a fixed point, so one solution is $x(t) = 0$ for all t. Also

$$\int \frac{dx}{x^{1/3}} = \int dt \qquad \Rightarrow \qquad \frac{3}{2}x^{2/3} = t + C$$

Initial data ensures $C = 0$, so $x(t) = (2t/3)^{3/2}$. The co-existence of solutions (nonuniqueness) is due to the fact that $x^{1/3}$ is not continuously differentiable.

Theorem 2.2.5 *Consider the initial value problem (IVP)*

$$\dot{x} = f(x), \qquad x(0) = x_0.$$

Suppose that f is continuously differentiable on an open interval $I \subset \mathbb{R}$ with $x_0 \in I$. Then the IVP has a solution $x(t)$ on some interval $(-\tau, \tau)$ about $t = 0$ and the solution is unique.

Example 2.2.6

$$\dot{x} = 1 + x^2, \qquad x(0) = 0.$$

$$\int \frac{dx}{1 + x^2} = \int dt \qquad \Rightarrow \qquad \tan^{-1} x = t + C.$$

Initial data ensures $C = 0$, so $x = \tan t$ and only exists for $-\pi/2 < t < \pi/2$. This is an example of *blow-up*, where $x(t)$ reaches ∞ in finite time.

As we have seen, all flows on a line either approach a fixed point or diverge to $\pm\infty$. In essence overshoot and damped oscillations can never occur in a first order system. It is never possible to have periodic motion for the first order system $\dot{x} = f(x)$, $x \in \mathbb{R}$, $f : \mathbb{R} \to \mathbb{R}$. Flow on a circle may lead to periodic motion, however, and this will be considered later.

2.2.4 Potentials

For a dynamical system $\dot{x} = f(x)$, the potential $V(x)$ is defined by

$$f(x) = -\frac{dV}{dx}.$$

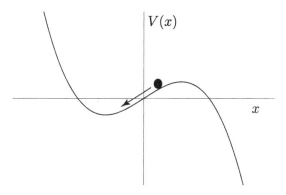

2.2.5 Numerical simulation of dynamical systems

This is a vast subject in its own right (see Chapter 5). It is a powerful means of obtaining insight into the behaviour of nonlinear dynamical systems by obtaining approximate solutions to problems that are analytically intractable.

Euler's method Given a dynamical systems $\dot{x} = f(x)$ and a known solution at time t we approximate the solution at time $t+\Delta t$ by $x(t+\Delta t)$, where

$$x(t + \Delta t) = x(t) + f(x(t))\Delta t.$$

Introducing the notation $x_n = x(t_n)$, where $t_n = n\Delta t$, we have the following iterative scheme

$$x_{n+1} = x_n + f(x_n)\Delta t.$$

Improved Euler method One problem with the Euler method is that it estimates the derivative $\dot{x}$ only at the left end of the time-interval between t_n and t_{n+1}. A better approach is to use the average derivative across this interval. First construct the estimate $\tilde{x}_{n+1} \approx x_n + f(x_n)\Delta t$ and then take the average of $f(x_n)$ and $f(\tilde{x}_n)$ and use that to construct the next step in the iterative scheme. The improved Euler method is then given by

$$\tilde{x}_n = x_n + f(x_n)\Delta t \qquad \text{(the trial step),}$$

$$x_{n+1} = x_n + \frac{1}{2}[f(x_n) + f(\tilde{x}_n)]\Delta t.$$

This scheme tends to a smaller error $E = |x(t_n) - x_n|$ for a given step-size Δt. For the (first-order) Euler method $E \propto \Delta t$, whilst for the (second order) improved Euler method $E \propto (\Delta t)^2$.

An accurate and commonly used scheme is the so-called fourth-order **Runge–Kutta scheme:**

$$x_{n+1} = x_n + \Delta t \left[\frac{1}{6}k_1(n) + \frac{1}{3}k_2(n) + \frac{1}{3}k_3(n) + \frac{1}{6}k_4(n) \right],$$

where

$$k_1(n) = f(x_n),$$
$$k_2(n) = f(x_n + \frac{1}{2}\Delta t k_1(n)),$$
$$k_3(n) = f(x_n + \frac{1}{2}\Delta t k_2(n)),$$
$$k_4(n) = f(x_n + \Delta t k_3(n)).$$

A number of free software packages for numerical analysis of dynamical systems exists, for example:

XPPAUT (http://www.math.pitt.edu/~bard/xpp/xpp.html)
PyDSTool (http://www.ni.gsu.edu/~rclewley/PyDSTool)
AUTO (http://indy.cs.concordia.ca/auto/)
MATCONT (http://www.scholarpedia.org/article/matcont).

2.2.6 Bifurcations

The qualitative structure of a flow can change as a parameter is varied. These qualitative changes are called *bifurcations* and the parameter values at which they occur are called *bifurcation* points.

Saddle-node bifurcation
$$\dot{x} = \mu - x^2$$

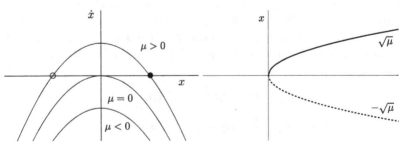

Transcritical bifurcation
$$\dot{x} = \mu x - x^2$$

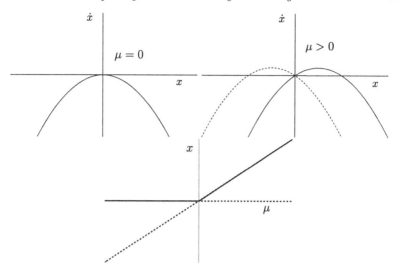

Example 2.2.7

$$\dot{x} = r \ln x + x - 1$$

Fixed point at $x = 1$. Let $u = x - 1$, then

$$\dot{u} = r \ln(1 + u) + u \approx r\left(u - \frac{u^2}{2} + \cdots\right) + u$$
$$= (r+1)u - \frac{1}{2}ru^2$$

Rescale $(v = (r/2)u)$:

$$\dot{v} = (r+1)v - v^2.$$

By a near identity change of coordinates we have found the *normal form* for the dynamics (valid close to the bifurcation point).

Pitchfork bifurcation: supercritical

$$\dot{x} = \mu x - x^3$$

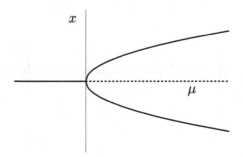

Pitchfork bifurcation: subcritical
$$\dot{x} = \mu x + x^3$$

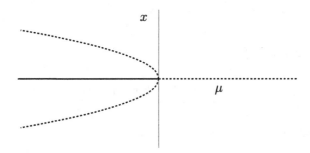

Example 2.2.8

$$\dot{x} = \mu x + x^3 - x^5$$

Fixed points

$$x = 0 \qquad \text{and} \qquad -(\mu + x^2) + x^4 = 0$$

with the roots:

$$x^2 = \frac{1 \pm \sqrt{1 + 4\mu}}{2}.$$

If $\mu > 0$ then $x^2 = (1 + \sqrt{1 + 4\mu})/2$: total of three fixed points. If $-1/4 < \mu < 0$, $x^2 = (1 \pm \sqrt{1 + 4\mu})/2$: total of five fixed points. Define $\mu_c = -1/4$.

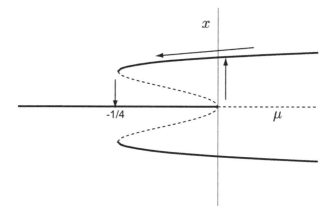

1. In the range $\mu_c < \mu < 0$ there co-exist three stable fixed points (and two unstable). There is *multi-stability* and initial conditions determine the final state.
2. The bifurcation at μ_c is a saddle-node bifurcation.
3. The system exhibits hysteresis.

Cusp singularity

Consider the following system with two parameters, $\dot{x} = h + \mu x - x^3$.

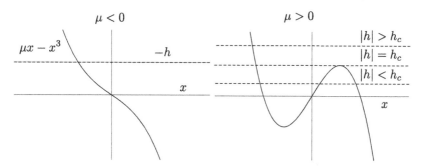

Critical case: horizontal line is tangent to min or max of $f(x) = \mu x - x^3$. Local max/min at $x = \pm\sqrt{\mu/3}$.

$$h_c(\mu) = \frac{2\mu}{3}\sqrt{\frac{\mu}{3}}$$

At $h = \pm h_c(\mu)$ there is a saddle-node bifurcation. There are two bifurcation curves $\pm h_c(\mu)$.

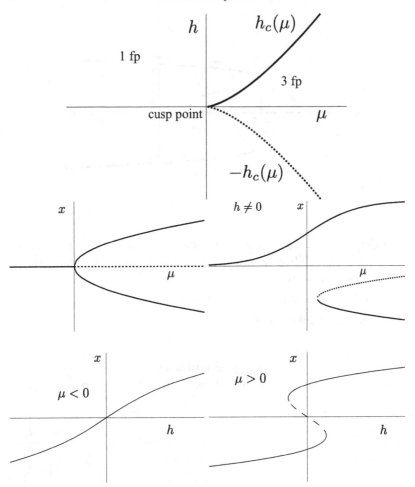

Example 2.2.9 Budworm population dynamics:

$$\dot{N} = RN\left(1 - \frac{N}{K}\right) - \frac{BN^2}{A^2 + N^2}, \qquad A, B, R > 0.$$

The budworm population $N(t)$ grows logistically (first term) in the absence of predators. The second term describes mortality due to predation (mainly by birds). Non-dimensionalise: $x = N/A$.

$$\frac{A}{B}\frac{dx}{dt} = \frac{R}{B}Ax\left(1 - \frac{Ax}{K}\right) - \frac{x^2}{1 + x^2} \equiv f(x).$$

Introduce

$$\tau = \frac{Bt}{A}, \qquad r = \frac{RA}{B}, \qquad k = \frac{K}{A}$$

so that

$$\frac{dx}{d\tau} = rx\left(1 - \frac{x}{k}\right) - \frac{x^2}{1 + x^2}.$$

Fixed points at

$$\bar{x} = 0, \qquad r\left(1 - \frac{\bar{x}}{k}\right) - \frac{\bar{x}}{1 + \bar{x}^2} = 0.$$

Linearisation:

$$f'(x) = r - \frac{2rx}{k} - \frac{2x}{(1 + x^2)^2},$$

so $f'(0) = r > 0$, so $\bar{x} = 0$ is unstable. Other roots may be found graphically by finding the intercepts of $x/(1 + x^2)$ and $r(1 - x/k)$:

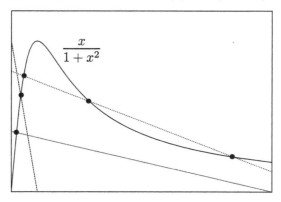

Hence there can be either one, two or three interceptions depending upon the choice of (k, r). For example when there are three fixed points $c > b > a > 0$, then since $\bar{x} = 0$ is unstable, a is stable, b is unstable and c is stable. We compute the details of the bifurcation in the following manner.

A saddle-node occurs when $r(1 - x/k)$ intersects $x/(1 + x^2)$ tangentially. Thus we require $\bar{x}$ given by $f(\bar{x}) = 0$ and

$$\frac{d}{dx}\left[r\left(1 - \frac{x}{k}\right)\right] = \frac{d}{dx}\left[\frac{x}{1 + x^2}\right]$$

or that

$$-\frac{r}{k} = \frac{1 - x^2}{(1 + x^2)^2}, \qquad x = \bar{x}.$$

This gives us:

$$r = \frac{2\bar{x}^3}{(1+\bar{x}^2)^2}, \qquad k = \frac{2\bar{x}^3}{\bar{x}^2 - 1}.$$

Since $k > 0$, we require $x > 1$. The bifurcation curve is defined by $(r(\bar{x}), k(\bar{x}))$.

2.2.7 Flows on the circle

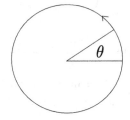

Basic model of an oscillator:

$$\dot{\theta} = f(\theta), \qquad \theta \in [0, 2\pi),$$

where $f(\theta) = f(\theta + 2\pi)$.

Uniform oscillator

$$\dot{\theta} = \omega, \qquad \theta = \theta_0 + \omega t$$

Period: $T = 2\pi/\omega$.

Non-uniform oscillator

$$\dot{\theta} = \omega - a\sin\theta$$

Consider here $\omega > 0$, $a \geq 0$ (similar results for negative ω and a).

- If $a < \omega$: non-uniform flow which is fastest at $\theta = -\pi/2$ and slowest at $\pi/2$. When a is only slightly less than ω the system takes a long time to pass through the bottleneck at $\theta = \pi/2$, after which it quickly traverses the rest of the circle.
- If $a > \omega$: there exists a stable–unstable pair of fixed points at $\sin^{-1}[\omega/a]$ born via a saddle-node bifurcation. Oscillations do not exist.

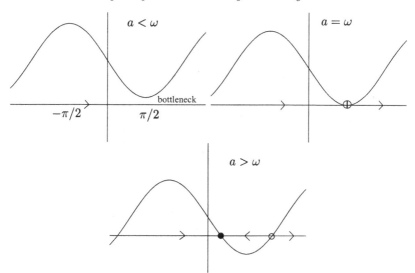

Period:

$$T = \int \mathrm{d}t = \int_0^{2\pi} \frac{\mathrm{d}t}{\mathrm{d}\theta} \mathrm{d}\theta = \int_0^{2\pi} \frac{\mathrm{d}\theta}{\omega - a\sin\theta} = \frac{2\pi}{\sqrt{\omega^2 - a^2}}$$

Close to $a = \omega$

$$T = \frac{2\pi}{\sqrt{\omega + a}} \frac{1}{\sqrt{\omega - a}} \approx \frac{\sqrt{2}\pi}{\sqrt{\omega}} \frac{1}{\sqrt{\omega - a}}.$$

2.3 Linear systems in $\mathbb{R}^2$

Consider the (autonomous) differential equation

$$\frac{\mathrm{d}\boldsymbol{x}}{\mathrm{d}t} \equiv \dot{\boldsymbol{x}} = \boldsymbol{A}\boldsymbol{x}, \qquad \boldsymbol{x} \in \mathbb{R}^2,$$

where A is a 2×2 constant matrix.

2.3.1 Harmonic oscillator

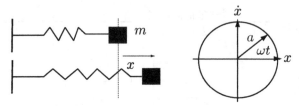

According to classical theory a simple harmonic oscillator is a particle of mass m moving under the action of a force $F = -kx$ (Hooke's law). Newton's laws of motion take the form

$$m\ddot{x} = -kx \quad \text{or} \quad \ddot{x} + \omega^2 x = 0, \quad \text{where } \omega = \sqrt{\frac{k}{m}}.$$

The general solution to this differential equation is of the form

$$x(t) = A \cos \omega t + B \sin \omega t$$

which represents an oscillatory motion of angular frequency ω. The constants of integration A and B are determined by the initial conditions for x and $\dot{x}$, where

$$\dot{x}(t) = -A\omega \sin \omega t + B\omega \cos \omega t$$

so that $x(0) = A$ and $\dot{x}(0) = B\omega$. An easy way to imagine the geometry of simple harmonic motion is to write the equations of motion as a second-order system. Introduce $v = \dot{x}$, then

$$\dot{x} = v,$$
$$\dot{v} = -\omega^2 x.$$

There is a fixed point at $(x, v) = (0, 0)$. Combining the above we have

$$\frac{dv}{dx} = -\omega^2 \frac{x}{v}.$$

After integrating this separable ODE we have

$$v^2 + \omega^2 x^2 = \text{constant}.$$

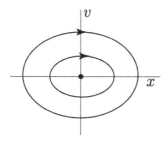

Reminder: Consider the matrix A

$$A = \begin{pmatrix} a_{11} & a_{12} \\ a_{21} & a_{22} \end{pmatrix}$$

• The *trace* and *determinant* of the matrix A

$$\mathrm{Tr}\,(A) = a_{11} + a_{22}$$
$$\det(A) = a_{11}a_{22} - a_{21}a_{12}$$

• Singularity: the matrix A is singular if $\det(A) = 0$
• Identity matrix

$$I = \begin{pmatrix} 1 & 0 \\ 0 & 1 \end{pmatrix}$$

Example 2.3.1 Consider the system

$$\dot{x} = Ax, \qquad A = \begin{pmatrix} a & 0 \\ 0 & -1 \end{pmatrix}.$$

Matrix multiplication yields

$$\dot{x} = ax,$$
$$\dot{y} = -y.$$

Since these two equations are uncoupled they can be solved separately

$$x(t) = x_0 e^{at},$$
$$y(t) = y_0 e^{-t}.$$

• **Stable nodes**: (i) $a < -1$ and (ii) $-1 < a < 0$

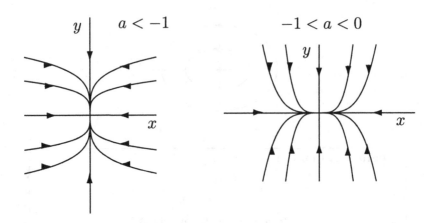

- **Star:** $a = -1$

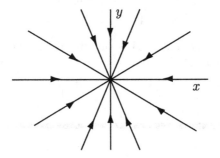

- **Saddle point:** $a > 0$

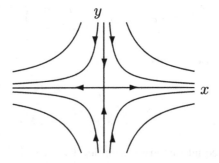

The y-axis is called the *stable manifold* of the saddle point x^*: the set of initial conditions x_0 such that $x(t) \to x^*$ as $t \to \infty$. The x-axis is called the *unstable manifold* of the saddle point x^*: the set of initial conditions x_0 such that $x(t) \to x^*$ as $t \to -\infty$.

- **Line of fixed points:** $a = 0$

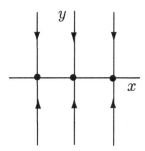

2.3.2 Classification of fixed points

Consider

$$\dot{\boldsymbol{x}} = \boldsymbol{A}\boldsymbol{x}, \qquad \boldsymbol{A} = \begin{bmatrix} a & b \\ c & d \end{bmatrix}.$$

Try a solution of the form

$$\boldsymbol{x} = \mathrm{e}^{\lambda t} \boldsymbol{v}.$$

This leads to the linear homogeneous equation

$$\boldsymbol{A}\boldsymbol{v} = \lambda \boldsymbol{v},$$

where $\boldsymbol{v}$ is an *eigenvector* of $\boldsymbol{A}$ with corresponding *eigenvalue* λ. For the system above to have a non-trivial solution we require that

$$\boxed{\det(\boldsymbol{A} - \lambda \boldsymbol{I}) = 0}$$

which is called the *characteristic equation*. Here $\boldsymbol{I}$ is the 2×2 identity matrix. Substituting the components of A into the characteristic equation gives

$$\lambda^2 - (a + d)\lambda + (ad - bc) = 0$$

or

$$\lambda^2 - \mathrm{Tr}\,(\boldsymbol{A})\,\lambda + \det(\boldsymbol{A}) = 0$$

so that

$$\boxed{\lambda_{\pm} = \frac{1}{2}\left[\mathrm{Tr}\,(\boldsymbol{A}) \pm \sqrt{(\mathrm{Tr}\,(\boldsymbol{A}))^2 - 4\det(\boldsymbol{A})} \right]}$$

The general solution for $\boldsymbol{x}(t)$:

$$x(t) = c_1 \mathrm{e}^{\lambda_1 t} \boldsymbol{v}_1 + c_2 \mathrm{e}^{\lambda_2 t} \boldsymbol{v}_2.$$

Exercise Solve the initial value problem

$$\dot{x} = x + y, \qquad \dot{y} = 4x - 2y, \qquad (x_0, y_0) = (2, -3)$$

If $\lambda_{1,2}$ are complex ($\lambda_{1,2} = \alpha \pm i\omega$), the fixed point is either a *centre* or a *spiral*. Since $\boldsymbol{x}(t)$ involves linear combinations of $\mathrm{e}^{\alpha \pm i\omega}$, $\boldsymbol{x}(t)$ is a combination of terms involving $\mathrm{e}^{\alpha t} \cos(\omega t)$ and $\mathrm{e}^{\alpha t} \sin(\omega t)$ (by Euler's formula, $\mathrm{e}^{i\omega t} = \cos(\omega t) + i \sin(\omega t)$):

- If $\alpha < 0 \Rightarrow$ stable focus (or stable spiral).
- If $\alpha > 0 \Rightarrow$ unstable focus (or unstable spiral).
- If $\alpha = 0 \Rightarrow$ a centre (periodic solution with period $T = 2\pi/\omega$), marginally stable.

Here we classify the different types of behaviour according to the values of $\mathrm{Tr}\,(\boldsymbol{A})$ and $\det(\boldsymbol{A})$:

- $\lambda_{\pm}$ are real if $\mathrm{Tr}\,(\boldsymbol{A})^2 > 4 \det(\boldsymbol{A})$.
- Real eigenvalues have the same sign if $\det(\boldsymbol{A}) > 0$ and are positive if $\mathrm{Tr}\,(\boldsymbol{A}) > 0$ (negative if $\mathrm{Tr}\,(\boldsymbol{A}) < 0$) – **stable and unstable nodes**.
- Real eigenvalues have opposite signs if $\det(\boldsymbol{A}) < 0$ – **saddle node**.
- Eigenvalues are complex if $\mathrm{Tr}\,(\boldsymbol{A})^2 < 4 \det(\boldsymbol{A})$ – **focus**.

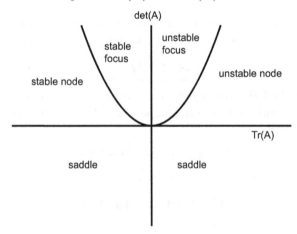

2.4 Linear systems in $\mathbb{R}^n$

Consider the differential equation

$$\dot{x} = Ax, \qquad x \in \mathbb{R}^n,$$

where A is an $n \times n$ constant matrix. Given the initial condition $x(0) = x_0$, the solution is

$$x(t) = e^{tA} x_0, \quad e^{tA} = \sum_{k=0}^{\infty} \frac{t^k}{k!} A^k. \tag{2.1}$$

To check this, use

$$\frac{d}{dt} e^{tA} = \sum_{k=1}^{\infty} \frac{t^{k-1}}{(k-1)!} A^k = A e^{tA}.$$

Thus

$$\frac{dx(t)}{dt} = \frac{d}{dt} e^{tA} x_0 = A e^{tA} x_0 = A x(t)$$

The solution (2.1) also allows one to solve the inhomogeneous equation

$$\dot{x} = A x + g(t).$$

Multiplying both sides by e^{-tA} gives

$$\frac{d}{dt} \left[e^{-tA} x(t) \right] = e^{-tA} g(t).$$

Integrating with respect to t then gives

$$e^{-tA} x(t) - x_0 = \int_0^t e^{-t' A} g(t') dt'$$

or

$$x(t) = e^{tA} x_0 + e^{tA} \int_0^t e^{-t' A} g(t') dt'.$$

2.4.1 Normal forms

Consider the linear change of variables $x = P y$, where P is an $n \times n$ invertible matrix $(\det(P) \neq 0)$. Then if $\dot{x} = A x$

$$\dot{y} = P^{-1} \dot{x} = P^{-1} A x = P^{-1} A P y.$$

Choosing P such that $\Lambda = P^{-1} A P$ is a diagonal matrix we have that

$$\dot{y} = \Lambda y.$$

If $x(0) = x_0$ then $y(0) = P^{-1} x_0$. In the new coordinates the solution is

$$y(t) = e^{t\Lambda} y_0.$$

Transforming back to original coordinates:

$$x(t) = P y(t) = P e^{t\Lambda} y_0 = P e^{t\Lambda} P^{-1} x_0. \tag{2.2}$$

Comparing equations (2.1) and (2.2) implies that

$$e^{tA} = Pe^{t\Lambda}P^{-1}.$$

Strategy: choose a matrix P such that Λ takes a form which allows us to calculate $e^{t\Lambda}$ and hence e^{tA}. The matrix Λ is then called a Normal Form whose particular structure depends on the eigenvalues of A.

Real distinct eigenvalues

Suppose that A has n distinct eigenvalues $\lambda_1, \ldots, \lambda_n$ with corresponding eigenvectors e_i so that

$$Ae_i = \lambda_i e_i.$$

Let $P = [e_1, \ldots, e_n]$ be the matrix with the eigenvectors of A as columns. Since the eigenvectors are real and linearly-independent, $\det(P) \neq 0$. Thus

$$AP = [Ae_1, \ldots, Ae_n] = [\lambda_1 e_1, \ldots, \lambda_n e_n] = [e_1, \ldots, e_n]\text{diag}(\lambda_1, \ldots, \lambda_n)$$
$$= P\text{diag}(\lambda_1, \ldots, \lambda_n).$$

Hence for real distinct eigenvalues $\Lambda = \text{diag}(\lambda_1, \ldots, \lambda_n)$. It follows that

$$e^{tA} = P\text{diag}(e^{\lambda_1 t}, \ldots, e^{\lambda_n t})P^{-1}.$$

Example 2.4.1 $A = \begin{pmatrix} -2 & 1 \\ 0 & 2 \end{pmatrix}$.

Characteristic equation is $\det(A - \lambda I_2) = 0 \Rightarrow (\lambda + 2)(\lambda - 2) = 0$. Then

$$\lambda_1 = -2, \quad e_1 = \begin{pmatrix} 1 \\ 0 \end{pmatrix}, \quad \lambda_2 = 2, \quad e_2 = \begin{pmatrix} 1 \\ 4 \end{pmatrix},$$

$$P = \begin{pmatrix} 1 & 1 \\ 0 & 4 \end{pmatrix}, \quad P^{-1} = \frac{1}{4}\begin{pmatrix} 4 & -1 \\ 0 & 1 \end{pmatrix},$$

and

$$e^{tA} = P\begin{pmatrix} e^{-2t} & 0 \\ 0 & e^{2t} \end{pmatrix}P^{-1} = \begin{pmatrix} e^{-2t} & \frac{1}{4}(e^{2t} - e^{-2t}) \\ 0 & e^{2t} \end{pmatrix}.$$

Pair of complex eigenvalues

Consider a 2×2 matrix with a pair of complex eigenvalues $\rho \pm i\omega$. The associated complex eigenvector is q such that

$$Aq = (\rho + i\omega)q, \quad q \in \mathbb{C}^2.$$

Let $q = u + iv$, where $u, v \in \mathbb{R}^2$, and equate real and imaginary parts:

$$Au = \rho u - \omega v,$$
$$Av = \omega u + \rho v,$$

or

$$A[v, u] = [v, u] \begin{pmatrix} \rho & -\omega \\ \omega & \rho \end{pmatrix}.$$

Hence, set

$$P = [v, u] = [\mathrm{Im}(q), \mathrm{Re}(q)], \quad \Lambda = \begin{pmatrix} \rho & -\omega \\ \omega & \rho \end{pmatrix}$$

to see that

$$AP = P\Lambda, \quad \text{or} \quad \Lambda = P^{-1}AP.$$

Having obtained the normal form, we need to solve the system of equations

$$\dot{x} = \rho x - \omega y, \quad \dot{y} = \omega x + \rho y, \quad x, y \in \mathbb{R}.$$

Let $z = x + iy$. Then

$$\dot{z} = \dot{x} + i\dot{y} = (\rho + i\omega)z. \tag{2.3}$$

Introduce polar coordinates $z = re^{i\theta}$ ($x = r\cos\theta, y = r\sin\theta$). Then an equivalent form for $\dot{z}$ is

$$\dot{z} = \dot{r}e^{i\theta} + ir\dot{\theta}e^{i\theta}. \tag{2.4}$$

Comparing equations (2.3) and (2.4) we deduce that

$$\dot{r} + ir\dot{\theta} = (\rho + i\omega)r$$

which, on equating real and imaginary parts, yields

$$\dot{r} = \rho r, \quad \dot{\theta} = \omega.$$

Hence, we obtain the solution

$$r(t) = e^{\rho t}r_0, \quad \theta(t) = \omega t + \theta_0.$$

After writing $x(t) = r(t)\cos(\omega t + \theta_0)$ and $y(t) = r(t)\sin(\omega t + \theta_0)$ with $x_0 = r_0\cos\theta_0$ and $y_0 = r_0\sin\theta_0$, it follows that

$$\begin{pmatrix} x(t) \\ y(t) \end{pmatrix} = e^{\rho t} \begin{pmatrix} \cos\omega t & -\sin\omega t \\ \sin\omega t & \cos\omega t \end{pmatrix} \begin{pmatrix} x_0 \\ y_0 \end{pmatrix}.$$

Therefore, stability depends on the real part of the eigenvalue $\rho \pm iw$, i.e. $\mathrm{Re}(\rho \pm iw) = \rho$.

Example 2.4.2 $A = \begin{pmatrix} 2 & 1 \\ -2 & 0 \end{pmatrix}$.

Characteristic equation is $\det(A - \lambda I_2) = 0 \Rightarrow (\lambda - 2)\lambda + 2 = 0$. Then

$$\lambda = 1 + i, \quad q = \begin{pmatrix} 1 \\ -1 + i \end{pmatrix}, \quad \mathrm{Im}(q) = \begin{pmatrix} 0 \\ 1 \end{pmatrix}, \quad \mathrm{Re}(q) = \begin{pmatrix} 1 \\ -1 \end{pmatrix},$$

$$P = \begin{pmatrix} 0 & 1 \\ 1 & -1 \end{pmatrix}, \quad P^{-1} = \begin{pmatrix} 1 & 1 \\ 1 & 0 \end{pmatrix},$$

and

$$e^{tA} = e^t P \begin{pmatrix} \cos t & -\sin t \\ \sin t & \cos t \end{pmatrix} P^{-1} = e^t \begin{pmatrix} \cos t + \sin t & \sin t \\ -2\sin t & \cos t - \sin t \end{pmatrix}.$$

Degenerate eigenvalues

Suppose that A has p distinct eigenvalues $\lambda_1, \ldots, \lambda_p$, $p \leq n$. Then

$$\det(A - \lambda I_n) = \prod_{k=1}^{p} (\lambda - \lambda_k)^{n_k},$$

where $n_k \geq 1$ and $\sum_{k=1}^{p} n_k = n$. If all the eigenvectors are distinct then $p = n$ and $n_k = 1$ for all k. If $p < n$ then at least one $n_k > 1$ and the characteristic polynomial has repeated roots. The number n_k is called the multiplicity of λ_k.

Consider the two-dimensional case. Recall the Cayley–Hamilton theorem: the matrix A satisfies its own characteristic equation. Therefore, $(A - \lambda I_2)^2 x = 0$ for all $x \in \mathbb{R}^2$. There are then two possibilities:

1. $(A - \lambda I_2)x = 0$ for all $x \in \mathbb{R}^2 \Rightarrow \Lambda = \begin{pmatrix} \lambda & 0 \\ 0 & \lambda \end{pmatrix}$.

2. $(A - \lambda I_2)e_2 \neq 0$ for some vector $e_2 \neq 0$. Define $e_1 = (A - \lambda I_2)e_2$. Then $(A - \lambda I_2)e_1 = 0$ so that

$$Ae_1 = \lambda e_1, \quad Ae_2 = e_1 + \lambda e_2 \Rightarrow A[e_1, e_2] = [e_1, e_2] \begin{pmatrix} \lambda & 1 \\ 0 & \lambda \end{pmatrix}.$$

Hence, we may set

$$P = [e_1, e_2], \quad \Lambda = \begin{pmatrix} \lambda & 1 \\ 0 & \lambda \end{pmatrix}.$$

Consider the following system given in its normal form:

$$\dot{x} = \lambda x + y, \quad \dot{y} = \lambda y.$$

It can be solved as

$$x(t) = e^{\lambda t}(x_0 + ty_0), \quad y(t) = e^{\lambda t}y_0.$$

Here

$$\frac{dy}{dx} = \frac{y}{\lambda x + y}$$

and for $\lambda = 1$ the phase portrait is shown below.

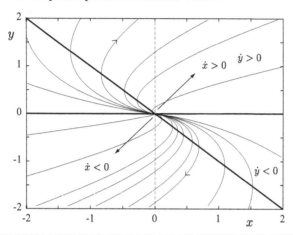

Solving linear systems

- Real eigenvalue λ $\Rightarrow$ $Ce^{\lambda t}$
- Real eigenvalue λ of multiplicity r $\Rightarrow$ $C_1 e^{\lambda t} + C_2 t e^{\lambda t} + \cdots + C_r t^{r-1} e^{\lambda t}$
- Pair of complex eigenvalues $\lambda = \rho \pm i\omega$ $\Rightarrow$ $e^{\rho t}(B \cos \omega t + C \sin \omega t)$
- Pair of complex eigenvalues $\lambda = \rho \pm i\omega$, each with multiplicity r $\Rightarrow$
  $e^{\rho t}(B_1 \cos \omega t + C_1 \sin \omega t + B_2 t \cos \omega t + C_2 t \sin \omega t + \cdots + B_r t^{r-1} \cos \omega t + C_r t^{r-1} \sin \omega t)$

2.5 Nonlinear systems in $\mathbb{R}^2$ (in $\mathbb{R}^n$)

We shall consider equations of the form

$$\dot{x}_1 = f_1(x_1, x_2),$$
$$\dot{x}_2 = f_2(x_1, x_2).$$

This system can be written in vector notation as

$$\dot{\mathbf{x}} = \mathbf{f}(\mathbf{x}),$$

where $\mathbf{x} = (x_1, x_2)$, $\mathbf{f}(\mathbf{x}) = (f_1(\mathbf{x}), f_2(\mathbf{x}))$. $\mathbf{x}$ represents a point in the phase plane, and $\dot{\mathbf{x}}$ is the velocity vector at that point.

Theorem 2.5.1 *Existence and uniqueness theorem (in $\mathbb{R}^n$):*
Suppose $\dot{x} = f(x)$ and $f : \mathbb{R}^n \to \mathbb{R}^n$ is continuously differentiable (i.e.
$\partial f_i / \partial x_j$, i, $j = 1, \ldots, n$ exist and are continuous for all x). Then there
exist $t_1 > 0$ and $t_2 > 0$ such that the solution with $x(t_0) = x_0$ exists and
is unique for all $t \in (t_0 - t_1, t_0 + t_2)$.

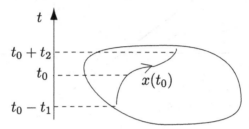

Phase-space and flows We refer to the local solution through x_0 as a *solution curve* or *trajectory*. Suppose that $\dot{x} = f(x)$, $x \in \mathbb{R}^n$, $f : \mathbb{R}^n \to \mathbb{R}^n$. We define a flow $\phi(x, t)$ such that $\phi(x, t)$ is the solution of the ODE at time t with initial value x_0 at $t = 0$. The solution $x(t)$ with $x(0) = x_0$ is now written as $\phi(x_0, t)$:

$$\frac{d\phi(x, t)}{dt} = f(\phi(x, t)), \qquad \phi(x, 0) = x_0.$$

1. $\phi(x, 0) = x_0$
2. $\phi(x, t + s) = \phi(\phi(x, t), s) = \phi(\phi(x, s), t)$

Then by varying initial condition x_0 we generate a family of trajectories called the *flow* generated by Φ.

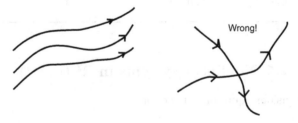

Note that uniqueness imples that trajectories cannot cross. An **equilibrium** or fixed point satisfies $\Phi(x, t) = x$ for all t. Thus $f(x) = 0$. An important feature of nonlinearities is that there can exist more than one (isolated) fixed point.

2.5.1 Stability

A fixed point x_0 is an attracting fixed point if all trajectories that start near x_0 approach it as $t \to \infty$. If x_0 attracts all trajectories it is called globally attracting.

A fixed point x_0 is **Lyapunov** (neutrally) stable if for all $\epsilon > 0$ there exists $\delta > 0$ such that $|x(0) - x_0| < \delta$ implies that $|x(t) - x_0| < \epsilon$ for all $t > 0$.

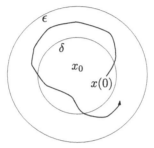

In other words, if a solution starts near an equilibrium x_0 then it stays near x_0 (for example harmonic oscillator).

A fixed point is **asymptotically stable** if it is Lyapunov stable and there exists $\delta > 0$ such that if $|x(0) - x_0| < \delta$ then $|x(t) - x_0| \to 0$ as $t \to \infty$.

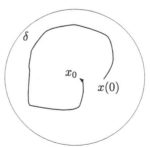

2.5.2 Linearisation

Consider the system

$$\dot{x} = f(x, y),$$
$$\dot{y} = g(x, y),$$

and suppose that (x^*, y^*) is a fixed point. Considering a small distur-
bance from the fixed point

$$u = x - x^*, \qquad v = y - y^*$$

we have (by Taylor series expansion)

$$\dot{u} = \dot{x} = f(u + x^*, v + y^*)$$

$$= f(x^*, y^*) + \left.\frac{\partial f}{\partial x}\right|_{(x^*, y^*)} \cdot u \; + \left.\frac{\partial f}{\partial y}\right|_{(x^*, y^*)} \cdot v + O(u^2, v^2, uv).$$

This leads to

$$\dot{u} = \left.\frac{\partial f}{\partial x}\right|_{(x^*, y^*)} \cdot u + \left.\frac{\partial f}{\partial y}\right|_{(x^*, y^*)} \cdot v + O(u^2, v^2, uv)$$

and similarly

$$\dot{v} = \left.\frac{\partial g}{\partial x}\right|_{(x^*, y^*)} \cdot u + \left.\frac{\partial g}{\partial y}\right|_{(x^*, y^*)} \cdot v + O(u^2, v^2, uv).$$

Hence

$$\begin{pmatrix} \dot{u} \\ \dot{v} \end{pmatrix} = A \begin{pmatrix} u \\ v \end{pmatrix} \quad - \text{ the linearised system,}$$

with

$$A = \begin{pmatrix} \frac{\partial f}{\partial x} & \frac{\partial f}{\partial y} \\ \frac{\partial g}{\partial x} & \frac{\partial g}{\partial y} \end{pmatrix}_{(x^*, y^*)} \quad - \text{ the } \mathbf{Jacobian\ matrix}.$$

Theorem 2.5.2 *Linear stability: Suppose that $\dot{x} = f(x)$ has an equilibrium at $x = 0$ and the linearisation $\dot{x} = Ax$. If A has no zero or purely
imaginary eigenvalues then the local stability of the fixed point (which is
called* **hyperbolic** *in this case) is determined by the linear system. In
particular, if all eigenvalues have a negative real part Re $(\lambda_i) < 0$ for all
$i = 1, \dots, n$ then the fixed point is asymptotically stable.*

Theorem 2.5.3 *Hartman–Grobman theorem: The local phase-
portrait near a hyperbolic fixed point is topologically equivalent to the
phase-portrait of the linearisation.*

Theorem 2.5.4 *Structural stability: A phase portrait is structurally
stable if its topology cannot be changed by an arbitrarily small perturbation to the vector field, i.e. a system is structurally stable if it is topologically equivalent to any ϵ-perturbation*

$$\dot{x} = f(x) + \epsilon p(x),$$

where $\epsilon \ll 1$ and p is smooth enough.

For example, the phase portrait of a saddle is structurally stable, but that of a centre is not: an arbitrarily small amount of damping converts the center to a spiral.

Exercise Consider the system

$$\dot{x} = -y + ax(x^2 + y^2),$$
$$\dot{y} = x + ay(x^2 + y^2),$$

where a is a parameter. Show that the linearised system incorrectly predicts that the origin is a centre for all values a. (Hint: rewite the system in polar coordinates $x = r\cos\theta$, $y = r\sin\theta$)

Example 2.5.5 The Lotka–Volterra model of population dynamics of two competing species. Assume (i) each species grows in the absence of the other with logistic growth ($\dot{x} = x(1 - x)$) and (ii) when both species are present they compete for food. Consider a particular model of rabbits (r) and sheep (s):

$$\dot{r} = r(3 - r - 2s) \equiv f(r, s),$$
$$\dot{s} = s(2 - r - s) \equiv g(r, s).$$

Fixed points are defined by $\dot{r} = \dot{s} = 0$. One finds $(\bar{r}, \bar{s}) = (0,0), (0,2), (3,0),$ $(1,1)$. To classify them we compute

$$A = \begin{bmatrix} \frac{\partial f}{\partial r} & \frac{\partial f}{\partial s} \\ \frac{\partial g}{\partial r} & \frac{\partial g}{\partial s} \end{bmatrix} = \begin{bmatrix} 3 - 2r - 2s & -2r \\ -s & 2 - r - 2s \end{bmatrix}$$

1. $(\bar{r}, \bar{s}) = (0,0)$

$$A = \begin{bmatrix} 3 & 0 \\ 0 & 2 \end{bmatrix}$$

The eigenvalues are both positive so $(0,0)$ is an unstable node. Trajectories leave the origin parallel to the eigenvector for $\lambda = 2$, i.e. tangential to $(0,1)$.

2. $(\bar{r}, \bar{s}) = (0,2)$

$$A = \begin{bmatrix} -1 & 0 \\ -2 & -2 \end{bmatrix}, \qquad \Lambda = \begin{bmatrix} -1 & 0 \\ 0 & -2 \end{bmatrix}$$

Hence $(0,2)$ is a stable node. Slow eigendirection is $(1,-2)$.

3. $(\bar{r}, \bar{s}) = (3, 0)$

$$A = \begin{bmatrix} -3 & -6 \\ 0 & -1 \end{bmatrix}, \qquad \Lambda = \begin{bmatrix} -3 & 0 \\ 0 & -1 \end{bmatrix}$$

Hence $(3, 0)$ is a stable node. Slow eigendirection is $(3, -1)$.

4. $(\bar{r}, \bar{s}) = (1, 1)$

$$A = \begin{bmatrix} -1 & -2 \\ 1 & -1 \end{bmatrix}, \qquad \Lambda = \begin{bmatrix} -1 + \sqrt{2} & 0 \\ 0 & -1 - \sqrt{2} \end{bmatrix}$$

Hence, $(1, 1)$ is a saddle.

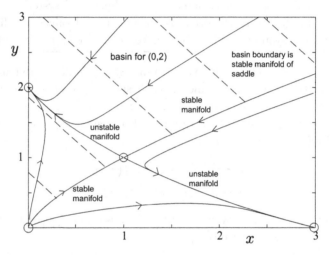

The above example nicely illustrates the notion of a **basin of attraction**. Given an attracting fixed point $\bar{x}$ we define its basin of attraction to be the set of initial conditions x_0 such that $x(t) \to \bar{x}$ as $t \to \infty$. For instance the basin of attraction for the node at $(3, 0)$ consists of all points lying below the stable manifold of the saddle. Because the stable manifold separates the basins of two nodes, it is called the **basin boundary**.

Theorem 2.5.6 *Lyapunov theorem: Suppose that x^* is a fixed point for the differential equation $\dot{x} = f(x)$, $x \in \mathbb{R}^n$. Then x^* is Lyapunov stable if there exists a (continuously differentiable) function $L(x)$ (called a Lyapunov function) with the following properties in some neighbourhood of x^*:*

1. *$L(x)$ and its partial derivatives are continuous;*
2. *$L(x) > 0$ for all $x \neq x^*$ and $L(x^*) = 0$;*

3. $\dot{L} \leq 0$ *for all* $x \neq x^*$.

Note that $\dot{L}$ is determined by the chain-rule

$$\dot{L} = \sum_i \frac{\partial L}{\partial x_i} \dot{x}_i = \sum_i \frac{\partial L}{\partial x_i} f(x_i).$$

Example 2.5.7 Show that $L(x,y) = x^2 + 4y^2$ is a Lyapunov function for

$$\dot{x} = -x + 4y, \qquad \dot{y} = -x - y^3.$$

The fixed point is at $(0,0)$.

1. $L(x,y)$ is continuously differentiable.
2. $L(x,y) > 0$, $L(0,0) = 0$.
3.

$$\dot{L} = \frac{\partial L}{\partial x}\dot{x} + \frac{\partial L}{\partial y}\dot{y} = -2x^2 - 8y^4 < 0.$$

Hence $L(x,y)$ is a Lyapunov function.

Sufficiently close to the fixed point, L forms a bowl and L decreases along trajectories.

2.6 Nonlinear oscillations

2.6.1 Limit cycles in $\mathbb{R}^2$

A limit cycle is an isolated closed trajectory (see diagram below).

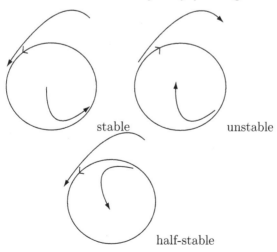

stable   unstable

half-stable

Limit cycles are often found in models that exhibit self-sustained oscillations. There are many example throughout the applied sciences: the beating of a heart, chemical reactions, daily (circadian) rhythms in human body temperature and hormone secretion, dangerous self-excited oscillations in bridges (Takoma Narrows) and airplane wings. Limit cycles are inherently a nonlinear phenomenon – a linear system can have closed orbits but they are not isolated.

Example 2.6.1 Consider the following two-dimensional system in polar coordinates:

$$\dot{r} = r(K - r^2), \qquad \dot{\theta} = 1.$$

For $K > 0$ there exists an unstable fixed point at the origin and a stable limit cycle at $r = \sqrt{K}$.

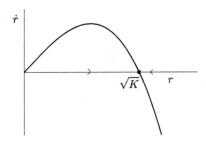

For $K < 0$ there is no limit cycle but only a stable fixed point at the origin. We can represent qualitative change in the dynamics as the parameter K varies using a **bifurcation diagram**.

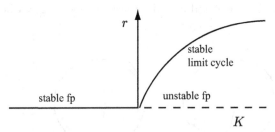

This is an example of a **super-critical Hopf bifurcation**. Since an arbitrarily small perturbation of the origin will produce a self-sustained oscillation the system is said to exhibit a **soft excitation**.

Example 2.6.2 Van der Pol oscillator

$$\ddot{x} + \mu(x^2 - 1)\dot{x} + x = 0,$$

where $\mu \geq 0$ is a parameter. Historically, this equation arose in connection with the nonlinear electrical circuits used in the first radios. This equation looks like a simple harmonic oscillator, but with a nonlinear damping term $\mu(x^2 - 1)\dot{x}$. This term acts like ordinary positive damping for $|x| > 1$; however, like negative damping for $|x| < 1 \Rightarrow$ it causes large-amplitude oscillations to decay, but it pumps them back up if they become too small.

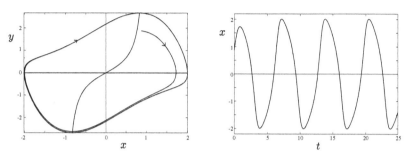

The Van der Pol equation has a unique, stable limit cycle for each $\mu > 0$.

Example 2.6.3 Consider the following example:

$$\dot{r} = \mu r + r^3 - r^5, \qquad \dot{\theta} = 1.$$

For $\mu < 0$ there exists a stable fixed point at the origin, an unstable limit cycle at $r = r_-$ and a stable limit cycle at $r = r_+$, where

$$r_\pm^2 = \frac{1}{2}\left[1 \pm \sqrt{1 + 4\mu}\right].$$

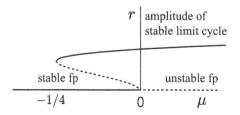

For $\mu > 0$ there exists an unstable fixed point at the origin and a stable limit cycle at $r = r_+$.

At $\mu = 0$ there is a **sub-critical Hopf bifurcation**. The system also exhibits hysteresis: once large amplitude oscillations have begun they cannot be turned off by bringing μ back to zero. These oscillations persist until $\mu_c = -1/4$ where the stable and unstable cycles collide and annihilate in a **saddle-node bifurcation of limit cycles**. When the system passes through the sub-critical Hopf bifurcation it jumps from the fixed point to a large amplitude oscillation (termed a **hard excitation**).

2.6.2 Poincaré—Bendixson theorem

It is generally difficult to establish the existence of a limit cycle. In two-dimensions one has the following useful theorem:

Theorem 2.6.4 *Suppose that there exists a bounded region D of phase-space such that any trajectory entering D cannot leave D. If there are no fixed points in D then there exists at least one periodic orbit in D.*

Typically, D will be an annular region with an unstable focus or node in the hole in the middle (so trajectories enter the inner boundary) and all trajectories cross the outer boundary inwards.

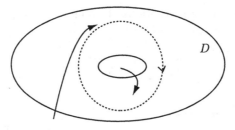

The standard trick to apply the Poincaré–Bendixson theorem is to construct a *trapping region* D, i.e. a closed connected set such that the vector field points inward everywhere on the boundary of D.

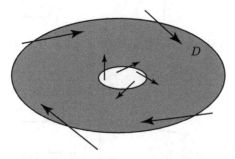

The Poincaré–Bendixson theorem tells us that the dynamics of planar systems is severely limited – if a trajectory is confined to a closed, bounded region that contains no fixed points, then the trajectory must eventually approach a closed orbit. There is no chaos for planar systems!

In higher-dimensional systems (in $\mathbb{R}^n$, $n \geq 3$) the Poincaré-Bendixson theorem no longer applies and trajectories may wander around forever in a bounded region without settling down to a fixed point or a closed orbit. In some cases, the trajectories are attracted to a complex geometric object called a *strange attractor*.

Example 2.6.5 In a fundamental biochemical process called glycolysis, living cells obtain energy by breaking down sugar. In yeast cells, for example, glycolysis can proceed in an oscillatory fashion, with concentrations of intermediate products varying periodically. A model of this process is given by

$$\dot{x} = -x + ay + x^2y \equiv f(x,y),$$
$$\dot{y} = b - ay - x^2y \equiv g(x,y),$$

where x and y are concentrations of ADP (adenosine phosphate) and F(6)P (Fructose-6 phosphate) and $a, b > 0$ are kinetic parameters. Construct a trapping region for this system.

First we find the nullclines ($f(x,y) = 0 = g(x,y)$) (see figure) and then

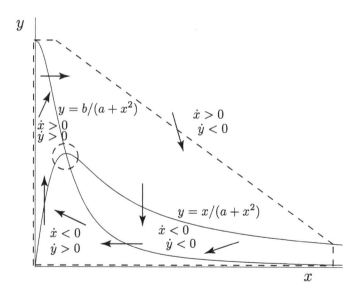

show that all trajectories are *inwards* in some region. To construct the
bounding region consider large x and y. Then $\dot{x} \approx x^2 y$ and $\dot{y} \approx -x^2 y$,
so $dy/dx \approx -1$ along trajectories. Hence, the vector field at large x is
parallel to the diagonal, which suggests comparing the sizes of $\dot{x}$ and
$-\dot{y}$. So, consider

$$\dot{x} - (-\dot{y}) = -x + ay + x^2 y + (b - ay - x^2 y) = b - x.$$

Hence $-\dot{y} > \dot{x}$ if $x > b$. This implies that the vector field points inward
on the diagonal line (of the above figure) because dy/dx is more negative
than -1 and therefore the vectors are steeper than the diagonal – we
have a trapping region! We must now find under those conditions which
make the fixed point unstable (so as to repel orbits). Linearisation:

$$A = \begin{bmatrix} -1 + 2xy & a + x^2 \\ -2xy & -(a + x^2) \end{bmatrix}.$$

Fixed point

$$\bar{x} = b, \qquad \bar{y} = \frac{b}{a + b^2}.$$

Determinant $\det(A) = a + b^2 > 0$ and

$$\text{Tr}(A) = -\frac{b^4 + (2a - 1)b^2 + (a + a^2)}{a + b^2}.$$

Hence the fixed point is unstable for $\text{Tr}(A) > 0$ and stable for $\text{Tr}(A) < 0$.
The border of stability $\text{Tr}(A) = 0$ occurs when

$$b^2 = \frac{1}{2}\left[1 - 2a \pm \sqrt{1 - 8a}\right].$$

Numerical integration shows that there is one stable limit cycle in the
parameter regime, which guarantees an unstable fixed point.

2.6.3 Relaxation oscillators

Consider the Van der Pol equation

$$\ddot{x} + \mu(x^2 - 1)\dot{x} + x = 0$$

for the special case that $\mu \gg 1$ (the strongly nonlinear limit). Using

$$\ddot{x} + \mu(x^2 - 1)\dot{x} = \frac{d}{dt}\left[\dot{x} + \mu(x^3/3 - x)\right]$$

and introducing

$$F(x) = \frac{x^3}{3} - x, \qquad w = \dot{x} + \mu F(x)$$

we may write

$$\dot{w} = \ddot{x} + \mu\dot{x}(x^2 - 1) = -x.$$

Hence, the Van der Pol system has a planar representation:

$$\dot{x} = w - \mu F(x),$$
$$\dot{w} = -x.$$

With the re-scaling $y = w/\mu$ we have

$$\dot{x} = \mu[y - F(x)],$$
$$\dot{y} = -\frac{1}{\mu}x.$$

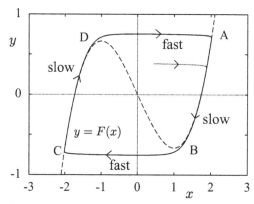

Suppose that the initial condition is not too close to the cubic nullcline, i.e. $y - F(x) \sim O(1)$. Then $|\dot{x}| \sim O(\mu) \gg 1$ and $|\dot{y}| \sim O(1/\mu) \ll 1$; hence the velocity is large in the horizontal direction and small in the vertical direction, so trajectories move horizontally. If the initial condition is above the cubic nullcline then $y - F(x) > 0$, so $\dot{x} > 0$; the trajectory moves sideways towards the right-hand branch of the nullcline. Once the trajectory gets so close that $y - F(x) \sim O(1/\mu^2)$, then $\dot{x}$ and $\dot{y}$ become comparable (both being $O(1/\mu)$). The trajectory crosses the nullcline vertically (see the figure on the previous page) and then moves slowly down the branch with a velocity $O(1/\mu)$, until it reaches the knee and can jump sideways.

The system has two widely separated time-scales. The jumps take a time $O(1/\mu)$ and the crawls a time $O(\mu)$. The period of oscillation can be approximated by the time spent on the slow branches:

$$T \approx \int_{t_A}^{t_B} \mathrm{d}t + \int_{t_C}^{t_D} \mathrm{d}t = 2 \int_{t_A}^{t_B} \mathrm{d}t \quad \text{by symmetry.}$$

On the slow branch $y = F(x)$, so

$$\dot{y} \approx \frac{\mathrm{d}y}{\mathrm{d}x}\dot{x} = F'(x)\dot{x} = (x^2 - 1)\dot{x}.$$

Using $\dot{y} = -x/\mu$ we have that $\dot{x} = -x/[\mu(x^2 - 1)]$, so

$$\mathrm{d}t \approx -\frac{\mu(x^2 - 1)}{x}\mathrm{d}x.$$

Now $x_A = 2$ and $x_B = 1$ (check this), so

$$T \approx 2 \int_2^1 -\frac{\mu}{x}(x^2 - 1)\mathrm{d}x = \mu[3 - 2\ln 2].$$

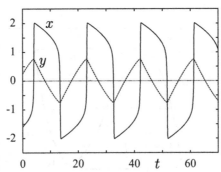

2.7 Perturbation methods

In this section we learn how to deal with weakly nonlinear oscillators of the form

$$\ddot{x} + x + \epsilon h(x, \dot{x}) = 0.$$

Two fundamental examples are the Van der Pol oscillator

$$\ddot{x} + x + \epsilon(x^2 - 1)\dot{x} = 0$$

and the Duffing equation

$$\ddot{x} + x + \epsilon x^3 = 0.$$

2.7.1 Regular perturbation theory and its failures

As a first approach we seek solutions to the equations of a weakly nonlinear oscillator in the form of a power series in ϵ. If $x(t, \epsilon)$ is a solution then we write

$$\boxed{x(t, \epsilon) = x_0(t) + \epsilon x_1(t) + \epsilon^2 x_2(t) + \cdots}$$

Consider as a simple example the weakly damped linear oscillator (exactly solvable)

$$\ddot{x} + 2\epsilon\dot{x} + x = 0, \qquad x(0) = 0, \quad \dot{x}(0) = 1.$$

The exact solution is

$$x(t, \epsilon) = (1 - \epsilon^2)^{-1/2}e^{-\epsilon t}\sin[(1 - \epsilon^2)^{1/2}t].$$

Using perturbation theory:

$$\left\{\frac{d^2}{dt^2} + 2\epsilon\frac{d}{dt} + 1\right\}(x_0 + \epsilon x_1 + \cdots) = 0.$$

Equating powers of ϵ gives

$$[\ddot{x}_0 + x_0] + \epsilon[\ddot{x}_1 + 2\dot{x}_0 + x_1] + O(\epsilon^2) = 0.$$

Since this holds for all sufficiently small ϵ the coefficients of each power of ϵ must vanish separately:

$$O(1) \qquad \ddot{x}_0 + x_0 = 0 \qquad\qquad x_0(0) = 0, \quad \dot{x}_0(0) = 1,$$
$$O(\epsilon) \qquad \ddot{x}_1 + 2\dot{x}_0 + x_1 = 0 \qquad x_1(0) = 0, \quad \dot{x}_1(0) = 0.$$

Solve the initial value problems in sequence. First

$$x_0(t) = \sin t.$$

Substituting into the equation for x_1 gives

$$\ddot{x}_1 + x_1 = -2\cos t$$

which has the solution $x_1(t) = -t \sin t$. This is called a *secular* term since it grows without bound as $t \to \infty$. Thus, according to perturbation theory

$$x(t, \epsilon) = \sin t - \epsilon t \sin t + O(\epsilon^2).$$

For fixed t this provides a convergent series expansion of the exact solution if ϵ is small enough ($\epsilon t \ll 1$). However, we are usually interested in the behaviour for fixed ϵ not fixed t. In that case we can only expect the perturbation approximation to work for $t \ll O(1/\epsilon)$. There are two major problems:

- The true solution exhibits two time-scales: a fast time $t \sim O(1)$ for the sinusoidal oscillation and a slow time $t \sim 1/\epsilon$ over which the amplitude decays. The perturbation approximation completely misrepresents the slow time-scale behaviour ($e^{-\epsilon t} \approx 1 - \epsilon t + O(\epsilon^2 t^2)$).
- The frequency of oscillations of the exact solution is $\omega = (1 - \epsilon^2)^{1/2} \approx 1 - \epsilon^2/2$, which is shifted slightly from the frequency $\omega = 1$ of the perturbation approximation. After a very long time $t \sim O(1/\epsilon^2)$ this frequency error will have a cumulative effect (super slow time-scale).

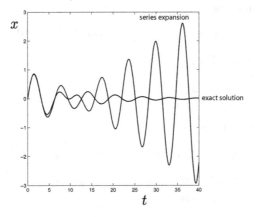

We would like an approximation that captures the behaviour for all t, or at least for large t.

2.7.2 Method of multiple scales (two-timing)

The previous example suggests that we should consider the series expansion

$$x(t, \epsilon) = x_0(\tau, T) + \epsilon x_1(\tau, T) + O(\epsilon^2),$$

where $\tau = t$ denotes the fast $O(1)$ time and $T = \epsilon t$ denotes the slow time. We treat τ and T as independent variables. Then the chain rule implies that

$$\dot{x} = \partial_\tau x + \epsilon \partial_T x.$$

Substituting the series solution and collecting powers of ϵ, we find

$$\dot{x} = \partial_\tau x_0 + \epsilon(\partial_T x_0 + \partial_\tau x_1) + O(\epsilon^2),$$
$$\ddot{x} = \partial_{\tau\tau} x_0 + \epsilon(\partial_{\tau\tau} x_1 + 2\partial_{T\tau} x_0) + O(\epsilon^2).$$

Example 2.7.1 Consider the weakly damped linear oscillator

$$\ddot{x} + 2\epsilon\dot{x} + x = 0.$$

Assuming the two-time expansion we have

$$\partial_{\tau\tau} x_0 + \epsilon(\partial_{\tau\tau} + 2\partial_{T\tau} x_0) + 2\epsilon\partial_\tau x_0 + x_0 + \epsilon x_1 + O(\epsilon^2) = 0.$$

Equating powers of ϵ yields

$$O(1) \quad \partial_{\tau\tau} x_0 + x_0 = 0,$$
$$O(\epsilon) \quad \partial_{\tau\tau} x_1 + 2\partial_{T\tau} x_0 + 2\partial_\tau x_0 + x_1 = 0.$$

The general solution for $x_0(\tau, T)$ is

$$x_0(\tau, T) = A(T)\sin\tau + B(T)\cos\tau.$$

Substituting into expression for x_1 gives

$$\partial_{\tau\tau} x_1 + x_1 = -2(\partial_{T\tau} x_0 + \partial_\tau x_0)$$
$$= -2(A' + A)\cos\tau + 2(B' + B)\sin\tau,$$

where $' \equiv d/dT$. Since we want to have an approximation free of secular terms we set the coefficients of the resonant terms to zero

$$A' + A = 0, \qquad B' + B = 0$$

so that

$$A(T) = A(0)e^{-T}, \qquad B(T) = B(0)e^{-T}.$$

Initial conditions determine $A(0)$ and $B(0)$ ($x(0) = 0$ and $\dot{x}(0) = 1$):

$$x(0) = x_0(0,0) + \epsilon x_1(0,0) + O(\epsilon^2) = 0,$$
$$\dot{x}(0) = \partial_\tau x_0(0,0) + \epsilon(\partial_T x_0(0,0) + \partial_\tau x_1(0,0)) + O(\epsilon^2) = 1,$$

i.e.

$$x_0(0,0) = 0, \qquad x_1(0,0) = 0,$$
$$\partial_\tau x_0(0,0) = 1, \qquad \partial_\tau x_1(0,0) + \partial_T x_0(0,0) = 0.$$

Therefore $A(0) = 1$ and $B(0) = 0$, so

$$x(t,\epsilon) = e^{-T} \sin \tau + O(\epsilon), \qquad x_0(t) = e^{-\epsilon t} \sin t + O(\epsilon).$$

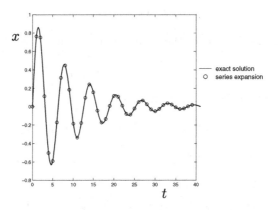

Exercise Use the method of multiple scales to show that the Van der Pol oscillator

$$\ddot{x} + x + \epsilon(x^2 - 1)\dot{x} = 0$$

has a stable limit cycle that is nearly circular with a radius $2 + O(\epsilon)$ and a frequency $\omega = 1 + O(\epsilon^2)$.

2.8 Coupled oscillators

Consider the model

$$\dot{\theta}_1 = \omega_1 + K_1 \sin(\theta_2 - \theta_1),$$
$$\dot{\theta}_2 = \omega_2 + K_2 \sin(\theta_1 - \theta_2).$$

In the uncoupled state ($K_1 = K_2 = 0$) we have $\theta_1(t) = \theta_1(0) + \omega_1 t$ and $\theta_2(t) = \theta_2(0) + \omega_2 t$ such that $d\theta_2/d\theta_1 = \omega_2/\omega_1$. If the slope is rational, $\omega_2/\omega_1 = p/q$, $p, q \in \mathbb{Z}$, then all trajectories lie on closed orbits of the torus (with coordinates (θ_1, θ_2)).

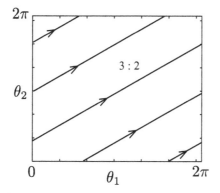

For irrational slopes the flow is said to be **quasiperiodic**. Each trajectory is dense on the torus (i.e. comes arbitrarily close to any given point).

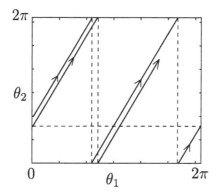

Introducing $\phi = \theta_1 - \theta_2$, the coupled system takes the form

$$\dot{\phi} = \omega_1 - \omega_2 - (K_1 + K_2)\sin(\phi).$$

There are two fixed points if $|\omega_1 - \omega_2| < K_1 + K_2$, defined by $\sin\phi^* = (\omega_1 - \omega_2)/(K_1 + K_2)$, and a saddle-node (tangent) bifurcation occurs when $|\omega_1 - \omega_2| = K_1 + K_2$. In this case $\dot{\phi} = 0$ so that $\dot{\theta}_1 = \dot{\theta}_2 =$ constant $= \omega^*$, where

$$\omega^* = \omega_2 + K_2 \sin\phi^* = \frac{K_1\omega_2 + K_2\omega_1}{K_1 + K_2}.$$

We may regard ω^* as a cooperative frequency that is an emergent property of the coupled system. When no cooperative frequency can be established the two oscillators cannot phase-lock (although they may still frequency-lock).

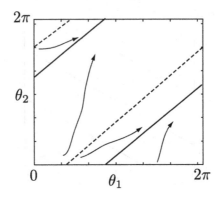

2.9 Poincaré maps

Poincaré maps are useful for studying the flows near a periodic orbit. Consider an n-dimensional system

$$\dot{x} = f(x), \quad x \in \mathbb{R}^n.$$

Let S be an $(n-1)$-dimensional surface (the 'surface of section'). S is required to be transverse to the flow, i.e. all trajectories starting on S flow through it (not parallel to it).

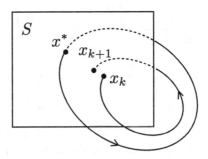

The **Poincaré map** is a mapping from S to itself, obtained by following trajectories from one intersection with S to the next. If $x_k \in S$ denotes the kth intersection, then the Poincaré map is defined by

$$x_{k+1} = P(x_k).$$

Suppose that x^* is a fixed point of P, i.e. $P(x^*) = x^*$. Then a trajectory starting at x^* returns to x^* after some time T, and is therefore a closed orbit for the original system $\dot{x} = f(x)$.

2.9.1 Linear stability of limit cycle

Consider a system

$$\dot{x} = f(x), \qquad x \in \mathbb{R}^n,$$

with a closed orbit. To ask whether the orbit is stable or not, we ask whether the corresponding fixed point x^* of the Poincaré map is stable. Consider $x^* + v_0$ in S, where v_0 is a perturbation. Then after the first return to S

$$x^* + v_1 = P(x^* + v_0) = P(x^*) + [DP(x^*)]v_0 + \text{small terms},$$

where $DP(x^*)$ is an $(n-1) \times (n-1)$ matrix called the **linearised Poincaré map** at x^*. Since $x^* = P(x^*)$, we have

$$v_1 = DP(x^*)v_0.$$

The stability criterion is expressed in terms of the eigenvalues λ_j of $DP(x^*)$: the closed orbit is linearly stable if and only if $|\lambda_j| < 1$ for all $j = 1, \ldots, n-1$.

(From the expression $v_k = \sum_{j=1}^{n-1} v_j(\lambda_j)^k e_j$, where e_j are eigenvectors and v_j are some scalars.) λ_j are called the **characteristic** or **Floquet multipliers** of the periodic orbit. In general, the characteristic multipliers can only be found by numerical integration.

Example 2.9.1

$$\dot{r} = r(1 - r^2), \qquad \dot{\theta} = 1.$$

Let S be the positive x-axis. Compute the Poincaré map and show that the system has a unique periodic orbit and determine its stability. Find the characteristic multipliers for the limit cycle.

Let r_0 be the initial condition on S. Since $\dot{\theta} = 1$, the first return to S occurs after a period $T = 2\pi$. Then $r_1 = P(r_0)$, where

$$\int_{r_0}^{r_1} \frac{dr}{r(1 - r^2)} = \int_0^{2\pi} dt = 2\pi,$$

so

$$r_1 = \left[1 + e^{-4\pi} \left(\frac{1}{r_0^2} - 1 \right) \right]^{-1/2}.$$

Therefore

$$P(r) = \left[1 + e^{-4\pi}\left(\frac{1}{r^2} - 1\right)\right]^{-1/2}.$$

We can show graphically that P has a unique stable fixed point at $r^* = 1$.

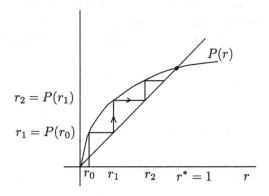

$$\frac{dP(r)}{dr} = e^{-4\pi}r^{-3}[1 + e^{-4\pi}(r^{-2} - 1)]^{-3/2}$$

$$\Rightarrow \quad \left.\frac{dP(r)}{dr}\right|_{r^*=1} = e^{-4\pi} \qquad - \text{ Floquet multiplier}$$

$$|e^{-4\pi}| < 1 \qquad \Rightarrow \qquad \text{the closed orbit is linearly stable.}$$

2.10 Introduction to chaos

2.10.1 Lorenz equations

Consider the Lorenz equations

$$\dot{x} = \sigma(y - x),$$
$$\dot{y} = rx - y - xz,$$
$$\dot{z} = xy - bz,$$

where σ, r, $b > 0$ are parameters. Ed Lorenz derived this three-dimensional system from a simplified model of convection rolls in the atmosphere (E. Lorenz (1963) Deterministic nonperiodic flow, *Journal of Atmospheric Sciences*, Vol. 20).

A mechanical model of the Lorenz equations was invented by W.

Malkus and L. Howard at MIT in the 1970s. The waterwheel simulator can be found at http://people.web.psi.ch/gassmann/waterwheel/WaterwheelLab.html.

Chaotic waterwheel on the campus of the University of Applied Sciences, CH-5200 Brugg, Switzerland (provided by Fritz Gassmann)

2.10.2 Properties of the Lorenz equations

- **Nonlinearity** – The Lorenz equations have two nonlinearities.
- **Symmetry** – If we replace $(x, y) \to (-x, -y)$, the equations stay the same. Hence if $(x(t), y(t), z(t))$ is a solution, so is $(-x(t), -y(t), z(t))$.
- **Volume contraction** – The Lorenz system is **dissipative**: volumes in phase-space contract under the flow.

  Consider any three-dimensional system $\dot{x} = f(x)$, $x \in \mathbb{R}^3$. Consider an arbitrary closed surface $S(t)$ of volume $V(t)$ in phase space. After time dt, S evolves into a new surface $S(t + dt)$. What is its volume $V(t + dt)$?

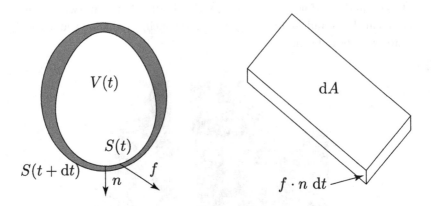

Let n denote the outward normal on S. Then $f \cdot n$ is the outward normal component of velocity (since f is the instantaneous velocity of the points). Therefore in time dt a patch of area dA sweeps out a volume $(f \cdot n \, dt)dA$. Hence

$$V(t + dt) = V(t) + \int_S (f \cdot n \, dt)dA$$

$$\Rightarrow \quad \dot{V} = \frac{V(t + dt) - V(t)}{dt} = \int_S f \cdot n \, dA.$$

By the divergence theorem we have

$$\dot{V} = \int_V \nabla \cdot f \, dV.$$

For the Lorenz system

$$\nabla \cdot f = \frac{\partial}{\partial x}[\sigma(y - x)] + \frac{\partial}{\partial y}[rx - y - xz] + \frac{\partial}{\partial z}(xy - bz) = -\sigma - 1 - b < 0.$$

Since the divergence is constant we have $\dot{V} = -(\sigma + 1 + b)V$ with solution $V(t) = V(0)e^{-(\sigma + 1 + b)t}$. Thus volumes in phase space shrink exponentially fast.

- The Lorenz equations cannot have repelling fixed points or repelling closed orbits, since repellers are incompatible with volume contraction. Thus all fixed points must be sinks or saddles and closed orbits (if they exist).

- **Fixed points** – There are two types of fixed points: $(x^*, y^*, z^*) = (0, 0, 0)$ for all values of the parameters and a symmetric pair of fixed points:

$$C^+ \equiv (x^* = y^* = \sqrt{b(r-1)}, \ z^* = r-1),$$
$$C^- \equiv (x^* = y^* = -\sqrt{b(r-1)}, \ z^* = r-1),$$

when $r > 1$. At $r = 1$ two fixed points coalesce with $(0,0,0)$.

- **Linear stability of the origin** – The linearisation at the origin gives

$$\dot{x} = \sigma(y-x),$$
$$\dot{y} = rx - y,$$
$$\dot{z} = -bz.$$

The equation for $z(t)$ is decoupled and shows that $z(t) \to 0$ exponentially fast. In addition,

$$\begin{pmatrix} \dot{x} \\ \dot{y} \end{pmatrix} = \begin{pmatrix} -\sigma & \sigma \\ r & -1 \end{pmatrix} \begin{pmatrix} x \\ y \end{pmatrix} \equiv A \begin{pmatrix} x \\ y \end{pmatrix},$$

with $\operatorname{Tr}(A) = -\sigma - 1 < 0$ and $\det(A) = \sigma(1-r)$.

- If $r > 1$ $\Rightarrow$ the origin is a saddle point because $\det(A) < 0$ (one outgoing and two incoming directions).
- If $r < 1$ $\Rightarrow$ all directions are incoming, $\operatorname{Tr}(A)^2 - 4\det(A) = (\sigma-1)^2 + 4\sigma r > 0$ and $(0,0,0)$ is a stable node.

- **Global stability of the origin** – We can show for $r < 1$ that the origin is globally stable (no limit cycles or chaos) by constructing a Lyapunov function. Consider $L(x,y,z) = x^2/\sigma + y^2 + z^2$. We have to show that if $r < 1$ and $(x,y,z) \neq (0,0,0)$, then $\dot{L} < 0$:

$$\frac{1}{2}\dot{L} = x\dot{x}/\sigma + y\dot{y} + z\dot{z} = -\left[x - \frac{r+1}{2}\right]^2 - \left[1 - \left(\frac{r+1}{2}\right)^2\right]y^2 - bz^2.$$

We can show that $\dot{L} = 0$ only at $(0,0,0)$, otherwise $\dot{V} < 0$. Therefore the origin is globally stable.

- **Stability of C^+ and C^-** – Assume $r > 1$ so that C^+ and C^- exist. We can find that they are linearly stable for

$$1 < r < r_H = \frac{\sigma(\sigma + b + 3)}{\sigma - b - 1}$$

(assuming $\sigma - b - 1 > 0$). The point $r = r_H$ is called the subcritical Hopf bifurcation.

When $r < r_H$ we observe a saddle cycle. This is a new type of unstable limit cycle that can arise in phase space of three or more dimensions (see the following diagram).

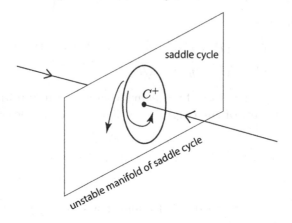

A saddle cycle.

The behaviour of the Lorenz system can be summarised in the following bifurcation diagram:

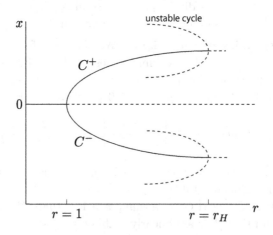

2.10.3 Chaos on a strange attractor

Lorenz used numerical integration to see what the trajectories would do in the long run. The parameters $\sigma = 10$, $b = 8/3$, $r = 28$ are used in the following diagrams. The motion is **aperiodic**.

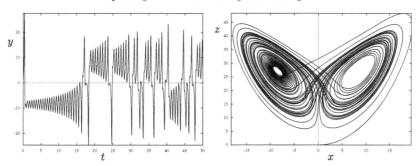

2.10.4 Exponential divergence

Consider $x(t)$ and $x(t)+\delta(t)$ with a vector of initial length $||\delta_0|| = 10^{-15}$.

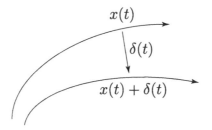

It can be found numerically that

$$||\delta(t)|| \propto ||\delta_0||e^{\lambda t},$$

where $\lambda = 0.9$.

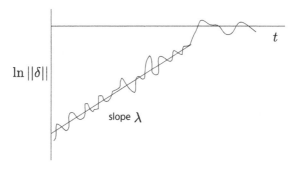

In general an n-dimensional system has n different Lyapunov exponents

$$\delta_k(t) \propto \delta_k(0)e^{\lambda_k t}.$$

Our λ is the largest Lyapunov exponent.

If a measures our tolerance, then our prediction becomes intolerable (when $||\delta(t)|| \geq a$) after a time

$$t \approx \frac{1}{\lambda} \ln \frac{a}{||\delta_0||}.$$

2.10.5 Defining chaos

Chaos is aperiodic long-term behaviour in a deterministic system that exhibits sensitive dependence on initial conditions.

1. *Aperiodic long-term behaviour* means that there are trajectories which do not settle down to fixed points, periodic orbits, or quasi-periodic orbits as $t \to \infty$.
2. *Deterministic* means no noise.
3. *Sensitive dependence on initial conditions* means that nearby trajectories separate exponentially fast (the Lyapunov exponent $\lambda > 0$).

2.10.6 Defining attractor and strange attractor

An attractor Λ is a closed set A with the following properties:

1. *A is an invariant set*: any trajectory $x(t)$ that starts in A stays in A for all time.
2. *A attracts an open set of initial conditions*: there is an open set U containing A such that if $x(0) \in U$, then the distance from $x(t)$ to A tends to zero as $t \to \infty$. This means that A attracts all trajectories that start sufficiently close to it. The largest such U is called the *basin of attraction* of A.
3. *A is minimal*: there is no proper subset of A that satisfies conditions 1 and 2.

Attractors with positive Lyapunov exponents are called strange attractors, and trajectories are called chaotic if at least one Lyapunov exponent is positive (i.e. there is sensitive dependence upon initial conditions). In a strange chaotic attractor the positive Lyapunov exponent indicates exponential spreading within the attractor in the direction transverse to the flow and the negative exponent indicates exponential contraction onto the attractor.

In nonlinear systems it is possible for more than one attractor to exist. To which attractor a trajectory ends up in depends upon the initial conditions. The closure of the set of initial conditions which approach a given attractor is called a basin of attraction. In many nonlinear systems the boundary between basins is **not** smooth and has a **fractal** structure.

2.10.7 Lorenz map

The main idea behind this is to show that z_n should predict z_{n+1}. The Lorenz equations can be integrated for a long time, then the local maxima of $z(t)$ are measured and plotted z_{n+1} vs z_n to get the Lorenz map $z_{n+1} = f(z_n)$.

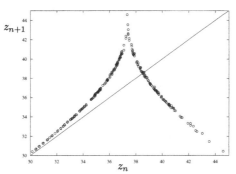

Exploring parameter space

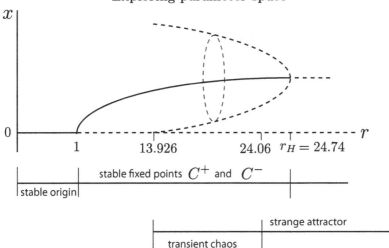

Bifurcation diagram of the Lorenz system.

Using parameters $\sigma = 10$, $b = 8/3$, $r = 21$ gives us the following graphs:

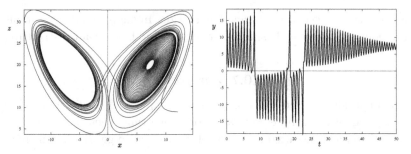

Using parameters $\sigma = 10$, $b = 8/3$, $r = 350$ gives us the following graphs:

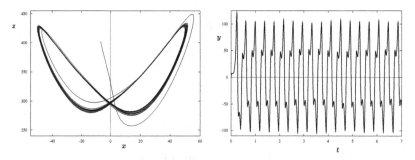

2.10.8 Tests for chaos

- Sensitive dependence on initial conditions – nearby trajectories separate exponentially fast (the Lyapunov exponent $\lambda > 0$)

- Another way to classify dynamical behaviour relies upon spectral analysis. Suppose that the time evolution of a dynamical system is represented by the time variation of some function $f(t)$ of its dynamical variables. Using the notion of a Fourier transform we may express $f(t)$ in terms of a continuum of frequency components:

$$ f(t) = \int_{-\infty}^{\infty} a(\omega)e^{-i\omega t}\mathrm{d}\omega, \qquad a(\omega) = \frac{1}{2\pi} \int_{-\infty}^{\infty} f(t)e^{i\omega t}\mathrm{d}t. $$

From the Fourier transform $a(\omega) \in \mathbb{C}$ we define the *power spectrum*

$$ S(\omega) = |a(\omega)|^2. $$

It is this quantity that is often used to classify experimental data as being chaotic or quasi-periodic.

2.11 One-dimensional maps

Here we study a class of dynamical systems in which time is *discrete* rather than continuous (i.e. difference equations or iterated maps).

Consider the one-dimensional map

$$x_{n+1} = f(x_n),$$

where f is a smooth function from the real line to itself. The sequence $x_0, x_1, x_2, \ldots$ is called the orbit starting from x_0. Maps are useful in various ways:

- Tools for analysing differential equations (e.g. Poincaré maps, the Lorenz map).
- Models of natural phenomena (where discrete time is better to be considered, e.g. digital electronics, in parts of economics and finance theory).
- Simple examples of chaos (maps show a much wilder behaviour than differential equations).

2.11.1 Fixed points and linear stability

If $f(x^*) = x^*$, then x^* is a fixed point. The orbit remains at x^* for all future iterations ($x_n = x^* \Rightarrow x_{n+1} = f(x_n) = f(x^*) = x^*$).

To determine the stability of x^*, we consider a nearby orbit $x_n = x^* + \eta_n$. Then we have

$$x^* + \eta_{n+1} = f(x^* + \eta_n) = f(x^*) + f'(x^*)\eta_n + O(\eta_n^2).$$

This equation reduces to the equation of the linearised map

$$\eta_{n+1} = f'(x^*)\eta_n$$

with **multiplier** $\lambda = f'(x^*)$. The solution of the linear map can be found explicitly by writing a few terms: $\eta_1 = \lambda\eta_0$, $\eta_2 = \lambda\eta_1 = \lambda^2\eta_0, \ldots, \eta_n = \lambda^n\eta_0$.

- If $|\lambda| = |f'(x^*)| < 1$, $\eta_n \to 0$ as $n \to \infty \Rightarrow x^*$ is **linearly stable**.
- If $|\lambda| > 1 \Rightarrow x^*$ is **unstable**.
- If $|\lambda| = 1 \Rightarrow$ **marginal** case (the neglected $O(\eta_n^2)$ terms determine the local stability).
- Fixed points with multiplier $\lambda = 0$ are called **superstable** (perturbations decay much faster).

Example 2.11.1 A cobweb for the map $x_{n+1} = \sin(x_n)$ helps to show that $x^* = 0$ is globally stable.

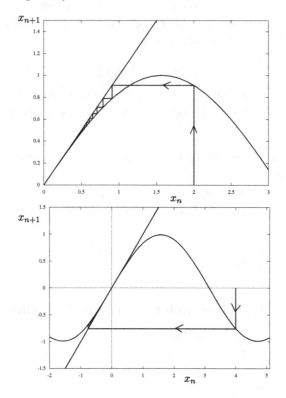

Example 2.11.2 Given $x_{n+1} = \cos(x_n)$ we can show that a typical orbit spirals into the fixed point $x^* = 0.739\ldots$ as $n \to \infty$ ($x = 0.739\ldots$ is the unique solution of $x = \cos(x)$).

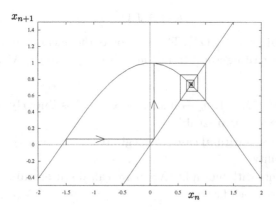

The spiral motion implies that x_n converges to x^* through damped oscillations (typically if $\lambda < 0$). If $\lambda > 0$ the convergence is monotonic.

2.11.2 Logistic map

Consider the **logistic map**

$$x_{n+1} = rx_n(1 - x_n),$$

which is a discrete time analog of the logistic equation for population growth studied earlier. $x_n \geq 0$ is a dimensionless measure of the population in the nth generation and $r \geq 0$ is the intrinsic growth rate. The graph of the logistic map is a parabola with a maximum value of $r/4$ at $x = 0.5$. Here we restrict the control parameter $0 \leq r \leq 4$ so that the equation maps the interval $0 \leq x \leq 1$ into itself.

If $r < 1$, $x_n \to 0$ as $n \to \infty$:

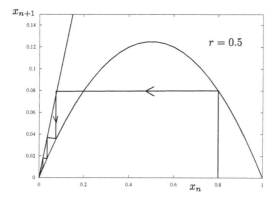

If $1 < r < 3$ the population grows and eventually reaches a nonzero steady state:

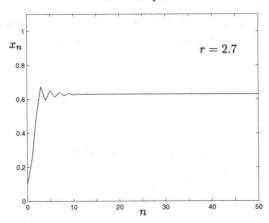

For larger r we observe oscillations in which x_n repeats every two iterations, i.e. a **period-2 cycle**:

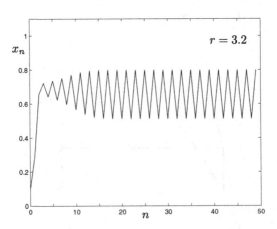

At still larger r, a cycle repeats every four generations, i.e. a **period-4 cycle**:

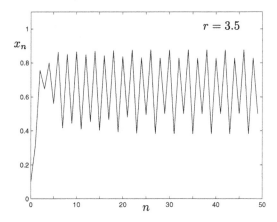

For many values of r, the sequence x_n never settles down to a fixed point or a periodic orbit, i.e. the long-term behaviour is aperiodic.

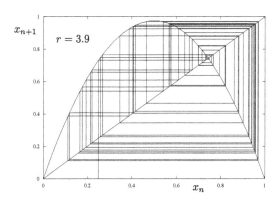

To see the long-term behaviour for all values of r at once, we can plot the **orbit diagram** (the system's attractor as a function of r):

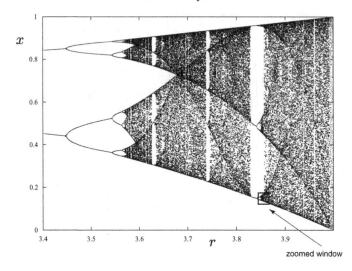

We observe a cascade of period doublings until at $r \approx 3.57$, where the map becomes chaotic. For $r > 3.57$ the orbit diagram reveals a mixture of order and chaos. The large periodic window beginning near $r \approx 3.83$ contains a stable period-3 cycle. A blow-up of part of the period-3 window is shown below (a copy of the orbit diagram reappears in miniature):

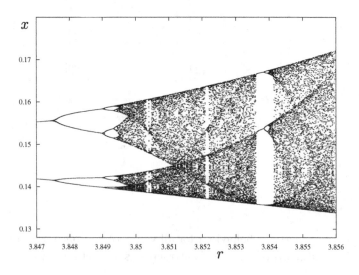

Some analysis of logistic map

The fixed points satisfy $x^* = f(x^*) = rx^*(1 - x^*)$. Hence $x^* = 0$ for all r and $x^* = 1 - 1/r$ for $r \geq 1$ (from the condition $0 \leq x^* \leq 1$). Stability depends on multiplier $f'(x^*) = r - 2rx^*$.

- If $f'(0) = r \Rightarrow x^*$ – stable if $r < 1$ and unstable if $r > 1$
- If $f'(1 - 1/r) = 2 - r \Rightarrow x^* = 1 - 1/r$ is stable if $|2 - r| < 1$, i.e. $1 < r < 3$ and unstable if $r > 3$.

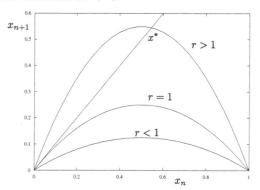

Then x^* bifurcates from the origin in a transcritical bifurcation at $r = 1$. As r increases beyond 1, the slope at x^* gets steeper. The critical slope $f'(x^*) = -1$ is attained when $r = 3$. The resulting bifurcation is called a **flip bifurcation** (often associated with period-doubling).

Here we will show that the logistic map has a 2-cycle for $r > 3$. A 2-cycle exists if and only if there are two points p and q such that $f(p) = q$ and $f(q) = p$. Equivalently, such a p must satisfy $f(f(p)) = p \Rightarrow p$ is a fixed point of the second-iterate map $f^2(x) \equiv f(f(x))$.

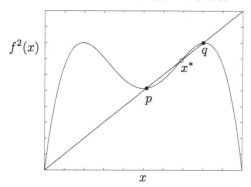

To find p and q we have to solve $f^2(x) = x$, i.e. $r^2 x(1-x)[1 - rx(1-x)] - x = 0$. Since the fixed points $x^* = 0$ and $x^* = 1 - 1/r$ are solutions of this equation we can reduce the equation to a quadratic one by factoring out the fixed points. Solving the resulting quadratic equation we get

$$p,\ q = \frac{r + 1 \pm \sqrt{(r - 3)(r + 1)}}{2r}.$$

For $r > 3$ the roots p and q are real and we have a 2-cycle. For $r < 3$ the roots are complex and a 2-cycle does not exist.

To analyse the stability of a cycle we can reduce the problem to a question about the stability of a fixed point. Both p and q are solutions of $f^2(x) = x \Rightarrow p$ and q are fixed points of the second-iterate map $f^2(x)$. The original 2-cycle is stable if p and q are stable fixed points. To determine whether p is a stable fixed point of f^2 we compute the multiplier

$$\lambda = \frac{\mathrm{d}}{\mathrm{d}x}(f(f(x)))\ |_{x=p} = f'(f(p))f'(p) = f'(q)f'(p).$$

The multiplier is the same at $x = q$. After carrying out the differentiations and substituting for p and q we obtain

$$\lambda = r(1 - 2q)r(1 - 2p) = 4 + 2r - r^2.$$

The 2-cycle is linearly stable if $|4 + 2r - r^2| < 1$, i.e. for $3 < r < 1 + \sqrt{6}$.

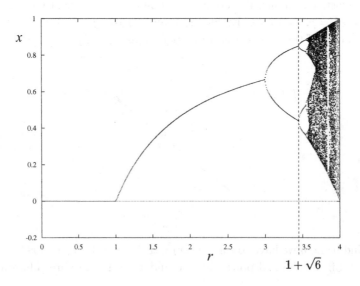

Lyapunov exponent

To be called *chaotic*, a system should also show sensitive dependence on initial conditions, in the sense that neighbouring orbits separate exponentially fast. The definition of the Lyapunov exponent for a chaotic differential equation can be extended to one-dimensional maps.

Given some initial condition x_0, consider a nearby point $x_0 + \delta_0$, where $\delta_0 \ll 1$. Let δ_n be the separation after n iterates. If $|\delta_n| \approx |\delta_0| e^{n\lambda}$, then λ is called the Lyapunov exponent. A positive Lyapunov exponent is a signature of chaos.

A more precise and computationally useful formula for λ can be derived. We note that $\delta_n = f^n(x_0 + \delta_0) - f^n(x_0)$. Then by taking logarithms we have

$$\lambda \approx \frac{1}{n} \ln \left| \frac{\delta_n}{\delta_0} \right| = \frac{1}{n} \ln \left| \frac{f^n(x_0 + \delta_0) - f^n(x_0)}{\delta_0} \right| = \frac{1}{n} \ln |(f^n)'(x_0)| \,,$$

in the limit $\delta_0 \to 0$. Using the chain rule we have

$$(f^n)'(x_0) = \prod_{i=0}^{n-1} f'(x_i)$$

and

$$\lambda \approx \frac{1}{n} \ln \left| \prod_{i=0}^{n-1} f'(x_i) \right| = \frac{1}{n} \sum_{i=0}^{n-1} \ln |f'(x_i)| \,.$$

Then the **Lyapunov exponent** for the orbit starting at x_0 is defined as

$$\lambda = \lim_{n \to \infty} \left[\frac{1}{n} \sum_{i=0}^{n-1} \ln |f'(x_i)| \right]$$

The graphs below show the Lyapunov exponent for the logistic map found numerically:

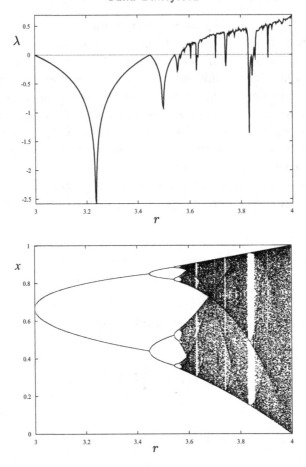

The bifurcation diagram of the logistic map $x_{n+1} = rx_n(1 - x_n)$ demonstrates the presence of the period-3 window near $3.8284\ldots \leq r \leq 3.8415\ldots$. The third-iterate map $f^3(x)$ is the key to understanding the birth of the period-3 cycle (note that the notation $f^3(x)$ here means $x_{n+3} = f^3(x_n)$). Any point p in a period-3 cycle repeats every three iterates, so such points satisfy $p = f^3(p)$, and are therefore fixed points of the third-iterate map. Consider $f^3(x)$ for $r = 3.835$:

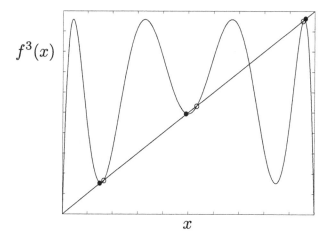

The black dots correspond to a stable period-3 cycle (we can see this by the slope) and the open dots correspond to an unstable 3-cycle (where the slope exceeds 1).

If we decrease r the graph changes shape and the marked intersections have vanished (see the following figure for $r = 3.8$):

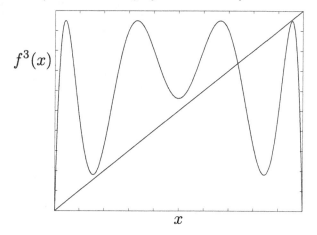

At some critical r the graph $f^3(x)$ must have become tangent to the diagonal (the stable and unstable period-3 cycle coalesce and annihilate in a **tangent bifurcation**).

2.11.3 Routes to chaos

If a nonlinear system has chaotic dynamics then it is natural to ask how this complexity develops as its parameters vary. For example, in the logistic map $x_{n+1} = rx_n(1 - x_n)$ it is easy to show that if $r = 1/2$ then there is a fixed point at $x = 0$ which attracts all solutions with initial values x_0 between 0 and 1, while if $r = 4$ the system is chaotic. How, then, does the transition to chaos occur as the parameter r varies? The identification and description of routes to chaos has important consequences for the interpretation of experimental and numerical observations of nonlinear systems. If an experimental system appears chaotic then it can be very difficult to determine whether the experimental data comes from a truly chaotic system, or if the results of the experiment are unreliable because there is too much external noise.

Period doubling

The period doubling route to chaos is found in the following logistic map:

$$x_{n+1} = f_r(x_n) \equiv rx_n(1 - x_n).$$

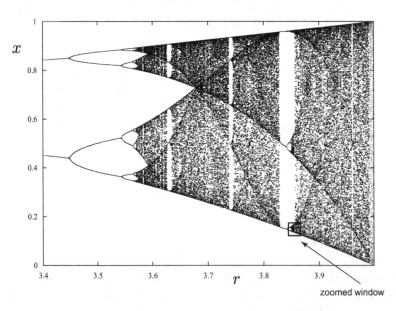

For small r, the attractor is always periodic and has period 2^n, with n increasing as r increases. Beyond some critical value $r = r_c$, with $r_c \approx 3.569946$, the attractor may be more complicated. This period-doubling cascade can be observed in many maps. In the logistic map, if this bifurcation occurs with $r = r_n$ then $r_n \to r_c$ geometrically as

$n \to \infty$, with

$$\lim_{n\to\infty} \frac{r_n - r_{n-1}}{r_{n+1} - r_n} = \delta \approx 4.66920.$$

Interestingly, for unimodal maps (smooth, concave down with single maxima, e.g. the logistic map, the sin map $r\sin(\pi x), \ldots$) this convergence rate is universal (although the constant r_c depends on the map).

Intermittency

By intermittency we mean the occurrence of a signal which alternates randomly between long regular (laminar) phases (intermissions) and relatively short irregular bursts. A mechanism for this was first proposed by Pomeau and Manneville in 1979 observed such behaviour by numerically solving the Lorenz model.

2.11.4 State space reconstruction

If the dynamical system can be modelled by a system of ODEs, then one can make an unambiguous identification between the state variables and the phase-space coordinates of the dynamical system. When the equations of motion are unknown the attractor has to be reconstructed from some measured time series $z(t)$. From Takens' embedding theorem we have the following: for almost every observable $z(t)$ and time delay τ, an m-dimensional portrait constructed from the vectors

$$\{z(t_0), z(t_0 + \tau), \ldots, z(t_0 + (m-1)\tau)\}$$

can have the same properties (the same Lyapunov exponents) as the original attractor. Strictly speaking, the phase portrait obtained by this procedure gives an embedding of the original manifold. The choice of delay τ is pretty arbitrary but the choice of the embedding dimension m is not. This presents us with some difficulties!

2.12 Fractals and fractal dimensions

A fractal is a complex geometric object with fine structure at arbitrarily small scales, perhaps with some degree of self-similarity. Fractals commonly appear in nature (see below).

Fractals in nature.

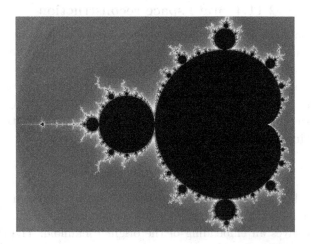

Mandelbrot set $z \to z^2 + c$, where c is a complex parameter.

Consider the example of the Cantor set. Start with the closed interval $S_0 = [0,1]$ and remove the (middle-third) interval $(1/3, 2/3)$. This leaves a pair of closed intervals, which we call S_1. Repeated middle-thirds removals gives rise to the Cantor set $C = S_\infty$.

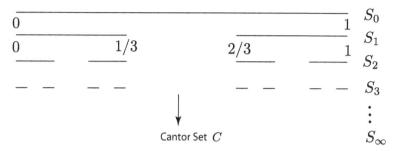

Cantor Set C

1. C has structure at arbitrarily small scales.
2. C is self-similar (e.g. the left half of S_2 is a scaled version of S_1).
3. C has noninteger dimension ($\ln 2 / \ln 3 \approx 0.63$).

Consider the Koch curve: start with a line segment S_0. To generate S_1 delete the middle third of S_0 and replace it with the other two sides of an equilateral triangle. Iterate the process to obtain the Koch curve $K = S_\infty$.

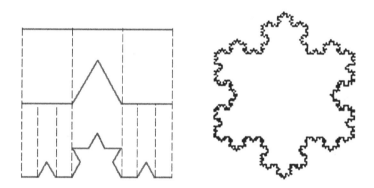

The length of K is infinite. To see this note that $L_1 = 4/3 L_0$ (because S_1 contains four segments each of length $L_0/3$). The length increases by a factor $4/3$ at each stage so that $L_n = (4/3)^n L_0 \to \infty$ as $n \to \infty$. Hence, every point is infinitely far from every other. This suggests that K is more than one-dimensional, possibly between 1 and 2. There are many definitions of fractal dimension (similarity, box, Hausdorff, ...).

Similarity dimension
Suppose that a self-similar set is composed of m copies of itself scaled down by a factor of r. Then the similarity dimension d is the exponent

defined by $m = r^d$, or equivalently,

$$d = \frac{\ln m}{\ln r}.$$

Box dimension

Let S be a subset of $\mathbb{R}^D$ and let $N(\epsilon)$ be the minimum number of D-dimensional boxes of side ϵ needed to cover S:

$$d_{\text{box}} = \lim_{\epsilon \to 0} \frac{\ln N(\epsilon)}{\ln(1/\epsilon)}.$$

For a smooth curve of length L, $N(\epsilon) \propto L/\epsilon$, so $d_{\text{box}} = 1$ as expected. For a planar region of area A bounded by a smooth curve, $N(\epsilon) \propto A/\epsilon^2$ and $d_{\text{box}} = 2$.

Example 2.12.1 Show that the box dimension of the Cantor set is $\ln 2 / \ln 3 \approx 0.63$.

Each S_n consists of 2^n intervals of length $(1/3)^n$, so if we pick $\epsilon = (1/3)^n$ we need 2^n of these intervals to cover the Cantor set. Hence, $N = 2^n$ when $\epsilon = (1/3)^n$. Since $\epsilon \to 0$ as $n \to \infty$ we find

$$d_{\text{box}} = \lim_{\epsilon \to 0} \frac{\ln N(\epsilon)}{\ln 1/\epsilon} = \frac{\ln 2^n}{\ln 3^n} = \frac{n \ln 2}{n \ln 3} = \frac{\ln 2}{\ln 3}.$$

Example 2.12.2 Show that the box dimension of the Koch curve is $\ln 4 / \ln 3 \approx 1.26$.

Each S_n consists of 4^n pieces of length $(1/3)^n L_0$. Ignoring the scale set by L_0 we have that

$$d_{\text{box}} = \lim_{\epsilon \to 0} \frac{\ln N(\epsilon)}{\ln 1/\epsilon} = \frac{\ln 4^n}{\ln 3^n} = \frac{n \ln 4}{n \ln 3} = \frac{\ln 4}{\ln 3}.$$

Correlation dimension

First we generate a set of very many points $\mathbf{x}_i$, $i = 1, \ldots, n$, on the attractor by letting the system evolve for a long time. Then fix a point $\mathbf{x}$ on the attractor A and let $N_{\mathbf{x}}(\epsilon)$ denote the number of points on A inside a ball of radius ϵ about $\mathbf{x}$. $N_{\mathbf{x}}(\epsilon)$ measures how frequently a typical trajectory visits an ϵ-neighborhood of $\mathbf{x}$. We vary ϵ and the number of points typically grows as a power law:

$$N_{\mathbf{x}}(\epsilon) \propto \epsilon^d.$$

We can average $N_{\mathbf{x}}(\epsilon)$ over many $\mathbf{x}$

$$C(\epsilon) \propto \epsilon^d,$$

where d is called the correlation dimension.

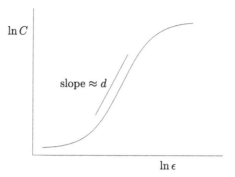

For example, the correlation dimension of the Lorenz attractor (for the standard parameter value $r = 28$, $\sigma = 10$, $b = 8/3$) can be found as $d_{\mathrm{corr}} = 2.05 \pm 0.01$.

2.13 More on bifurcations

2.13.1 Local bifurcations

Saddle-node

The saddle-node bifurcation is the basic mechanism for the creation and destruction of fixed points:

$$\dot{x} = \mu - x^2,$$
$$\dot{y} = -y.$$

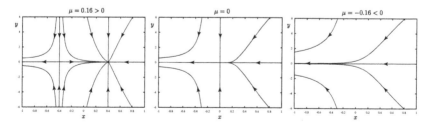

Consider a two-dimensional system

$$\dot{x} = f(x, y), \qquad \dot{y} = g(x, y)$$

that depends on a parameter μ. Then the nullclines can intersect for some value of μ. As μ varies the nullclines can pull away from each other (the fixed points approach each other and collide when $\mu = \mu_c$).

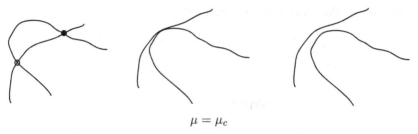

$$\mu = \mu_c$$

Example 2.13.1   Consider a genetic control system

$$\dot{x} = -ax + y$$

$$\dot{y} = \frac{x^2}{1 + x^2} - by$$

where x and y are proportional to the concentrations of the protein and the messenger *RNA*, a, $b > 0$.

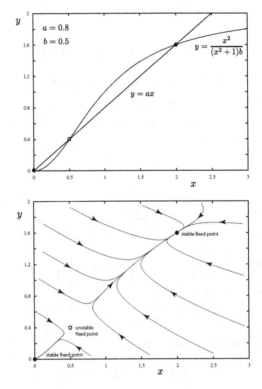

Transcritical and pitchfork

We can construct examples of transcritical and pitchfork bifurcations:

$$\dot{x} = \mu x - x^2, \qquad \dot{y} = -y \qquad \text{transcritical,}$$

$$\dot{x} = \mu x - x^3, \qquad \dot{y} = -y \qquad \text{supercritical,}$$

$$\dot{x} = \mu x + x^3, \qquad \dot{y} = -y \qquad \text{subcritical.}$$

Example 2.13.2 The supercritical pitchfork system for $\mu < 0$, $\mu = 0$ and $\mu > 0$.

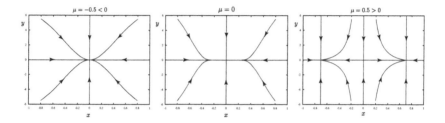

Hopf bifurcations

If the fixed point of a two-dimensional system is stable (for the fixed parameter μ), the eigenvalues λ_1, λ_2 must both lie in the left-half plane $\text{Re}(\lambda) < 0$. To destabilize the fixed point, we need one or both of the eigenvalues to cross into the right-half plane as μ varies.

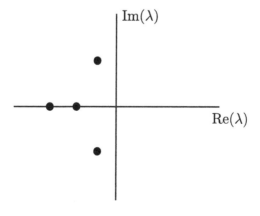

Supercritical HB

If the decay becomes slower and slower and finally changes to growth at a critical value μ_c, the equilibrium state will lose stability. The resulting motion is a small-amplitude, sinusoidal, limit cycle oscillation about the former steady state.

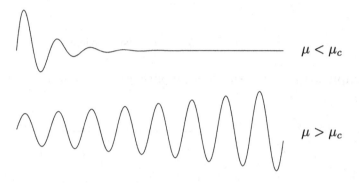

$$\mu < \mu_c$$

$$\mu > \mu_c$$

Example 2.13.3

$$\dot{r} = \mu r - r^3, \qquad \dot{\theta} = \omega + br^2.$$

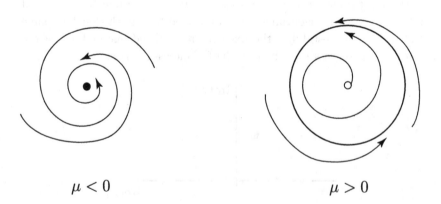

$$\mu < 0 \qquad\qquad\qquad \mu > 0$$

Normally, the eigenvalues would follow a curvy path and cross the imaginary axis with nonzero slope:

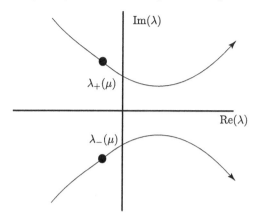

Subcritical HB

The subcritical case is always much more dramatic, and potentially dangerous in engineering applications. After the bifurcation the trajectories must jump to a distant attractor, which may be a fixed point, another limit cycle, infinity, or a chaotic attractor (in three and higher dimensions).

Example 2.13.4

$$\dot{r} = \mu r + r^3 - r^5, \qquad \dot{\theta} = \omega + br^2.$$

A subcritical HB occurs at $\mu = 0$, where the unstable cycle shrinks to zero amplitude. For $\mu > 0$ the large amplitude limit cycle is suddenly the only attractor.

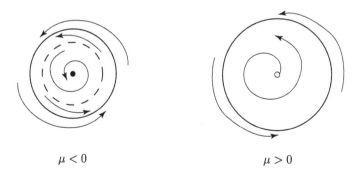

$\mu < 0$ $\qquad\qquad\qquad\qquad\qquad$ $\mu > 0$

2.13.2 Global bifurcations of cycles

Saddle-node bifurcation of cycles

A bifurcation in which two limit cycles coalesce and annihilate. An example occurs in the system

$$\dot{r} = \mu r + r^3 - r^5, \qquad \dot{\theta} = \omega + br^2.$$

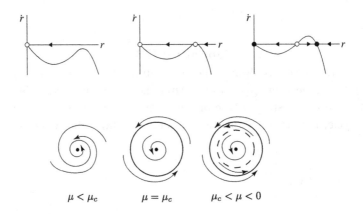

$$\mu < \mu_c \qquad\qquad \mu = \mu_c \qquad\qquad \mu_c < \mu < 0$$

Homoclinic bifurcation

In this bifurcation part of a limit cycle moves closer and closer to a saddle point. At the bifurcation the cycle touches the saddle point and becomes a homoclinic orbit. For example, consider the system

$$\dot{x} = y, \qquad \dot{y} = \mu y + x - x^2 + xy.$$

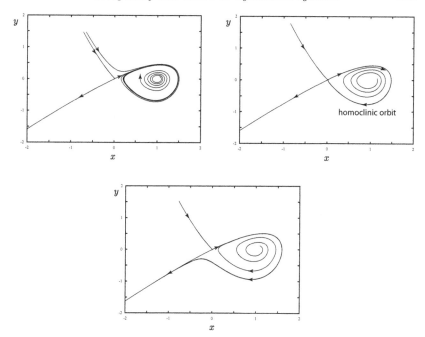

The period of oscillations tends to infinity at the homoclinic bifurcation.

Heteroclinic bifurcation

In the phase portrait of a dynamical system, a *heteroclinic orbit* (sometimes called a heteroclinic connection) is a path in phase space which joins two different equilibrium points. (Note: if the equilibrium points at the start and end of the orbit are the same, the orbit is a homoclinic orbit.) Consider the continuous dynamical system:

$$\dot{x} = f(x).$$

Suppose there are equilibria at $x = x_0$ and $x = x_1$. Then a solution $\phi(t)$ is a heteroclinic orbit from x_0 to x_1 if

$$\phi(t) \to x_0 \qquad \text{as} \qquad t \to -\infty$$

and

$$\phi(t) \to x_1 \qquad \text{as} \qquad t \to +\infty.$$

This implies that the orbit is contained in the stable manifold of x_1 and the unstable manifold of x_0.

124

For example, the phase portrait of the pendulum equation $\ddot{x} + \sin x = 0$:

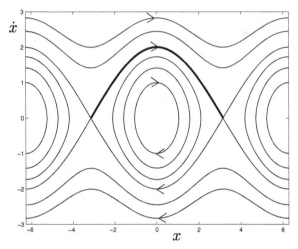

The highlighted curve shows the heteroclinc orbit from $(x, \dot{x}) = (-\pi, 0)$ to $(x, \dot{x}) = (\pi, 0)$. This orbit corresponds with the pendulum starting upright, making one revolution through its lowest position, and ending upright again.

A heteroclinic cycle is an invariant set in the phase space of a dynamical system. It is a topological circle of equilibrium points and connecting heteroclinic orbits.

A heteroclinic bifurcation is a global bifurcation involving a heteroclinic cycle.

Illustrative bibliography

Alligood, K. T., Sauer, T. D. and Yorke, J. A. (1996). *Chaos: An Introduction to Dynamical Systems*. Springer.

Boccara, N. (2004). *Modeling Complex Systems*. Springer.

Glendinning, P. (1994). *Stability, Instability and Chaos*. Cambridge University Press.

Strogatz, S. H. (2000). *Nonlinear Dynamics and Chaos*. Westview Press.

3
Interacting stochastic particle systems
Stefan Grosskinsky

Abstract

Interacting particle systems (IPS) are probabilistic mathematical models of complex phenomena involving a large number of interrelated components. There are numerous examples within all areas of natural and social sciences, such as traffic flow on motorways or communication networks, opinion dynamics, spread of epidemics or fires, genetic evolution, reaction diffusion systems, crystal surface growth, financial markets, etc. The central question is to understand and predict emergent behaviour on macroscopic scales, as a result of the microscopic dynamics and interactions of individual components. Qualitative changes in this behaviour depending on the system parameters are known as collective phenomena or phase transitions and are of particular interest.

In IPS the components are modelled as particles confined to a lattice or some discrete geometry. But applications are not limited to systems endowed with such a geometry, since continuous degrees of freedom can often be discretized without changing the main features. So depending on the specific case, the particles can represent cars on a motorway, molecules in ionic channels, or prices of asset orders in financial markets (see Chapter 6), to name just a few examples. In principle, such systems often evolve according to well-known laws, but in many cases microscopic details of motion are not fully accessible. Due to the large system size these influences on the dynamics can be approximated as effective random noise with a certain postulated distribution. The actual origin of the noise, which may be related to chaotic motion (see Chapter 2) or thermal interactions (see Chapter 4), is usually ignored. On this level the statistical description in terms of a random process where particles move and interact according to local stochastic rules is an appropriate mathematical model. It is neither possible nor required to keep track of every single particle. One is rather interested in predicting measurable quantities which correspond to expected values of certain observables, such as the growth rate of the crystalline surface or the flux of cars on the motorway. Although describing the system only on a mesoscopic level as explained above, stochastic particle systems are usually referred to as microscopic models and we stick to this convention. On a macroscopic scale, a continuum description of systems with a large number of particles is given by coarse-grained density fields, evolving in time according to a partial differential equation. The form of this equation depends on the particular application, and its mathematical connection to microscopic particle models is one of the fundamental questions in complexity science.

The focus of this chapter is not on detailed models of real world phenomena, but on simple paradigmatic IPS that capture the main features of real complex systems. Several such models have been introduced in the seminal paper [1]. They allow for a detailed mathematical analysis leading to a deeper understanding of the fundamental principles of emergence and collective phenomena. The chapter provides an introduction into the well-developed mathematical theory for the description of the dynamics of IPS, and involving graphical representations and an analytic description using semigroups and generators. Since the external conditions for real systems are often constant or slowly varying with respect to the system dynamics, observations are typically available in a time stationary situation. This is described by the invariant measures of the stochastic particle systems which are thus of major interest in their analysis, and are introduced at the end of Section 3.1 which covers the basic theory. Later sections provide a more detailed discussion of several examples of basic IPS, and probabilistic techniques used for their analysis such as coupling, duality and relative entropy methods are introduced. The second main aspect of the chapter, also covered in Sections 3.2 to 3.4, is to get acquainted with different types of collective phenomena in complex systems, which are also discussed in Chapters 1 and 4. We will discuss their intimate relationship with symmetries and conservation laws of the dynamics, and make connections to the classical theory of phase transitions in statistical mechanics.

Necessary prerequisites for the reader are a basic knowledge in undergraduate mathematics, in particular probability theory and stochastic processes. For the latter, discrete time Markov chains are sufficient since

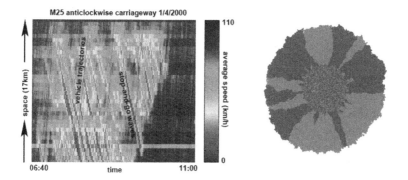

Figure 3.1 Left: Traffic data gathered from inductance loops showing car speeds on the M25 London orbital motorway averaged over 1 minute intervals (taken from [2]). Right: segregation pattern of a neutral genetic marker in a simple simulation of microbial species growing from a mixed population of circular shape (taken from [3], similar to experiments in [4]).

the concept of continuous-time processes will be introduced in Section 3.1. Acquaintance with measure theoretic concepts and basic functional analysis is helpful but not necessary.

Before we get immersed in mathematics let us quickly discuss two real world examples of recent studies, to illustrate the motivation and the origin of the processes with which we will work. The left-hand side of Figure 3.1 shows colour-coded data for car speed on the M25 London orbital motorway as a function of space and time. The striped patterns of low speed correspond to spurious stop-and-go waves during rush hour. Often there is no obvious external cause, such as an accident or lane closure, so this pattern has to be interpreted as an intrinsic collective phenomenon emerging from the interactions of cars on a busy road. A minimal mathematical description of this situation in terms of IPS would be to take a one-dimensional lattice $\Lambda = \mathbb{Z}$ (or a subset thereof), and at each site $x \in \Lambda$ denote the presence or absence of a car with an occupation number $\eta(x) = 1$ or 0, respectively. So the state space of our mathematical model is given by the set $\{0,1\}^{\Lambda}$ denoting all possible configurations $\eta = \big(\eta(x)\big)_{x \in \Lambda}$. In terms of dynamics, we only want to model normal traffic on a single lane road without car crashes or overtaking. So cars are allowed to proceed one lattice site to the right, with a given rate,[1] provided that the site in front is not occupied by another car. The rate may

---

[1] The concept of 'rate' and exact mathematical formulations of the dynamics will be introduced in Section 3.1.

depend on the surrounding configuration of cars (e.g. number of empty sites ahead), and relatively simple choices depending only on three or four neighbouring sites can already lead to interesting patterns and the emergence of stop-and-go waves. There are numerous such approaches in the literature known as cellular automata models for traffic flow, see e.g. [5] and references therein. The defining features of this process in terms of IPS are that no particles are created or destroyed (conservation of the number of particles) and that there is at most one particle per site (exclusion rule). Processes with both or only the first property will be discussed in detail in Sections 3.2 and 3.3, respectively. The right of Figure 3.1 shows a simple minded simulation of segregation patterns of a microbial species grown from a mixed initial population of circular shape [3], similar to experiments in [4]. Each individual appears either red or green (depicted in dark or light grey, respectively) as a result of a neutral genetic marker that affects only the colour. A possible IPS model of this system has a state space $\{0, R, G\}^\Lambda$, where the lattice is now two dimensional, say $\Lambda = \mathbb{Z}^2$ for simplicity, and 0 represents an empty site, R the presence of a red and G of a green individual. The dynamics can be modelled by letting each individual split into two ($R \to 2R$ or $G \to 2G$) with a given rate, and then place the offspring on an empty neighbouring site. If there is no empty neighbouring site the reproduction rate is zero (or equivalently the offspring is immediately killed). Therefore we have two equivalent species competing for the same resource (empty sites), and spatial segregation is a result of the fact that once the red particles died out in a certain region due to fluctuations, all the offspring is descending from green ancestors. Note that, in contrast to the first example, the number of particles in this model is not conserved. The simplest such process to model extinction or survival of a single species is called the contact process, and is discussed in detail in Section 3.4.

3.1 Basic theory

3.1.1 Markov processes

The *state space* X of a stochastic process is the set of all *configurations* which we typically denote by η or ζ. For interacting particle systems the state space is of the form $X = S^\Lambda$ where the *local state space* $S \subseteq \mathbb{Z}$ is a finite subset of the integers, such as $S = \{0, 1\}$ or $S = \{0, 1, 2\}$ to indicate e.g. the local occupation number $\eta(x) \in S$ for all $x \in \Lambda$. Λ is any countable set such as a regular lattice or the vertex set of a graph. We often do not explicitly specify the edge set or connectivity structure of Λ but, unless stated otherwise, we always assume it to be strongly connected to avoid degeneracies, i.e. any pair of points in Λ is connected (along a possibly directed path).

The particular structure of the state space is not essential to define a Markov process in general, what is essential for that is that X is a compact metric space. Spaces of the above form have that property w.r.t. the product topology.[2] The metric structure of X allows us to properly define regular sample paths and continuous functions, and set up a measurable structure, whereas compactness becomes important only later in connection with stationary distributions.

A *continuous-time stochastic process* $(\eta_t : t \geq 0)$ is then a family of random variables η_t taking values in X. This can be characterized by a probability measure $\mathbb{P}$ on the *canonical path space*

$$D[0, \infty) = \{\eta_. : [0, \infty) \to X \text{ càdlàg}\}. \tag{3.1}$$

By convention, this is the set of right continuous functions with left limits (càdlàg). The elements of $D[0, \infty)$ are the *sample paths* $t \mapsto \eta_t \in X$, written shortly as $\eta_.$ or $\boldsymbol{\eta}$. For an IPS with $X = S^\Lambda$, $\eta_t(x)$ denotes the occupation of site x at time t.

Note that as soon as $|S| > 1$ and Λ is infinite the state space $X = S^\Lambda$ is uncountable. But even if X itself is countable (e.g. for finite lattices), the path space is always uncountable due to the continuous-time $t \in \mathbb{R}$. Therefore we have to think about measurable structures on $D[0, \infty)$ and the state space X in the following. The technical details of this are not really essential for the understanding but we include them here for com-

---

[2] Why is $X = S^\Lambda$ a compact metric space?
The discrete topology σ_x on the local state space S is simply given by the power set, i.e. all subsets are 'open'. The choice of the metric does not influence this and is therefore irrelevant for that question. The *product topology* σ on X is then given by the smallest topology such that all the canonical projections $\eta(x) : X \to S$ (occupation at a site x for a given configuration η) are continuous (pre-images of open sets are open). That means that σ is generated by sets

$$\eta(x)^{-1}(U) = \{\eta : \eta(x) \in U\}, \quad U \subseteq S,$$

which are called *open cylinders*. Finite intersections of these sets

$$\{\eta : \eta(x_1) \in U_1, \ldots, \eta(x_n) \in U_n\}, \quad n \in \mathbb{N}, \ U_i \subseteq S$$

are called *cylinder sets* and any open set on X is a (finite or infinite) union of cylinder sets. Clearly S is compact since it is finite, and by *Tychonoff's theorem* any product of compact topological spaces is compact (w.r.t. the product topology). This holds for any countable lattice or vertex set Λ.

pleteness. The metric on X provides us with a generic topology of open sets generating the Borel σ-algebra, which we take as the measurable structure on X. Now, let $\mathcal{F}$ be the smallest σ-algebra on $D[0,\infty)$ such that all the mappings $\eta. \mapsto \eta_s$ for $s \geq 0$ are measurable w.r.t. $\mathcal{F}$. That means that every path can be evaluated or observed at arbitrary times s, i.e.

$$\{\eta_s \in A\} = \{\eta. | \eta_s \in A\} \in \mathcal{F} \qquad (3.2)$$

for all measurable subsets $A \in X$. This is certainly a reasonable minimal requirement for $\mathcal{F}$. If $\mathcal{F}_t$ is the smallest σ-algebra on $D[0,\infty)$ relative to which all the mappings $\eta. \mapsto \eta_s$ for $s \leq t$ are measurable, then $(\mathcal{F}_t : t \geq 0)$ provides a natural filtration for the process. The filtered space $\big(D[0,\infty), \mathcal{F}, (\mathcal{F}_t : t \geq 0)\big)$ serves as a generic choice for the probability space of a stochastic process.

Definition 3.1.1 A *(homogeneous) Markov process* on X is a collection $(\mathbb{P}^\zeta : \zeta \in X)$ of probability measures on $D[0,\infty)$ with the following properties:

(a) $\mathbb{P}^\zeta\big(\eta. \in D[0,\infty) : \eta_0 = \zeta\big) = 1$ for all $\zeta \in X$,
    i.e. $\mathbb{P}^\zeta$ is normalized on all paths with initial condition $\eta_0 = \zeta$.
(b) $\mathbb{P}^\zeta(\eta_{t+.} \in A | \mathcal{F}_t) = \mathbb{P}^{\eta_t}(A)$ for all $\zeta \in X$, $A \in \mathcal{F}$ and $t > 0$,
    (Markov property).
(c) The mapping $\zeta \mapsto \mathbb{P}^\zeta(A)$ is measurable for every $A \in \mathcal{F}$.

Note that the Markov property as formulated in (b) implies that the process is (time-)homogeneous, since the law $\mathbb{P}^{\eta_t}$ does not have an explicit time dependence. Markov processes can be generalized to be inhomogeneous (see e.g. [6]), but we will concentrate only on homogeneous processes. The condition in (c) allows us to consider processes with general initial distributions $\mu \in \mathcal{M}_1(X)$ via

$$\mathbb{P}^\mu := \int_X \mathbb{P}^\zeta \mu(d\zeta). \qquad (3.3)$$

When we do not want to specify the initial condition for the process we will often only write $\mathbb{P}$.

3.1.2 Continuous-time Markov chains and graphical representations

Throughout this section let X be a countable set. Markov processes on X are called *Markov chains*. They can be understood without a path space description on a more basic level, by studying the time evolution of distributions $p_t(\eta) := \mathbb{P}(\eta_t = \zeta)$ (see e.g. [6] or [7]). The dynamics of continuous-time Markov chains can be characterized by *transition rates* $c(\zeta, \zeta') \geq 0$, which have to be specified for all $\zeta, \zeta' \in X$. For a given process $(\mathbb{P}^\zeta : \zeta \in X)$ the rates are defined via

$$\mathbb{P}^\zeta(\eta_t = \zeta') = c(\zeta, \zeta') \, t + o(t) \quad \text{as } t \searrow 0 \quad \text{for} \quad \zeta \neq \zeta', \quad (3.4)$$

and represent probabilities per unit time. We will not go into the details here of why the linearization in (3.4) is valid for small times t. It can be shown under the assumption of uniform continuity of $t \mapsto \mathbb{P}^\zeta(\eta_t = \zeta')$ as $t \searrow 0$, which is also called *strong continuity* (see e.g. [8], Section 19). This is discussed in more detail for general Markov processes in Section 3.1.4. We will see in the next subsection how a given set of rates determines the path measures of a process. Now we would like to get an intuitive understanding of the time evolution and the role of the transition rates. For a process with $\eta_0 = \zeta$, we denote by

$$W_\zeta := \inf\{t \geq 0 : \eta_t \neq \zeta\} \quad (3.5)$$

the *holding time* in state ζ. Its distribution is related to the total exit rate out of state ζ,

$$c_\zeta := \sum_{\zeta' \neq \zeta} c(\zeta, \zeta'). \quad (3.6)$$

We assume in the following that $c_\zeta < \infty$ for all $\zeta \in X$ (which is only a restriction if X is infinite). As shown below, this ensures that the process has a well-defined waiting time in each state c_ζ, which is essential to construct the dynamics locally in time. To have well-defined global dynamics for all $t \geq 0$ we also have to exclude that the chain explodes,[3] which is ensured by a uniform bound

$$\bar{c} = \sup_{\zeta \in X} c_\zeta < 0. \quad (3.7)$$

If $c_\zeta = 0$, ζ is an *absorbing state* and $W_\zeta = \infty$ almost surely (a.s.).

Proposition 3.1.2 *If $c_\zeta \in (0, \infty)$, $W_\zeta \sim Exp(c_\zeta)$ and* $\mathbb{P}^\zeta(\eta_{W_\zeta} = \zeta') = c(\zeta, \zeta')/c_\zeta.$

---

[3] Explosion means that the Markov chain exhibits infinitely many jumps in finite time. For more details see e.g. [7], Section 2.7.

Proof As W_ζ has the *'loss of memory'* property

$$\mathbb{P}^\zeta(W_\zeta > s + t | W_\zeta > s) = \mathbb{P}^\zeta(W_\zeta > s + t | \eta_s = \zeta)$$
$$= \mathbb{P}^\zeta(W_\zeta > t), \qquad (3.8)$$

the distribution of the holding time W_ζ does not depend on how much time the process has already spent in state ζ. Thus

$$\mathbb{P}^\zeta(W_\zeta > s + t, \ W_\zeta > s) = \mathbb{P}^\zeta(W_\zeta > s + t)$$
$$= \mathbb{P}^\zeta(W_\zeta > s) \, \mathbb{P}^\zeta(W_\zeta > t). \qquad (3.9)$$

This is the functional equation for an exponential and implies that

$$\mathbb{P}^\zeta(W_\zeta > t) = e^{\lambda t} \ \text{(with initial condition } \mathbb{P}^\zeta(W_\zeta > 0) = 1). \qquad (3.10)$$

The exponent is given by

$$\lambda = \frac{d}{dt}\mathbb{P}^\zeta(W_\zeta > t)\Big|_{t=0} = \lim_{t \searrow 0} \frac{\mathbb{P}^\zeta(W_\zeta > t) - 1}{t} = -c_\zeta, \qquad (3.11)$$

since with (3.4) and (3.6)

$$\mathbb{P}^\zeta(W_\zeta > 0) = 1 - \mathbb{P}^\zeta(\eta_t \neq \zeta) + o(t) = 1 - c_\zeta t + o(t). \qquad (3.12)$$

Now, conditioned on a jump occurring in the time interval $[t, t + h)$ we have

$$\mathbb{P}^\zeta(\eta_{t+h} = \zeta' | t \leq W_\zeta < t + h) = \mathbb{P}^\zeta(\eta_h = \zeta' | W_\zeta < h)$$
$$= \frac{\mathbb{P}^\zeta(\eta_h = \zeta')}{\mathbb{P}^\zeta(W_\zeta < h)} \to \frac{c(\zeta, \zeta')}{c_\zeta} \qquad (3.13)$$

as $h \searrow 0$, using the Markov property and L'Hopital's rule with (3.4) and (3.11). With right-continuity of paths, this implies the second statement. $\qquad \square$

We summarize some important properties of exponential random variables, the proof of which can be found in any standard textbook. Let $W_1, W_2, \ldots$ be a sequence of independent exponentials $W_i \sim Exp(\lambda_i)$. Then $\mathbb{E}(W_i) = 1/\lambda_i$ and

$$\min\{W_1, \ldots, W_n\} \sim Exp\Big(\sum_{i=1}^n \lambda_i\Big). \qquad (3.14)$$

The sum of iid (independent, identically distributed) exponentials with $\lambda_i = \lambda$ is Γ-distributed, i.e.

$$\sum_{i=1}^n W_i \sim \Gamma(n, \lambda) \quad \text{with PDF} \quad \frac{\lambda^n w^{n-1}}{(n-1)!} e^{-\lambda w}. \qquad (3.15)$$

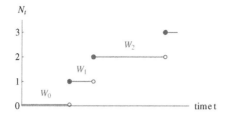

Figure 3.2 Sample path (càdlàg) of a Poisson process with holding times $W_0, W_1, \ldots$.

Example The *Poisson process* $(N_t : t \geq 0)$ with rate $\lambda > 0$ (short $PP(\lambda)$) is a Markov chain with $X = \mathbb{N} = \{0, 1, \ldots\}$, $N_0 = 0$ and $c(n, m) = \lambda\,\delta_{n+1,m}$. A sample path is illustrated in Figure 3.2.

With iid random variables $W_i \sim Exp(\lambda)$ we can write $N_t = \max\{n : \sum_{i=1}^n W_i \leq t\}$. This implies

$$\mathbb{P}(N_t = n) = \mathbb{P}\Big(\sum_{i=1}^n W_i \leq t < \sum_{i=1}^{n+1} W_i\Big)$$

$$= \int_0^t \mathbb{P}\Big(\sum_{i=1}^n W_i = s\Big)\,\mathbb{P}(W_{n+1} > t - s)\,ds$$

$$= \int_0^t \frac{\lambda^n s^{n-1}}{(n-1)!}\,e^{-\lambda s}\,e^{-\lambda(t-s)}\,ds = \frac{(\lambda t)^n}{n!}e^{-\lambda t}, \qquad (3.16)$$

so $N_t \sim Poi(\lambda t)$ has a Poisson distribution. Alternatively a Poisson process can be characterized by the following.

Proposition 3.1.3 $(N_t : t \geq 0) \sim PP(\lambda)$ *if and only if it has stationary, independent increments, i.e.*

$$N_{t+s} - N_s \sim N_t - N_0 \quad and$$
$$N_{t+s} - N_s \text{ independent of } (N_u : u \leq s), \qquad (3.17)$$

and for each t, $N_t \sim Poi(\lambda t)$.

Proof By the loss of memory property and (3.16) increments have the distribution

$$N_{t+s} - N_s \sim Poi(\lambda t) \quad \text{for all} \quad s \geq 0, \qquad (3.18)$$

and are independent of N_s, which is enough together with the Markov property.

The other direction follows by deriving the jump rates from the properties in (3.17) and the Poisson law using (3.4). □

Remember that for independent Poisson variables $Y_1, Y_2, \ldots$ with $Y_i \sim Poi(\lambda_i)$ we have $\mathbb{E}(Y_i) = Var(Y_i) = \lambda_i$ and

$$\sum_{i=1}^{n} Y_i \sim Poi\left(\sum_{i=1}^{n} \lambda_i\right). \tag{3.19}$$

With Proposition 3.1.3 this immediately implies that adding a finite number of independent Poisson processes $(N_t^i : t \geq 0) \sim PP(\lambda_i)$, $i = 1, \ldots, n$ results in a Poisson process, i.e.

$$M_t = \sum_{i=1}^{n} N_t^i \quad \Rightarrow \quad (M_t : t \geq 0) \sim PP\left(\sum_{i=1}^{n} \lambda_i\right). \tag{3.20}$$

Example A continuous-time simple random walk $(\eta_t : t \geq 0)$ on $X = \mathbb{Z}$ with jump rates p to the right and q to the left is given by

$$\eta_t = R_t - L_t \text{ where } (R_t : t \geq 0) \sim PP(p), \ (L_t : t \geq 0) \sim PP(q). \tag{3.21}$$

The process can be constructed by the following graphical representation:

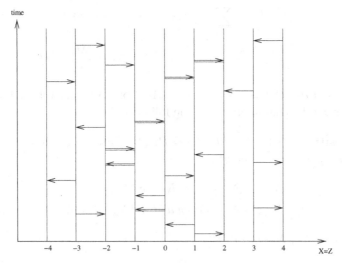

In each column the arrows $\rightarrow\sim PP(p)$ and $\leftarrow\sim PP(q)$ are independent Poisson processes. Together with the initial condition, the trajectory of the process shown in grey is then uniquely determined. An analogous construction is possible for a general Markov chain, which is a continuous-time random walk on X with jump rates $c(\zeta, \zeta')$. In this way we can also construct interacting random walks and more general IPS, as is shown in the next section.

Note that the restriction $c_\zeta < \infty$ for all $\zeta \in X$ excludes e.g. random walks on $X = \mathbb{Z}$ which move non-locally and jump to any site with rate $c(\zeta, \zeta') = 1$. In the graphical construction for such a process there would not be a well-defined first jump event and the path could not be constructed. However, as long as the rates are summable, such as

$$c(\zeta, \zeta') = (\zeta - \zeta')^{-2} \quad \text{for all } \zeta, \zeta' \in \mathbb{Z}, \tag{3.22}$$

we have $c_\zeta < \infty$, and the basic properties of adding Poisson processes or taking minima of exponential random variables extend to infinitely many. So the process is well defined and the path can be constructed in the graphical representation.

3.1.3 Three basic IPS

For the IPS introduced in this section the state space is of the form $X = \{0, 1\}^\Lambda$, *particle configurations* $\eta = (\eta(x) : x \in \Lambda)$. $\eta(x) = 1$ means that there is a particle at site x and if $\eta(x) = 0$ site x is empty. The lattice Λ can be any countable set, typical examples we have in mind are regular lattices $\Lambda = \mathbb{Z}^d$, subsets of those, or the vertex set of a given graph.

As noted before, if Λ is infinite X is uncountable, so we are not dealing with Markov chains in this section. But for the processes we consider the particles move/interact only locally and one at a time, so a description with jump rates still makes sense. More specifically, for a given $\eta \in X$ there are only countably many η' for which $c(\eta, \eta') > 0$. Define the configurations η^x and $\eta^{xy} \in X$ for $x \neq y \in \Lambda$ by

$$\eta^x(z) = \begin{cases} \eta(z), & z \neq x \\ 1 - \eta(x), & z = x \end{cases} \quad \text{and} \quad \eta^{xy}(z) = \begin{cases} \eta(z), & z \neq x, y, \\ \eta(y), & z = x, \\ \eta(x), & z = y, \end{cases} \tag{3.23}$$

so that η^x corresponds to creation/annihilation of a particle at site x and η^{xy} to motion of a particle between x and y. Then following standard notation we write for the corresponding jump rates

$$c(x, \eta) = c(\eta, \eta^x) \quad \text{and} \quad c(x, y, \eta) = c(\eta, \eta^{xy}). \tag{3.24}$$

All other jump rates including e.g. multi-particle interactions or simultaneous motion are zero.

Definition 3.1.4 Let $p(x, y) \geq 0$, $x, y \in \Lambda$, be the transition rates of an irreducible continuous-time random walk on Λ. The *exclusion process* (EP) on X is then characterized by the jump rates

$$c(x, y, \eta) = p(x, y)\eta(x)(1 - \eta(y)), \quad x, y \in \Lambda, \qquad (3.25)$$

where particles only jump to empty sites (exclusion interaction). If Λ is a regular lattice and $p(x, y) > 0$ only if x and y are nearest neighbours, the process is called *simple EP* (SEP). If in addition $p(x, y) = p(y, x)$ for all $x, y \in \Lambda$ it is called *symmetric SEP* (SSEP) and otherwise *asymmetric SEP* (ASEP).

Note that the presence of a direct connection (or directed edge) (x, y) is characterized by $p(x, y) > 0$, and irreducibility of $p(x, y)$ is equivalent to Λ being strongly connected. Particles only move and are not created or annihilated, therefore the number of particles in the system is conserved in time. In general, such IPS are called *lattice gases*. The ASEP in one dimension $d = 1$ is one of the most basic and most studied models in IPS and nonequilibrium statistical mechanics (see e.g. [9] and references therein), and a common quick way of defining it is

$$10 \xrightarrow{p} 01, \quad 01 \xrightarrow{q} 10, \qquad (3.26)$$

where particles jump to the right (left) with rate p (q). Variants and extensions of exclusion processes are used to model all kinds of transport phenomena, including for instance traffic flow (see e.g. [9, 10] and references therein).

The graphical construction given on the next page is analogous to the single particle process given above, with the additional constraint of the exclusion interaction. We will discuss exclusion processes in more detail in Section 3.2. Exclusion is of course not the only possible interaction between random walkers, and we will discuss a different example with a simpler zero-range interaction in Section 3.3.

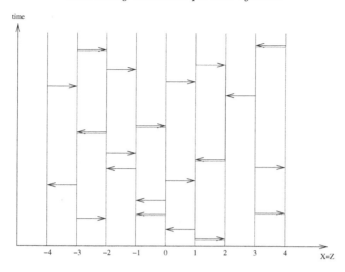

Definition 3.1.5 The *contact process* (CP) on X is characterized by the jump rates

$$c(x, \eta) = \begin{cases} 1, & \eta(x) = 1 \\ \lambda \sum_{y \sim x} \eta(y), & \eta(x) = 0 \end{cases}, \quad x \in \Lambda. \tag{3.27}$$

Particles can be interpreted as infected sites which recover with rate 1 and are infected independently with rate $\lambda > 0$ by particles on connected sites $y \sim x$.

In contrast to the EP, the CP does not have a conserved quantity like the number of particles, but it does have an absorbing state $\eta \equiv 0$, since there is no spontaneous infection. A compact notation for the CP is

$$1 \xrightarrow{1} 0, \quad 0 \to 1 \quad \text{with rate} \quad \lambda \sum_{y \sim x} \eta(x). \tag{3.28}$$

The graphical construction below now contains a third independent Poisson process $\times \sim PP(1)$ on each line marking the recovery events. The infection events are marked by the independent $PP(\lambda)$ Poisson processes $\to$ and $\leftarrow$.

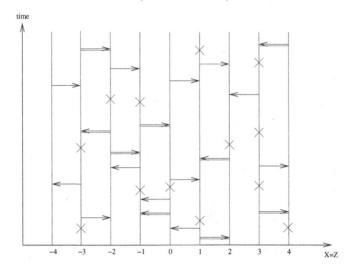

The CP and related models have applications in population dynamics and the spread of infectious diseases/viruses, etc. (see e.g. [11] and references therein).

Definition 3.1.6 Let $p(x,y) \geq 0$, $x,y \in \Lambda$ be irreducible transition rates on Λ as for the EP. The *linear voter model* (VM) on X is characterized by the jump rates

$$c(x,\eta) = \sum_{y \in \Lambda} p(x,y)\Big(\eta(x)\big(1 - \eta(y)\big) + \big(1 - \eta(x)\big)\eta(y)\Big), \quad x \in \Lambda. \quad (3.29)$$

0 and 1 can be interpreted as two different opinions, and a site x adopts the opinion of site y with rate $p(x,y)$ independently for all connected sites with different opinion.

Note that the voter model is symmetric under flipping occupation numbers, i.e.

$$c(x,\eta) = c(x,\zeta) \quad \text{if} \quad \zeta(x) = 1 - \eta(x) \quad \text{for all x} \in \Lambda. \quad (3.30)$$

Consequently it has two absorbing states $\eta \equiv 0,1$, which correspond to fixation of one of the opinions. For the general (non-linear) voter model the jump rates $c(x,\eta)$ can be any function that exhibits the symmetry (3.30), no spontaneous change of opinion and monotonicity, i.e. for $\eta(x) = 0$ we have

$$c(x,\eta) = 0 \quad \text{if} \quad \sum_{y \sim x} \eta(y) = 0,$$

$$c(x,\eta) \geq c(x,\zeta) \quad \text{if} \quad \eta(y) \geq \zeta(y) \quad \text{for all} \quad y \sim x, \quad (3.31)$$

with corresponding symmetric rules for $\eta(x) = 1$. This model and its generalizations have applications in opinion dynamics and formation of cultural beliefs (see e.g. [12] and references therein).

3.1.4 Semigroups and generators

Let X be a compact metric space and denote by

$$C(X) = \{f : X \to \mathbb{R} \text{ continuous}\} \tag{3.32}$$

the set of real-valued continuous functions. This is a Banach space with sup-norm $\|f\|_\infty = \sup_{\eta \in X} |f(\eta)|$, since by compactness of X, $\|f\|_\infty < \infty$ for all $f \in C(X)$. Functions f can be regarded as observables, and we are interested in their time evolution rather than the evolution of the full distribution. This is not only mathematically easier to formulate, but also more relevant in most applications. The full detail on the state of the process is typically not directly accessible, but is approximated by a set of measurable quantities in the spirit of $C(X)$ (but admittedly often much smaller than $C(X)$). And moreover, by specifying $\mathbb{E}(f(\eta_t))$ for all $f \in C(X)$ we have completely characterized the distribution of the process at time t, since $C(X)$ is dual to the set $\mathcal{M}_1(X)$ of all probability measures on X.[4]

Definition 3.1.7 For a given process $(\eta_t : t \geq 0)$ on X, for each $t \geq 0$ we define the operator

$$S(t) : C(X) \to C(X) \quad \text{by} \quad (S(t)f)(\zeta) := \mathbb{E}^\zeta f(\eta_t). \tag{3.33}$$

In general, $f \in C(X)$ does not imply $S(t)f \in C(X)$, but all the processes we consider have this property and are called *Feller processes*.

Proposition 3.1.8 Let $(\eta_t : t \geq 0)$ be a Feller process on X. Then the family $(S(t) : t \geq 0)$ is a **Markov semigroup**, i.e.

(a) $S(0) = Id$,   *(identity at $t = 0$)*
(b) $t \mapsto S(t)f$ is right-continuous for all $f \in C(X)$,   *(right-continuity)*
(c) $S(t + s)f = S(t)S(s)f$ for all $f \in C(X)$, $s, t \geq 0$,

                              (semigroup/Markov property)
(d) $S(t)\mathbb{1} = \mathbb{1}$ for all $t \geq 0$,   *(conservation of probability)*
(e) $S(t)f \geq 0$ for all non-negative $f \in C(X)$.   *(positivity)*

---

[4] The fact that probability measures on X can by characterized by expected values of functions on the dual $C(X)$ is a direct consequence of the Riesz representation theorem (see e.g. [13], Theorem 2.14).

Proof (a) $S(0)f(\zeta) = \mathbb{E}^\zeta\big(f(\eta_0)\big) = f(\zeta)$ since $\eta_0 = \zeta$, which is equivalent to (a) of Definition 3.1.1.

(b) for fixed $\eta \in X$ right-continuity of $t \mapsto S(t)f(\eta)$ (a mapping from $[0, \infty)$ to $\mathbb{R}$) follows directly from right-continuity of η_t and continuity of f. Right-continuity of $t \mapsto S(t)f$ (a mapping from $[0, \infty)$ to $C(X)$) w.r.t. the sup-norm on $C(X)$ requires us to show uniformity in η, which is more involved (see e.g. [14], Chapter IX, Section 1).

(c) follows from the Markov property of η_t (Definition 3.1.1(c))

$$S(t + s)f(\zeta) = \mathbb{E}^\zeta f(\eta_{t+s}) = \mathbb{E}^\zeta\Big(\mathbb{E}^\zeta\big(f(\eta_{t+s}|\mathcal{F}_t)\big)\Big) = \mathbb{E}^\zeta\Big(\mathbb{E}^{\eta_t}\big(f(\tilde{\eta}_s)\big)\Big)$$
$$= \mathbb{E}^\eta\big((S(s)f)(\eta_t)\big) = S(t)S(s)f(\zeta), \qquad (3.34)$$

where $\tilde{\eta} = \eta_{t+}$. denotes the path of the process started at time t.

(d) $S(t)1 = \mathbb{E}^\eta(1) = \mathbb{E}^\eta\big(1_{\eta_t}(X)\big) = 1$ since $\eta_t \in X$ for all $t \geq 0$ (conservation of probability).

(e) is immediate by definition. $\qquad\square$

Remarks Note that (b) implies in particular $S(t)f \to f$ as $t \to 0$ for all $f \in C(X)$, which is usually called *strong continuity* of the semigroup (see e.g. [8], Section 19). Furthermore, $S(t)$ is also *contractive*, i.e. for all $f \in C(X)$

$$\big\|S(t)f\big\|_\infty \leq \big\|S(t)|f|\big\|_\infty \leq \|f\|_\infty\big\|S(t)1\big\|_\infty = \|f\|_\infty, \qquad (3.35)$$

which follows directly from conservation of probability (d). Strong continuity and contractivity imply that $t \mapsto S(t)f$ is actually uniformly continuous for all $t > 0$. Using also the semigroup property (c) we have for all $t > \epsilon > 0$ and $f \in C(X)$

$$\big\|S(t)f - S(t - \epsilon)f\big\|_\infty = \big\|S(t - \epsilon)\big(S(\epsilon)f - f\big)\big\|_\infty$$
$$\leq \big\|S(\epsilon)f - f\big\|_\infty, \qquad (3.36)$$

which vanishes for $\epsilon \to 0$ and implies left-continuity in addition to right-continuity (b).

Theorem 3.1.9 *Suppose $(S(t) : t \geq 0)$ is a Markov semigroup on $C(X)$. Then there exists a unique (Feller) Markov process $(\eta_t : t \geq 0)$ on X such that*

$$\mathbb{E}^\zeta f(\eta_t) = S(t)f(\zeta) \quad \text{for all } f \in C(X),\ \zeta \in X \text{ and } t \geq 0. \quad (3.37)$$

Proof See [15] Theorem I.1.5 and references therein.

The semigroup $(S(t) : t \geq 0)$ describes the time evolution of expected values of observables f on X for a given Markov process. It provides a full representation of the process which is dual to the path measures $(\mathbb{P}^\zeta : \zeta \in X)$.

For a general initial distribution $\mu \in \mathcal{M}_1(X)$ the path measure (3.3) is $\mathbb{P}^\mu = \int_X \mathbb{P}^\zeta \mu(d\zeta)$. Thus

$$\mathbb{E}^\mu f(\eta_t) = \int_X \big(S(t)f\big)(\zeta)\,\mu(d\zeta) = \int_X S(t)f\,d\mu \text{ for all } f \in C(X). \quad (3.38)$$

Definition 3.1.10 For a process $(S(t) : t \geq 0)$ with initial distribution μ we denote by $\mu S(t) \in \mathcal{M}_1(X)$ the *distribution at time t*, which is uniquely determined by

$$\int_X f\,d[\mu S(t)] := \int_X S(t)f\,d\mu \quad \text{for all } f \in C(X). \quad (3.39)$$

The notation $\mu S(t)$ is a convention from functional analysis, where we write

$$\langle S(t)f, \mu \rangle := \int_X S(t)f\,d\mu = \langle f, S(t)^*\mu \rangle = \langle f, \mu S(t) \rangle. \quad (3.40)$$

The distribution μ is in fact evolved by the adjoint operator $S(t)^*$, which can also be denoted by $S(t)^*\mu = \mu S(t)$. The fact that $\mu S(t)$ is uniquely specified by (3.39) is again a consequence of the Riesz representation theorem (see e.g. [13], Theorem 2.14).

Since $(S(t) : t \geq 0)$ has the semigroup structure given in Proposition 3.1.8(c), in analogy with the proof of Proposition 3.1.2 we expect that it has the form of an exponential generated by the linearization $S'(0)$, i.e.

$$S(t) = \exp(tS'(0)) = Id + S'(0)\,t + o(t), \quad \text{with} \quad S(0) = Id, \quad (3.41)$$

which is made precise in the following.

Definition 3.1.11 The *generator* $\mathcal{L} : D_\mathcal{L} \to C(X)$ for the process $(S(t) : t \geq 0)$ is given by

$$\mathcal{L}f := \lim_{t \searrow 0} \frac{S(t)f - f}{t} \quad \text{for } f \in D_\mathcal{L}, \quad (3.42)$$

where the *domain* $D_\mathcal{L} \subseteq C(X)$ is the set of functions for which the limit exists.

The limit in (3.42) is to be understood w.r.t. the sup-norm $\|.\|_\infty$ on $C(X)$. In general, $D_\mathcal{L} \subsetneq C(X)$ is a proper subset for processes on infinite lattices, and we will see later that this is in fact the case even for the simplest examples SEP and CP we introduced above.

Proposition 3.1.12   *$\mathcal{L}$ as defined above is a **Markov generator**, i.e.*

(a) $1 \in D_{\mathcal{L}}$ *and* $\mathcal{L}1 = 0$,   *(conservation of probability)*
(b) *for $f \in D_{\mathcal{L}}$, $\lambda \geq 0$:* $\min_{\zeta \in X} f(\zeta) \geq \min_{\zeta \in X}(f - \lambda \mathcal{L}f)(\zeta)$,
                                                        (positivity)
(c) *$D_{\mathcal{L}}$ is dense in $C(X)$ and the range $\mathcal{R}(Id - \lambda \mathcal{L}) = C(X)$ for suffi-ciently small $\lambda > 0$.*

Proof (a) is immediate from the definition (3.42) and $S(t)1 = 1$, the rest is rather technical and can be found in [15], Section I.2 and in references therein.

Theorem 3.1.13   **(Hille–Yosida)** *There is a one-to-one correspon-dence between Markov generators and semigroups on $C(X)$, given by (3.42) and*

$$S(t)f := \lim_{n \to \infty} \left(Id - \frac{t}{n}\mathcal{L}\right)^{-n} f \quad \text{for } f \in C(X), \ t \geq 0. \quad (3.43)$$

Furthermore, for $f \in D_{\mathcal{L}}$ also $S(t)f \in D_{\mathcal{L}}$ for all $t \geq 0$ and

$$\frac{d}{dt} S(t)f = S(t)\mathcal{L}f = \mathcal{L}S(t)f, \quad (3.44)$$

*called the **forward** and **backward equation**, respectively.*

Proof See [15], Theorem I.2.9. and references therein.

Remarks Properties (a) and (b) in Proposition 3.1.12 are related to conservation of probability $S(t)1 = 1$ and positivity of the semigroup (see Proposition 3.1.8). By taking closures a bounded linear operator is uniquely determined by its values on a dense set. So property (c) in Proposition 3.1.12 ensures that the semigroup $S(t)$ is uniquely defined via (3.43) for all $f \in C(X)$, and that $Id - \frac{t}{n}\mathcal{L}$ is actually invertible for n large enough, as is required in the definition. The fact that $D_{\mathcal{L}}$ is dense in $C(X)$ is basically the statement that $t \mapsto S(t)$ is indeed differ-entiable at $t = 0$, confirming the intuition (3.41). This can be proved as a consequence of strong continuity of the semigroup.

Given that $S(t)f$ is the unique solution to the backward equation

$$\frac{d}{dt} u(t) = \mathcal{L}\,u(t) \quad \text{with initial condition} \quad u(0) = f, \quad (3.45)$$

one often writes $S(t) = e^{t\mathcal{L}}$ in analogy to scalar exponentials as indicated in (3.41).

It can be shown that the $\mathbb{R}$-valued process $f(\eta_t) - S(t)f(\eta_0)$ is a martingale (see e.g. [8, 16]). As an alternative to the Hille–Yosida approach, the process $(\mathbb{P}^\zeta : \zeta \in X)$ can be characterized as a unique solution to the martingale problem for a given Markov generator $\mathcal{L}$ (see [15], Sections I.5 and I.6).

Connection to Markov chains

The forward and backward equation, as well as the role of the generator and semigroup are in complete (dual) analogy to the theory of continuous-time Markov chains, where the Q-matrix generates the time evolution of the distribution at time t (see e.g. [7], Section 2.1). The approach we introduced above is more general and can of course describe the time evolution of Markov chains with countable X. With jump rates $c(\eta, \eta')$ the generator can be computed directly using (3.4) for small $t \searrow 0$,

$$S(t)f(\eta) = \mathbb{E}^\eta\big(f(\eta_t)\big) = \sum_{\eta' \in X} \mathbb{P}^\eta(\eta_t = \eta')\, f(\eta')$$

$$= \sum_{\eta' \neq \eta} c(\eta, \eta')\, f(\eta')\, t + f(\eta)\Big(1 - \sum_{\eta' \neq \eta} c(\eta, \eta')t\Big) + o(t). \tag{3.46}$$

With the definition (3.42) this yields

$$\mathcal{L}f(\eta) = \lim_{t \searrow 0} \frac{S(t)f - f}{t} = \sum_{\eta' \in X} c(\eta, \eta')\big(f(\eta') - f(\eta)\big). \tag{3.47}$$

Example For the simple random walk with state space $X = \mathbb{Z}$ we have

$$c(\eta, \eta + 1) = p \quad \text{and} \quad c(\eta, \eta - 1) = q, \tag{3.48}$$

while all other transition rates vanish. The generator is given by

$$\mathcal{L}f(\eta) = p\big(f(\eta + 1) - f(\eta)\big) + q\big(f(\eta - 1) - f(\eta)\big), \tag{3.49}$$

and in the symmetric case $p = q$ it is proportional to the discrete Laplacian.

In general, since the state space X for Markov chains is not necessarily compact, we have to restrict ourselves to bounded continuous functions f. A more detailed discussion of conditions on f for (3.47) to be a convergent sum for Markov chains can be found in Section 3.1.6. For IPS with (possibly uncountable) $X = \{0,1\}^\Lambda$ we can formally write down similar expressions for the generator. For a lattice gas (e.g. SEP) we have

$$\mathcal{L}f(\eta) = \sum_{x,y \in \Lambda} c(x, y, \eta)\big(f(\eta^{xy}) - f(\eta)\big) \qquad (3.50)$$

and for pure reaction systems like the CP or the VM

$$\mathcal{L}f(\eta) = \sum_{x \in \Lambda} c(x, \eta)\big(f(\eta^x) - f(\eta)\big). \qquad (3.51)$$

For infinite lattices Λ convergence of the sums is an issue and we have to find a proper domain $D_{\mathcal{L}}$ of functions f for which they are finite.

Definition 3.1.14 For $X = S^\Lambda$ with $S \subseteq \mathbb{N}$, $f \in C(X)$ is a *cylinder function* if there exists a finite subset $\Delta_f \subseteq \Lambda$ such that

$$f(\eta) = f(\zeta) \text{ for all } \eta, \zeta \in X \text{ with } \eta(x) = \zeta(x) \text{ for all } x \in \Delta_f, \quad (3.52)$$

i.e. f depends only on a finite set of coordinates of a configuration. We write $C_0(X) \subseteq C(X)$ for the set of all cylinder functions.

Examples The indicator function $\mathbb{1}_\eta$ is in general not a cylinder function (only on finite lattices), whereas the local particle number $\eta(x)$ or the product $\eta(x)\eta(x + y)$ are. These functions are important observables, and their expectations correspond to *local densities*

$$\rho(t, x) = \mathbb{E}^\mu\big(\eta_t(x)\big) \qquad (3.53)$$

and *two-point correlation functions*

$$\rho(t, x, x + y) = \mathbb{E}^\mu\big(\eta_t(x)\eta_t(x + y)\big). \qquad (3.54)$$

For $f \in C_0(X)$ the sum (3.51) contains only finitely many non-zero terms, so converges for any given η. However, we need $\mathcal{L}f$ to be finite w.r.t. the sup-norm of our Banach space $\big(C(X), \|\cdot\|_\infty\big)$. To assure this, we also need to impose some regularity conditions on the jump rates. For simplicity we will assume them to be of finite range as explained below. This is much more than is necessary, but it is easy to work with and fulfilled by all the examples we consider. Basically the independence of cylinder functions f and jump rates c on coordinates x outside a finite range $\Delta \subseteq \Lambda$ can be replaced by a weak dependence on coordinates $x \notin \Delta$ decaying with increasing Δ (see e.g. [15], Sections I.3 and VIII.0 for a more general discussion).

Definition 3.1.15 The jump rates of an IPS on $X = \{0, 1\}^\Lambda$ are said to be of *finite range* $R > 0$ if for all $x \in \Lambda$ there exists a finite $\Delta \subseteq \Lambda$ with $|\Delta| \leq R$ such that

$$c(x, \eta^z) = c(x, \eta) \quad \text{for all } \eta \in X \text{ and } z \notin \Delta. \tag{3.55}$$

in case of a pure reaction system. For a lattice gas the same should hold for the rates $c(x, y, \eta)$ for all $y \in \Lambda$, with the additional requirement

$$\left| \{ y \in \Lambda : c(x, y, \eta) > 0 \} \right| \leq R \quad \text{for all } \eta \in X \text{ and } x \in \Lambda. \tag{3.56}$$

Proposition 3.1.16 *Under the condition of finite range jump rates,* $\|\mathcal{L}f\|_\infty < \infty$ *for all* $f \in C_0(X)$. *Furthermore, the operators* $\mathcal{L}$ *defined in (3.50) and (3.51) are uniquely defined by their values on* $C_0(X)$ *and are Markov generators in the sense of Proposition 3.1.12.*

Proof Consider a pure reaction system with rates $c(x, \eta)$ of finite range R. Then for each $x \in \Lambda$, $c(x, \eta)$ assumes only a finite number of values (at most 2^R), and therefore $\bar{c}(x) = \sup_{\eta \in X} c(x, \eta) < \infty$. Then we have for $f \in C_0(X)$, depending on coordinates in $\Delta_f \subseteq \Lambda$,

$$\|\mathcal{L}f\|_\infty \leq 2\|f\|_\infty \sup_{\eta \in X} \sum_{x \in \Delta_f} c(x, \eta) \leq 2\|f\|_\infty \sum_{x \in \Delta_f} \sup_{\eta \in X} c(x, \eta)$$

$$\leq 2\|f\|_\infty \sum_{x \in \Delta_f} \bar{c}(x) < \infty, \tag{3.57}$$

since the last sum is finite with finite summands. A similar computation works for lattice gases.

The proof of the second statement is more involved, see e.g. [15], Theorem I.3.9. Among many other points, this involves choosing a proper metric such that $C_0(X)$ is dense in $C(X)$, which is not the case for the one induced by the sup-norm. $\square$

Generators are linear operators and Proposition 3.1.12 then implies that the sum of two or more generators is again a Markov generator (modulo technicalities regarding domains, which can be substantial in more general situations than ours, see e.g. [8]). In that way we can define more general processes, e.g. a sum of (3.50) and (3.51) could define a contact process with nearest-neighbour particle motion. In general such mixed processes are called *reaction-diffusion processes* and are important in applications, e.g. in chemistry or material science [12]. They will not be covered in these notes where we concentrate on developing the mathematical theory for the most basic models.

3.1.5 Stationary measures and reversibility

Definition 3.1.17 A measure $\mu \in \mathcal{M}_1(X)$ is *stationary* or *invariant* if $\mu S(t) = \mu$ or, equivalently, $\forall f \in C(X)$

$$\int_X S(t)f \, d\mu = \int_X f \, d\mu \quad \text{or,} \quad \mu\big(S(t)f\big) = \mu(f). \qquad (3.58)$$

The set of all invariant measures of a process is denoted by $\mathcal{I}$. A measure μ is called *reversible* if

$$\mu\big(fS(t)g\big) = \mu\big(gS(t)f\big) \quad \text{for all } f, g \in C(X). \qquad (3.59)$$

To simplify notation here and in the following we use the standard notation $\mu(f) = \int_X f \, d\mu$ for integration. This is also the expected value w.r.t. the measure μ, but we use the symbol $\mathbb{E}$ only for expectations on path space w.r.t. the measure $\mathbb{P}$.

Taking $g = 1$ in (3.59) we see that every reversible measure is also stationary. Stationarity of μ implies that

$$\mathbb{P}^\mu(\eta_. \in A) = \mathbb{P}^\mu(\eta_{t+.} \in A) \quad \text{for all } t \geq 0, \ A \in \mathcal{F}, \qquad (3.60)$$

using the Markov property (Definition 3.1.1(c)) with notation (3.3) and (3.58). Using $\eta_t \sim \mu$ as initial distribution, the definition of a stationary process can be extended to negative times on the path space $D(-\infty, \infty)$. If μ is also reversible, this implies

$$\mathbb{P}^\mu(\eta_{t+.} \in A) = \mathbb{P}^\mu(\eta_{t-.} \in A) \quad \text{for all } t \geq 0, \ A \in \mathcal{F}, \qquad (3.61)$$

i.e. the process is time-reversible. More details on this are given at the end of this section.

Proposition 3.1.18 *Consider a Feller process on a compact state space X with generator $\mathcal{L}$. Then*

$$\mu \in \mathcal{I} \quad \Leftrightarrow \quad \mu(\mathcal{L}f) = 0 \quad \text{for all } f \in C_0(X), \qquad (3.62)$$

and similarly

$$\mu \text{ is reversible} \quad \Leftrightarrow \quad \mu(f\mathcal{L}g) = \mu(g\mathcal{L}f) \quad \text{for all } f, g \in C_0(X). \qquad (3.63)$$

Proof The correspondence between semigroups and generators in the Hille–Yosida theorem is given in terms of limits in (3.42) and (3.43). By strong continuity of $S(t)$ in $t = 0$ and restricting to $f \in C_0(X)$ we can rewrite both conditions as

$$\mathcal{L}f := \lim_{n \to \infty} \underbrace{\frac{S(1/n)f - f}{1/n}}_{:=g_n} \quad \text{and} \quad S(t)f := \lim_{n \to \infty} \underbrace{\left(Id + \frac{t}{n}\mathcal{L}\right)^n f}_{:=h_n}. \qquad (3.64)$$

Now $\mu \in \mathcal{I}$ implies that for all $n \in \mathbb{N}$

$$\mu\big(S(1/n)f\big) = \mu(f) \quad \Rightarrow \quad \mu(g_n) = 0. \tag{3.65}$$

Then we have

$$\mu(\mathcal{L}f) = \mu\Big(\lim_{n \to \infty} g_n \Big) = \lim_{n \to \infty} \mu(g_n) = 0, \tag{3.66}$$

by bounded (or dominated) convergence, since g_n converges in $\big(C(X), \|.\|_\infty\big)$ as long as $f \in C_0(X)$, X is compact and $\mu(X) = 1$.

On the other hand, if $\mu(\mathcal{L}f) = 0$ for all $f \in C_0(X)$, we have by linearity

$$\mu(h_n) = \mu\bigg(\Big(Id + \frac{t}{n}\mathcal{L}\Big)^n f \bigg) = \sum_{k=0}^{n} \binom{n}{k} \frac{t^k}{n^k} \mu(\mathcal{L}^k f) = \mu(f) \tag{3.67}$$

using the binomial expansion, where only the term with $k = 0$ contributes with $\mathcal{L}^0 = Id$. This is by assumption since

$$\mu(\mathcal{L}^k f) = \mu\big(\mathcal{L}(\mathcal{L}^{k-1} f)\big) = 0 \tag{3.68}$$

and $\mathcal{L}^{k-1} f \in C_0(X)$. Then the same limit argument as above (3.66) implies $\mu\big(S(t)f\big) = \mu(f)$.

This finishes the proof of (3.62). A completely analogous argument works for the equivalence (3.63) on reversibility. $\square$

It is well known for Markov chains that on a finite state space there exists at least one stationary distribution (see Section 3.1.6). For IPS compactness of the state spaces X ensures a similar result.

Theorem 3.1.19 *For every Feller process with compact state space X we have:*

(a) $\mathcal{I}$ is non-empty, compact and convex.
(b) Suppose the weak limit $\mu = \lim_{t \to \infty} \pi S(t)$ exists for some initial distribution $\pi \in \mathcal{M}_1(X)$, i.e.

$$\pi S(t)(f) = \int_X S(t)f \, d\pi \to \mu(f) \quad \text{for all } f \in C(X), \tag{3.69}$$

then $\mu \in \mathcal{I}$.

Proof (a) Convexity of $\mathcal{I}$ follows directly from two basic facts. First, a convex combination of two probability measures $\mu_1, \mu_2 \in \mathcal{M}_1(X)$ is again a probability measure, i.e.

$$\nu := \lambda \mu_1 + (1 - \lambda)\mu_2 \in \mathcal{M}_1(X) \quad \text{for all } \lambda \in [0, 1]. \tag{3.70}$$

Second, the stationarity condition (3.62) is linear, i.e. if $\mu_1, \mu_2 \in \mathcal{I}$ then so is ν since

$$\nu(\mathcal{L}f) = \lambda\mu_1(\mathcal{L}f) + (1 - \lambda)\mu_2(\mathcal{L}f) = 0 \quad \text{for all } f \in C(X). \quad (3.71)$$

$\mathcal{I}$ is a closed subset of $\mathcal{M}_1(X)$ if we have

$$\mu_1, \mu_2, \ldots \in \mathcal{I}, \ \mu_n \to \mu \text{ weakly, } \text{implies} \ \mu \in \mathcal{I}. \quad (3.72)$$

But this is immediate by weak convergence, since for all $f \in C(X)$

$$\mu_n(\mathcal{L}f) = 0 \quad \text{for all } n \in \mathbb{N} \quad \Rightarrow \quad \mu(\mathcal{L}f) = \lim_{n\to\infty} \mu_n(\mathcal{L}f) = 0. \quad (3.73)$$

Under the topology of weak convergence $\mathcal{M}_1(X)$ is compact since X is compact,[5] and therefore also $\mathcal{I} \subseteq \mathcal{M}_1(X)$ is compact since it is a closed subset of a convex set.

Non-emptyness: by compactness of $\mathcal{M}_1(X)$ there exists a convergent subsequence of $\pi S(t)$ for every $\pi \in \mathcal{M}_1(X)$. With (b) the limit is in $\mathcal{I}$.
(b) Let $\mu := \lim_{t\to\infty} \pi S(t)$. Then $\mu \in \mathcal{I}$ since for all $f \in C(X)$,

$$\begin{aligned}
\mu(S(s)f) &= \lim_{t\to\infty} \int_X S(s)f \, d[\pi S(t)] = \lim_{t\to\infty} \int_X S(t)S(s)f \, d\pi \\
&= \lim_{t\to\infty} \int_X S(t+s)f \, d\pi = \lim_{t\to\infty} \int_X S(t)f \, d\pi \\
&= \lim_{t\to\infty} \int_X f \, d[\pi S(t)] = \mu(f). \quad (3.74)
\end{aligned}$$

$\square$

Remark By the Krein Milman theorem (see e.g. [17], Theorem 3.23), compactness and convexity of $\mathcal{I} \subseteq \mathcal{M}_1(X)$ implies that $\mathcal{I}$ is the closed convex hull of its extreme points $\mathcal{I}_e$, which are called **extremal invariant measures**. Every invariant measure can therefore be written as a convex combination of members of $\mathcal{I}_e$, so the extremal measures are the ones we need to find for a given process.

Definition 3.1.20 A Markov process with semigroup $(S(t) : t \geq 0)$ is *ergodic* if

(a) $\mathcal{I} = \{\mu\}$ is a singleton, and   (unique stationary measure)
(b) $\lim_{t\to\infty} \pi S(t) = \mu$ for all $\pi \in \mathcal{M}_1(X)$.   (convergence to equilibrium)

Phase transitions are related to the breakdown of ergodicity and in particular to non-uniqueness of stationary measures. This can be the result

---

[5] For more details on weak convergence see e.g. [18], Section 2.

of the presence of absorbing states (e.g. CP), or of spontaneous symmetry breaking/breaking of conservation laws (e.g. SEP or VM) as is discussed later. On finite lattices, IPS are Markov chains which are known
to have a unique stationary distribution under reasonable assumptions
of non-degeneracy (see Section 3.1.6). Therefore, from a mathematical
point of view, phase transitions occur only in infinite systems. Infinite
systems are often interpreted/studied as limits of finite systems, which
show traces of a phase transition by divergence or non-analytic behaviour
of certain observables. In terms of applications, infinite systems are approximations or idealizations of real systems which may be large but are
always finite, so results have to interpreted with care.

There is a well-developed mathematical theory of phase transitions for
reversible systems provided by the framework of Gibbs measures (see e.g.
[19]). But for IPS, which are in general non-reversible, the notion of phase
transitions is not unambiguous, and we will try to get an understanding
by looking at several examples.

Further remarks on reversibility

We have seen before that a stationary process can be extended to negative times on the path space $D(-\infty, \infty)$. A time-reversed stationary
process is again a stationary Markov process and the time evolution is
given by adjoint operators as explained in the following.

Let $\mu \in \mathcal{M}_1(X)$ be the stationary measure of the process $(S(t) : t \geq 0)$
and consider

$$L^2(X, \mu) = \left(f \in C(X) : \mu(f^2) < \infty \right) \tag{3.75}$$

the set of test functions square integrable w.r.t. μ. With the inner product $\langle f, g \rangle = \mu(fg)$ the closure of this (w.r.t. the metric given by the
inner product) is a Hilbert space, and the generator $\mathcal{L}$ and the $S(t)$,
$t \geq 0$ are bounded linear operators on $L^2(X, \mu)$. They are uniquely defined by their values on $C(X)$, which is a dense subset of the closure of
$L^2(X, \mu)$. Therefore they have an *adjoint* operator $\mathcal{L}^*$ and $S(t)^*$, respectively, uniquely defined by

$$\langle S(t)^* f, g \rangle = \mu(gS(t)^* f) = \mu(fS(t)g) = \langle f, S(t)g \rangle, \tag{3.76}$$

for all $f, g \in L^2(X, \mu)$ and analogously for $\mathcal{L}^*$. Note that the adjoint
operators on the self-dual Hilbert space $L^2(X, \mu)$ are not the same as
the adjoints mentioned in (3.40) on $\mathcal{M}_1(X)$ (dual to $C(X)$), which evolve
the probability measures. To compute the action of the adjoint operator

note that for all $g \in L^2(X, \mu)$

$$
\mu(gS(t)^*f) = \int_X fS(t)g \, d\mu = \mathbb{E}^\mu\big(f(\eta_0) \, g(\eta_t)\big) = \mathbb{E}^\mu\Big(\mathbb{E}\big(f(\eta_0)\big|\eta_t\big)g(\eta_t)\Big)
$$

$$
= \int_X \mathbb{E}\big(f(\eta_0)\big|\eta_t = \zeta\big)g(\zeta)\,\mu(d\zeta)
$$

$$
= \mu\Big(g\,\mathbb{E}\big(f(\eta_0)\big|\eta_t = \cdot\big)\Big), \tag{3.77}
$$

where the identity between the first and second line is due to μ being the stationary measure. Since this holds for all g it implies that

$$
S(t)^*f(\eta) = \mathbb{E}\big(f(\eta_0)\big|\eta_t = \eta\big), \tag{3.78}
$$

so the adjoint operator describes the evolution of the time-reversed process. Similarly, it can be shown that the adjoint generator $\mathcal{L}^*$ is actually the generator of the adjoint semigroup $(S(t)^* : t \geq 0)$. This includes some technicalities with domains of definition, see e.g. [20] and references therein. The process is time-reversible if $\mathcal{L} = \mathcal{L}^*$ and therefore reversibility is equivalent to $\mathcal{L}$ and $S(t)$ being *self-adjoint* as in (3.59) and (3.63).

3.1.6 Simplified theory for Markov chains

For Markov chains the state space X is countable, but not necessarily compact, think e.g. of a random walk on $X = \mathbb{Z}$. Therefore we have to restrict the construction of the semigroups to bounded continuous functions

$$
C_b(X) := \big\{f : X \to \mathbb{R} \text{ continuous and bounded}\big\}. \tag{3.79}
$$

In particular cases a larger space could be used, but the set $C_b(X)$ of bounded observables is sufficient to uniquely characterize the distribution of the of the Markov chain.[6] Note that if X is compact (e.g. for finite state Markov chains or for all IPS considered in Section 3.1.4), then $C_b(X) = C(X)$. The domain of the generator (3.47)

$$
\mathcal{L}f(\eta) = \sum_{\eta' \neq \eta} c(\eta, \eta')\big(f(\eta') - f(\eta)\big) \tag{3.80}
$$

for a Markov chain is then given by the full set of observables $\mathcal{D}_\mathcal{L} = C_b(X)$. This follows from the uniform bound $c_\eta \leq \bar{c}$ (3.7) on the jump

---

[6] cf. weak convergence of distributions, which is usually defined via expected values of $f \in C_b(X)$ (see e.g. [8], Chapter 4).

rates, since for every $f \in C_b(X)$

$$\|\mathcal{L}f\|_\infty = \sup_{\eta \in X} \mathcal{L}f(\eta) \le 2\|f\|_\infty \sup_{\eta \in X} \sum_{\eta' \in X} c(\eta, \eta')$$

$$= 2\|f\|_\infty \sup_{\eta \in X} c_\eta < \infty. \tag{3.81}$$

In particular, indicator functions $f = \mathbb{1}_\eta : X \to \{0, 1\}$ are always in $C_b(X)$ and we have

$$\int_X S(t) f \, d\mu = [\mu S(t)](\eta) =: p_t(\eta) \tag{3.82}$$

for the distribution at time t with $p_0(\eta) = \mu(\eta)$. Using this and (3.80) we get for the right-hand side of the backward equation (3.45) for all $\eta \in X$

$$\int_X \mathcal{L}S(t)\mathbb{1}_\eta d\mu = \sum_{\zeta \in X} \mu(\zeta) \sum_{\zeta' \in X} c(\zeta, \zeta') \big(S(t)\mathbb{1}_\eta(\zeta') - S(t)\mathbb{1}_\eta(\zeta) \big)$$

$$= \sum_{\zeta \in X} [\mu S(t)](\zeta) \Big(c(\zeta, \eta) - \mathbb{1}_\eta(\zeta) \sum_{\zeta' \in X} c(\zeta, \zeta') \Big)$$

$$= \sum_{\zeta \in X} p_t(\zeta) \, c(\zeta, \eta) - p_t(\eta) \sum_{\zeta' \in X} c(\eta, \zeta'), \tag{3.83}$$

where we use the convention $c(\zeta, \zeta) = 0$ for all $\zeta \in X$. In summary we get

$$\frac{d}{dt} p_t(\eta) = \sum_{\eta' \ne \eta} \big(p_t(\eta') \, c(\eta', \eta) - p_t(\eta) \, c(\eta, \eta') \big), \quad p_0(\eta) = \mu(\eta). \tag{3.84}$$

This is called the *master equation*, with intuitive gain and loss terms into state η on the right-hand side. It makes sense only for countable X, and in that case it is actually equivalent to (3.45), since the indicator functions form a basis of $C_b(X)$.

Analogous to the master equation (and using the same notation), we can get a meaningful relation for Markov chains by inserting the indicator function $f = \mathbb{1}_\eta$ in the stationarity condition (3.62). This yields with (3.80)

$$\mu(\mathcal{L}\mathbb{1}_\eta) = \sum_{\eta' \ne \eta} \big(\mu(\eta') \, c(\eta', \eta) - \mu(\eta) \, c(\eta, \eta') \big) = 0 \text{ for all } \eta \in X, \tag{3.85}$$

so that μ is a stationary solution of the master equation (3.84). A short computation yields

$$\mu\big(\mathbb{1}_\eta \mathcal{L} \mathbb{1}_{\eta'}\big) = \sum_{\zeta \in X} \mu(\zeta) \mathbb{1}_\eta(\zeta) \sum_{\xi \in X} c(\zeta,\xi)\big(\mathbb{1}_{\eta'}(\xi) - \mathbb{1}_{\eta'}(\zeta)\big)$$

$$= \mu(\eta)\, c(\eta,\eta'), \tag{3.86}$$

again using $c(\zeta,\zeta) = 0$ for all $\zeta \in X$. So inserting $f = \mathbb{1}_\eta$ and $g = \mathbb{1}_{\eta'}$ for $\eta' \neq \eta$ into the reversibility condition (3.63) on both sides we get

$$\mu(\eta')\, c(\eta',\eta) = \mu(\eta)\, c(\eta,\eta') \quad \text{for all } \eta, \eta' \in X, \eta \neq \eta', \tag{3.87}$$

which are called *detailed balance relations*. So if μ is reversible, every individual term in the sum (3.85) vanishes. On the other hand, not every solution of (3.85) has to fulfill (3.87), i.e. there are stationary measures which are not reversible. The detailed balance equations are typically easy to solve for μ, so if reversible measures exist they can be found as solutions of (3.87).

Examples Consider the simple random walk on the torus $X = \mathbb{Z}/L\mathbb{Z}$, moving with rate p to the right and q to the left. The uniform measure $\mu(\eta) = 1/L$ is an obvious solution to the stationary master equation (3.85). However, the detailed balance relations are only fulfilled in the symmetric case $p = q$. For the simple random walk on the infinite state space $X = \mathbb{Z}$ the constant solution cannot be normalized, and in fact (3.85) does not have a normalized solution.

Another important example is a *birth–death chain* with state space $X = \mathbb{N}$ and jump rates

$$c(\eta, \eta+1) = \alpha, \quad c(\eta+1, \eta) = \beta \quad \text{for all } \eta \in \mathbb{N}. \tag{3.88}$$

In this case the detailed balance relations have the solution

$$\mu(\eta) = (\alpha/\beta)^\eta. \tag{3.89}$$

For $\alpha < \beta$ this can be normalized, yielding a stationary, reversible measure for the process.

In particular, not every Markov chain has a stationary distribution. If X is finite there exists at least one stationary distribution, as a direct result of the Perron–Frobenius theorem in linear algebra (see e.g. [6], Section 6.6). For general countable (possibly infinite) state space X, existence of a stationary measure is equivalent to positive recurrence of the Markov chain (cf. [7], Section 3.5).

What about uniqueness of stationary distributions?

Definition 3.1.21 A Markov chain $(\mathbb{P}^\eta : \eta \in X)$ is called *irreducible*, if for all $\eta, \eta' \in X$

$$\mathbb{P}^\eta(\eta_t = \eta') > 0 \quad \text{for some } t \geq 0. \tag{3.90}$$

So an irreducible Markov chain can sample the whole state space, and it can be shown that this implies that it has at most one stationary distribution (cf. [7], Section 3.5). For us most important is the following statement on ergodicity as defined in Definition 3.1.20.

Proposition 3.1.22 *An irreducible Markov chain with finite state space X is ergodic.*

Proof Again a result of linear algebra, in particular the Perron–Frobenius theorem: The generator can be understood as a finite matrix $c(\eta, \eta')$, which has eigenvalue 0 with unique eigenvector μ. All other eigenvalues λ_i have negative real part, and the so-called *spectral gap*

$$\gamma := -\inf_i Re(\lambda_i) \tag{3.91}$$

determines the speed of convergence to equilibrium. □

Remarks The spectrum of the generator plays a similar role also for general Markov processes and IPS. The spectral gap is often hard to calculate, useful estimates can be found for reversible processes (see e.g. [16], Appendix 3 and also [20]).

Note that irreducibility is sufficient, but not necessary for uniqueness of the stationary measure or ergodicity of the process. In general, the state space can be decomposed into communicating subsets (classes) within which every state can be reached from every other state. For finite state space a class is called recurrent, if it can be reached from any state $\eta \in S$ (see [7], Section 1.5 for a more detailed discussion). Uniqueness of the recurrent class is then necessary and sufficient for ergodicity, and the stationary measure will concentrate on that class. For irreducible processes this class is the full state space, but it does not have to be. For example, for the contact process on a finite lattice the absorbing state $\eta \equiv 0$ is the unique recurrent class and the process is ergodic, whereas the voter model has two absorbing states $\eta \equiv 0, 1$ and is therefore not ergodic.

3.2 The asymmetric simple exclusion process

As given in Definition 3.1.4 an *exclusion process* (EP) has state space $X = \{0,1\}^\Lambda$ on a lattice Λ. The process is characterized by the generator

$$\mathcal{L}f(\eta) = \sum_{x,y\in\Lambda} c(x,y,\eta)\big(f(\eta^{xy}) - f(\eta)\big) \qquad (3.92)$$

with jump rates

$$c(x,y,\eta) = p(x,y)\,\eta(x)\big(1 - \eta(y)\big). \qquad (3.93)$$

$p(x,y)$ are irreducible transition rates of a single random walker on Λ. For the simple EP (SEP) Λ is a regular lattice such as $\mathbb{Z}^d$ and $p(x,y) = 0$ whenever x and y are not nearest neighbours. In this chapter we focus on results and techniques that apply to the asymmetric SEP (ASEP) as well as to the symmetric SEP (SSEP). For the latter there are more detailed results available based on reversibility of the process (see e.g. [15], Section VIII.1).

3.2.1 Stationary measures and conserved quantities

Definition 3.2.1   For a function $\rho : \Lambda \to [0,1]$, ν_ρ is a *product measure* on X if for all $k \in \mathbb{N}$, $x_1,\ldots,x_k \in \Lambda$ mutually different and $n_1,\ldots,n_k \in \{0,1\}$

$$\nu_\rho\big(\eta(x_1) = n_1,\ldots,\eta(x_k) = n_k\big) = \prod_{i=1}^{k} \nu^1_{\rho(x_i)}\big(\eta(x_i) = n_i\big), \quad (3.94)$$

where the *single-site marginals* are given by

$$\nu^1_{\rho(x_i)}\big(\eta(x_i) = 1\big) = \rho(x_i) \text{ and } \nu^1_{\rho(x_i)}\big(\eta(x_i) = 0\big) = 1 - \rho(x_i). \quad (3.95)$$

Remark In other words under ν_ρ the $\eta(x)$ are independent Bernoulli random variables $\eta(x) \sim Be\big(\rho(x)\big)$ with *local density* $\rho(x) = \nu\big(\eta(x)\big)$. The above definition can readily be generalized to non-Bernoulli product measures (see e.g. Section 3.3).

Theorem 3.2.2   *(a) Suppose $p(\cdot,\cdot)/C$ is doubly stochastic for some $C \in (0,\infty)$, i.e.*

$$\sum_{y'\in\Lambda} p(x,y') = \sum_{x'\in\Lambda} p(x',y) = C \quad \text{for all } x,y \in \Lambda, \qquad (3.96)$$

then $\nu_\rho \in \mathcal{I}$ for all constants $\rho \in [0,1]$ (uniform density).

(b) If $\lambda : \Lambda \to [0, \infty)$ *fulfils* $\lambda(x) p(x, y) = \lambda(y) p(y, x),$

    then $\nu_\rho \in \mathcal{I}$ *with density* $\rho(x) = \dfrac{\lambda(x)}{1 + \lambda(x)}, \quad x \in \Lambda.$

Proof For stationarity we have to show that $\nu_\rho(\mathcal{L}f) = 0$ for all $f \in C_0(X)$. This condition is linear in f and every cylinder function can be written as a linear combination of simple functions

$$f_\Delta(\eta) = \begin{cases} 1, & \eta(x) = 1 \text{ for each } x \in \Delta, \\ 0, & \text{otherwise,} \end{cases} \tag{3.97}$$

for $\Delta \subseteq \Lambda$ finite.[7] Therefore we have to check the stationarity condition only for such functions where we have

$$\nu_\rho(\mathcal{L}f_\Delta) = \sum_{x,y \in \Lambda} p(x, y) \int_X \eta(x) (1 - \eta(y)) (f_\Delta(\eta^{xy}) - f_\Delta(\eta)) \, d\nu_\rho. \tag{3.98}$$

For $x \neq y$ (we take $p(x, x) = 0$ for all $x \in \Lambda$) the integral terms in the sum look like

$$\int_X f_\Delta(\eta) \, \eta(x) (1 - \eta(y)) \, d\nu_\rho = \begin{cases} 0, & y \in \Delta, \\ (1 - \rho(y)) \prod_{u \in \Delta \cup \{x\}} \rho(u), & y \notin \Delta, \end{cases}$$

$$\int_X f_\Delta(\eta^{xy}) \, \eta(x) (1 - \eta(y)) \, d\nu_\rho = \begin{cases} 0, & x \in \Delta, \\ (1 - \rho(y)) \prod_{u \in \Delta \cup \{x\} \setminus \{y\}} \rho(u), & x \notin \Delta. \end{cases}$$

$$\tag{3.99}$$

This follows from the fact that the integrands take values only in $\{0, 1\}$ and the right-hand side is therefore the probability of the integrand being 1. Then re-arranging the sum we get

$$\nu_\rho(\mathcal{L}f_\Delta)$$
$$= \sum_{\substack{x \in A \\ y \notin A}} \left[\rho(y)(1 - \rho(x)) p(y, x) - \rho(x)(1 - \rho(y)) p(x, y) \right] \prod_{u \in A \setminus \{x\}} \rho(u). \tag{3.100}$$

Assumption of (b) is equivalent to

$$\frac{\rho(x)}{1 - \rho(x)} p(x, y) = \frac{\rho(y)}{1 - \rho(y)} p(y, x), \tag{3.101}$$

---

[7] Remember that cylinder functions depend only on finitely many coordinates and with local state space $\{0, 1\}$ therefore only take finitely many different values.

so the square bracket vanishes for all x, y in the sum (3.100). For $\rho(x) \equiv \rho$ in (a) we get

$$\nu_\rho(\mathcal{L}f_\Delta) = \rho^{|\Delta|}(1-\rho) \sum_{\substack{x \in \Delta \\ y \notin \Delta}} [p(y,x) - p(x,y)] = 0 \qquad (3.102)$$

due to $p(\cdot, \cdot)$ being proportional to a doubly-stochastic matrix.   □

For the ASEP (3.26) in one dimension with $\Lambda = \mathbb{Z}$ we have:

- Theorem 3.2.2(a) holds with $C = p + q$ and therefore $\nu_\rho \in \mathcal{I}$ for all $\rho \in [0,1]$. These measures have homogeneous density; they are reversible iff $p = q$, which is immediate from time-reversibility.
- Also Theorem 3.2.2(b) is fulfilled with $\lambda(x) = c\,(p/q)^x$ for all $c \geq 0$, since $c\,(p/q)^x\,p = c\,(p/q)^{x+1}\,q$. Therefore

$$\nu_\rho \in \mathcal{I} \quad \text{with} \quad \rho(x) = \frac{c(p/q)^x}{1 + c(p/q)^x} \quad \text{for all } c \geq 0. \quad (3.103)$$

For $p = q$ these measures are homogeneous and in fact the same ones we found above using Theorem 3.2.2(a). For $p \neq q$ the measures are not homogeneous and since e.g. for $p > q$ the density of particles (holes) is exponentially decaying as $x \to \pm\infty$ they concentrate on configurations such that

$$\sum_{x<0} \eta(x) < \infty \quad \text{and} \quad \sum_{x \geq 0} \big(1 - \eta(x)\big) < \infty. \qquad (3.104)$$

These are called *blocking measures* and turn out to be reversible also for $p \neq q$ (see [21]). Note that these measures are not translation invariant, but the dynamics of the ASEP is.

- To further understand the family of blocking measures, note that there are only countably many configurations with property (3.104), forming the disjoint union of

$$X_n = \Big\{ \eta : \sum_{x<n} \eta(x) = \sum_{x \geq n} \big(1 - \eta(x)\big) < \infty \Big\}, \quad n \in \Lambda. \qquad (3.105)$$

Whenever a particle crosses the bond $(n - 1, n)$ a hole crosses in the other direction, so the process cannot leave X_n and it is an invariant set for the ASEP. This is of course a consequence of the fact that no particles are created or destroyed. Conditioned on X_n, which is countable, the ASEP is an irreducible MC with unique stationary distribution $\nu_n := \nu_\rho(\cdot|X_n)$. Due to conditioning on X_n the distribution ν_n does actually not depend on ρ any more (cf. next section for a more detailed discussion). In [21] Liggett showed using couplings (cf. Section 3.4.2) that all extremal stationary measures of the ASEP in one dimension are

$$\mathcal{I}_e = \{\nu_\rho : \rho \in [0, 1]\} \cup \{\nu_n : n \in \mathbb{Z}\}. \tag{3.106}$$

To stress the role of the boundary conditions let us consider another example. For the ASEP on a one-dimensional torus $\Lambda_L = \mathbb{Z}/L\mathbb{Z}$ we have:

- Theorem 3.2.2(a) still applies so $\nu_\rho \in \mathcal{I}$ for all $\rho \in [0, 1]$. But part (b) no longer holds due to periodic boundary conditions, so there are no blocking measures. Under ν_ρ the total number of particles in the system is a binomial random variable

$$\Sigma_L(\eta) := \sum_{x \in \Lambda} \eta(x) \sim Bi(L, \rho) \quad \text{where}$$

$$\nu_\rho(\Sigma_L = N) = \binom{L}{N} \rho^N (1-\rho)^{L-N}. \tag{3.107}$$

Originating from statistical mechanics, the measures $\{\nu_\rho : \rho \in [0, 1]\}$ for the finite lattice Λ_L are called *grand-canonical measures/ensemble*.
- If we fix the number of particles at time 0, i.e. $\Sigma_L(\eta_0) = N$, we condition the ASEP on

$$X_{L,N} = \{\eta : \Sigma_L(\eta) = N\} \subsetneq X_L, \tag{3.108}$$

which is an invariant set since the number of particles is conserved by the dynamics. For each $N \in \mathbb{N}$, the process is irreducible on $X_{L,N}$ and $|X_{L,N}| = \binom{L}{N}$ is finite. Therefore it has a unique stationary measure $\pi_{L,N}$ on $X_{L,N}$ and the $\{\pi_{L,N} : N = 0, \ldots, L\}$ are called *canonical measures/ensemble*.

3.2.2 Symmetries and conservation laws

Definition 3.2.3 For a given Feller process $(S(t) : t \geq 0)$ a bounded[8] linear operator $T : C(X) \to C(X)$ is called a *symmetry*, if it commutes with the semigroup. So for all $t \geq 0$ we have $S(t)T = TS(t)$, i.e.

$$S(t)(Tf)(\eta) = T(S(t)f)(\eta), \quad \text{for all } f \in C(X), \ \eta \in X. \tag{3.109}$$

---

[8] $T : C(X) \to C(X)$ is bounded if there exists $B > 0$ such that for all $f \in C(X)$, $\|Tf\|_\infty \leq B\|f\|_\infty$.

Proposition 3.2.4 *For a Feller process with generator $\mathcal{L}$, a bounded linear operator $T : C(X) \to C(X)$ is a symmetry iff $\mathcal{L}T = T\mathcal{L}$, i.e.*

$$\mathcal{L}(Tf)(\eta) = T(\mathcal{L}f)(\eta), \quad \text{for all } f \in C_0(X). \tag{3.110}$$

We denote the set of all symmetries by $\mathcal{S}(\mathcal{L})$ or simply $\mathcal{S}$. The symmetries form a semigroup w.r.t. composition, i.e.

$$T_1, T_2 \in \mathcal{S} \quad \Rightarrow \quad T_1T_2 = T_1 \circ T_2 \in \mathcal{S}. \tag{3.111}$$

Proof The first part is similar to the proof of Proposition 3.1.18 on stationarity. For the second part, note that composition of operators is associative. Then for $T_1, T_2 \in \mathcal{S}$ we have

$$\mathcal{L}(T_1T_2) = (\mathcal{L}T_1)T_2 = (T_1\mathcal{L})T_2 = T_1(\mathcal{L}T_2) = (T_1T_2)\mathcal{L}, \tag{3.112}$$

so that $T_1T_2 \in \mathcal{S}$. $\qquad\square$

Proposition 3.2.5 *For a bijection $\tau : X \to X$ let $Tf := f \circ \tau$, i.e. $Tf(\eta) = f(\tau\eta)$ for all $\eta \in X$. Then T is a symmetry for the process $\big(S(t) : t \geq 0\big)$ iff*

$$S(t)(f \circ \tau) = \big(S(t)f\big) \circ \tau \quad \text{for all } f \in C(X). \tag{3.113}$$

*Such T (or equivalently τ) are called **simple symmetries**. Simple symmetries are invertible and form a group.*

Proof The first statement is immediate by the definition, T is bounded since $\|f \circ \tau\|_\infty = \|f\|_\infty$ and obviously linear.

In general, compositions of symmetries are symmetries according to Proposition 3.2.4, and if $\tau_1, \tau_2 : X \to X$ are simple symmetries then the composition $\tau_1 \circ \tau_2 : X \to X$ is also a simple symmetry. A simple symmetry τ is a bijection, so it has an inverse τ^{-1}. Then we have for all $f \in C(X)$ and all $t \geq 0$

$$\big(S(t)(f \circ \tau^{-1})\big) \circ \tau = S(t)(f \circ \tau^{-1} \circ \tau) = S(t)f, \tag{3.114}$$

since $\tau \in \mathcal{S}$. Composing with τ^{-1} leads to

$$\big(S(t)(f \circ \tau^{-1})\big) \circ \tau \circ \tau^{-1} = S(t)(f \circ \tau^{-1}) = \big(S(t)f\big) \circ \tau^{-1}, \tag{3.115}$$

so that τ^{-1} is also a simple symmetry. $\qquad\square$

Example For the ASEP on $\Lambda = \mathbb{Z}$ the translations $\tau_x : X \to X$ for $x \in \Lambda$, defined by

$$(\tau_x\eta)(y) = \eta(y - x) \quad \text{for all } y \in \Lambda \tag{3.116}$$

are simple symmetries. This can be easily seen since the jump rates are invariant under translations, i.e. we have for all $x, y \in \Lambda$

$$
\begin{aligned}
c(x, x+1, \eta) &= p\,\eta(x)\big(1 - \eta(x+1)\big) \\
&= p\,\eta(x+y-y)\big(1 - \eta(x+1+y-y)\big) \\
&= c(x+y, x+1+y, \tau_y \eta).
\end{aligned} \tag{3.117}
$$

An analogous relation holds for jumps to the left with rate $c(x, x-1, \eta) = q\eta(x)\big(1 - \eta(x-1)\big)$. Note that the family $\{\tau_x : x \in \Lambda\}$ forms a group. The same symmetry holds for the ASEP on $\Lambda_L = \mathbb{Z}/L\mathbb{Z}$ with periodic boundary conditions, where there are only L distinct translations τ_x for $x = 0, \ldots, L-1$ (since e.g. $\tau_L = \tau_0$, etc.). The argument using symmetry of the jump rates can be made more general.

Proposition 3.2.6 *Consider an IPS with jump rates $c(\eta, \eta')$ in general notation.[9] Then a bijection $\tau : X \to X$ is a simple symmetry iff*

$$
c(\eta, \eta') = c(\tau\eta, \tau\eta') \quad \text{for all } \eta, \eta' \in X. \tag{3.118}
$$

Proof Assuming the symmetry of the jump rates, we have for all $f \in C_0(X)$ and $\eta \in X$

$$
\begin{aligned}
\big(\mathcal{L}(Tf)\big)(\eta) = \big(\mathcal{L}(f \circ \tau)\big)(\eta) &= \sum_{\eta' \in X} c(\eta, \eta')\big(f(\tau\eta') - f(\tau\eta)\big) \\
&= \sum_{\eta' \in X} c(\tau\eta, \tau\eta')\big(f(\tau\eta') - f(\tau\eta)\big) \\
&= \sum_{\zeta' \in X} c(\tau\eta, \zeta')\big(f(\zeta') - f(\tau\eta)\big) \\
&= (\mathcal{L}f)(\tau\eta) = \big(T(\mathcal{L}f)\big)(\eta),
\end{aligned} \tag{3.119}
$$

where the identity in the second line just comes from relabeling the sum, which is possible since τ is bijective and the sum converges absolutely. On the other hand, $\mathcal{L}T = T\mathcal{L}$ implies that

$$
\sum_{\eta' \in X} c(\eta, \eta')\big(f(\tau\eta') - f(\tau\eta)\big) = \sum_{\eta' \in X} c(\tau\eta, \tau\eta')\big(f(\tau\eta') - f(\tau\eta)\big). \tag{3.120}
$$

Since this holds for all $f \in C_0(X)$ and $\eta \in X$ it uniquely determines that $c(\eta, \zeta) = c(\tau\eta, \tau\zeta)$ for all $\eta, \zeta \in X$ with $\eta \neq \zeta$. In fact, if there existed η, ζ for which this is not the case, we can plug $f = \mathbb{1}_{\tau\zeta}$ into (3.120) which yields a contradiction. For fixed η both sums then contain only a

---

[9] Remember that for fixed η there are only countably many $c(\eta, \eta') > 0$.

single term, so this is even possible on infinite lattices even though $\mathbb{1}_{\tau\zeta}$ is not a cylinder function.[10] $\square$

Proposition 3.2.7 *For an observable $g \in C(X)$ define the multiplication operator $T_g := g \, Id$ via*

$$T_g f(\eta) = g(\eta)\, f(\eta) \quad \text{for all } f \in C(X),\ \eta \in X. \qquad (3.121)$$

*Then T_g is a symmetry for the process $(\eta_t : t \geq 0)$ iff $g(\eta_t) = g(\eta_0)$ for all $t > 0$. In that case T_g (or equivalently g) is called a **conserved quantity**.*

Proof First note that T_g is linear and bounded since $\|f\|_\infty \leq \|g\|_\infty \|f\|_\infty$. If $g(\eta_t) = g(\eta_0)$ we have for all $t > 0$, $f \in C(X)$ and $\eta \in X$

$$\big(S(t)(T_g f)\big)(\eta) = \mathbb{E}^\eta\big(g(\eta_t)\, f(\eta_t)\big)$$
$$= g(\eta)\big(S(t)f\big)(\eta) = T_g\big(S(t)f\big)(\eta). \qquad (3.122)$$

On the other hand, if T_g is a symmetry the above computation implies that for all (fixed) $t > 0$

$$\mathbb{E}^\eta\big(g(\eta_t)\, f(\eta_t)\big) = \mathbb{E}^\eta\big(g(\eta)\, f(\eta_t)\big). \qquad (3.123)$$

Since this holds for all $f \in C(X)$ the value of $g(\eta_t)$ is uniquely specified by the expected values to be $g(\eta)$ since g is continuous (cf. argument in (3.120)). $\square$

Remarks If $g \in C(X)$ is a conserved quantity then so is $h \circ g$ for all $h : \mathbb{R} \to \mathbb{R}$ provided that $h \circ g \in C(X)$.

A subset $Y \subseteq X$ is called *invariant* if $\eta_0 \in Y$ includes $\eta_t \in Y$ for all $t > 0$. Then $g = \mathbb{1}_Y$ is a conserved quantity iff Y is invariant. In general, every level set

$$X_l = \{\eta \in X : g(\eta) = l\} \subseteq X \quad \text{for all } l \in \mathbb{R}, \qquad (3.124)$$

for a conserved quantity $g \in C(X)$ is invariant.

Examples For the ASEP on $\Lambda_L = \mathbb{Z}/L\mathbb{Z}$ the number of particles $\Sigma_L(\eta) = \sum_{x \in \Lambda_L} \eta(x)$ is conserved. The level sets of this integer valued function are the subsets

$$X_{L,N} = \big\{\eta : \Sigma_L(\eta) = N\big\} \quad \text{for } N = 0,\ldots,L, \qquad (3.125)$$

----

[10] So the function $\eta \mapsto \mathcal{L}\mathbb{1}_{\tau\zeta}(\eta)$ would in general not be well defined since it is given by an infinite sum for $\eta = \tau\zeta$. But here we are only interested in a single value for $\eta \neq \zeta$.

defined in (3.108). In particular, the indicator functions $\mathbb{1}_{X_{L,N}}$ are conserved quantities. Similar conservation laws exist for the ASEP on $\Lambda = \mathbb{Z}$ in connection with the blocking measures (3.105).

The most important result of this section is the connection between symmetries and stationary measures. For a measure μ and a symmetry T we define the measure μT via

$$(\mu T)(f) = \int_X f \, d\mu T := \int_X Tf \, d\mu = \mu(Tf) \quad \text{for all } f \in C(X), \quad (3.126)$$

which is analogous to the definition of $\mu S(t)$ in Definition 3.1.10.

Theorem 3.2.8 *For a Feller process $\big(S(t) : t \geq 0\big)$ with state space X we have*

$$\mu \in \mathcal{I}, \ T \in \mathcal{S} \quad \Rightarrow \quad \frac{1}{\mu T(X)} \, \mu T \in \mathcal{I}, \quad (3.127)$$

provided that the normalization $\mu T(X) \in (0, \infty)$.

Proof For $\mu \in \mathcal{I}$ and $T \in \mathcal{S}$ we have for all $t \geq 0$ and $f \in C(X)$

$$\begin{aligned}
(\mu T) S(t)(f) &= \mu\big(T\, S(t) f\big) = \mu\big(S(t) T f\big) \\
&= \mu S(t)(Tf) = \mu(Tf) = \mu T(f). \quad (3.128)
\end{aligned}$$

With $\mu T(X) \in (0, \infty)$, μT can be normalized and $\frac{1}{\mu T(X)} \mu T \in \mathcal{I}$. $\quad\square$

Remarks For $\mu \in \mathcal{I}$ it will often be the case that $\mu T = \mu$ so that μ is invariant under some $T \in \mathcal{S}$ and not every symmetry generates a new stationary measure. For ergodic processes $\mathcal{I} = \{\mu\}$ is a singleton, so μ has to respect all the symmetries of the process, i.e. $\mu T = \mu$ for all $T \in \mathcal{S}$.

If $T_g = g\, \mathrm{Id}$ is a conserved quantity, then $\mu T_g = g\, \mu$, i.e.

$$\mu T_g(Y) = \int_Y g(\eta)\, \mu(d\eta) \quad \text{for all measurable } Y \subseteq X. \quad (3.129)$$

So g is the *density* of μT_g w.r.t. μ and one also writes $g = \frac{d\mu T_g}{d\mu}$. This implies also that μT_g is *absolutely continuous* w.r.t. μ (short $\mu T_g \ll \mu$), which means that for all measurable Y, $\mu(Y) = 0$ implies $\mu T_g(Y) = 0$.[11]

For an invariant set $Y \subseteq X$ and the conserved quantity $g = \mathbb{1}_Y$ we have $\mu T_g = \mathbb{1}_Y \mu$. If $\mu(Y) \in (0, \infty)$ the measure of Theorem 3.2.8 can be

---

[11] In fact, absolute continuity and existence of a density are equivalent by the Radon–Nikodym theorem (see e.g. [8], Theorem 2.10).

written as a *conditional measure*

$$\frac{1}{\mu T_g(X)} \, \mu T_g = \frac{\mathbb{1}_Y}{\mu(Y)} \, \mu =: \mu(\cdot|Y) \tag{3.130}$$

concentrating on the set Y, since the normalization is $\mu T_g(X) = \mu(\mathbb{1}_Y) = \mu(Y)$.

Examples The homogeneous product measures ν_ρ, $\rho \in [0,1]$ are invariant under the translations τ_x, $x \in \Lambda$ for all translation invariant lattices with $\tau_x \Lambda = \Lambda$ such as $\Lambda = \mathbb{Z}$ or $\Lambda = \mathbb{Z}/L\mathbb{Z}$. But the blocking measures ν_n for $\Lambda = \mathbb{Z}$ are not translation invariant, and in fact $\nu_n = \nu_0 \circ \tau_{-n}$, so the family of blocking measures is generated from a single one by applying translations.

For $\Lambda_L = \mathbb{Z}/L\mathbb{Z}$ we have the invariant sets $X_{L,N}$ for a fixed number of particles $N = 0, \ldots, L$ as given in (3.108). Since the ASEP is irreducible on $X_{L,N}$ it has a unique stationary measure $\pi_{L,N}$ (see previous section). Using the above remark we can write $\pi_{L,N}$ as a conditional product measure ν_ρ (which is also stationary). For all $\rho \in (0,1)$ we have (by uniqueness of $\pi_{L,N}$)

$$\pi_{L,N} = \nu_\rho(\cdot|X_{L,N}) = \frac{\mathbb{1}_{X_{L,N}}}{\nu_\rho(X_{L,N})} \, \nu_\rho, \tag{3.131}$$

where $\nu_\rho(X_{L,N}) = \binom{L}{N}\rho^N(1-\rho)^{L-N}$ is binomial (see previous section).

Therefore we can compute explicitly

$$\pi_{L,N}(\eta) = \begin{cases} 0, & \eta \notin X_{L,N}, \\ \frac{\rho^N(1-\rho)^{L-N}}{\binom{L}{N}\rho^N(1-\rho)^{L-N}} = 1/\binom{L}{N}, & \eta \in X_{L,N}, \end{cases} \tag{3.132}$$

and $\pi_{L,N}$ is uniform on $X_{L,N}$ and in particular independent of ρ. We can write the grand-canonical product measures ν_ρ as convex combinations

$$\nu_\rho = \sum_{N=0}^{L} \binom{L}{N}\rho^N(1-\rho)^{L-N}\pi_{L,N}, \tag{3.133}$$

but this is not possible for the $\pi_{L,N}$ since they concentrate on irreducible subsets $X_{L,N} \subsetneq X_L$. Thus for the ASEP on $\Lambda_L = \mathbb{Z}/L\mathbb{Z}$ we have

$$\mathcal{I}_e = \{\pi_{L,N} : N = 0, \ldots, L\} \tag{3.134}$$

given by the canonical measures. So for each value of the conserved quantity Σ_L we have an extremal stationary measure and these are the only elements of $\mathcal{I}_e$. The latter follows from

$$X_L = \bigcup_{N=0}^{L} X_{L,N} \quad \text{and} \quad \text{irreducibility on each } X_{L,N}. \quad (3.135)$$

In fact, suppose that for some $\lambda \in (0,1)$ and $\mu_1, \mu_2 \in \mathcal{I}$

$$\pi_{L,N} = \lambda\mu_1 + (1-\lambda)\mu_2. \quad (3.136)$$

Then for all measurable $Y \subseteq X$ with $Y \cap X_{L,N} = \emptyset$ we have

$$0 = \pi_{L,N}(Y) = \lambda\mu_1(Y) + (1-\lambda)\mu_2(Y), \quad (3.137)$$

which implies that $\mu_1(Y) = \mu_2(Y) = 0$. So $\mu_1, \mu_2 \in \mathcal{I}$ concentrate on $X_{L,N}$ and thus $\mu_1 = \mu_2 = \pi_{L,N}$ by uniqueness of $\pi_{L,N}$ on $X_{L,N}$. So the conservation law provides a decomposition of the state space X_L into irreducible non-communicating subsets.

In general, taking into account all symmetries and conservation laws provides a full decomposition of the state space, and on each part concentrates a unique extremal stationary measure. This is the appropriate notion of *uniqueness of stationary measures* (cf. Definition 3.1.20) for systems with conserved quantities/symmetries. In general, a symmetry T is said to be *broken*, if there exists $\mu \in \mathcal{I}_e$ such that

$$\mu T(X) \in (0,\infty) \quad \text{and} \quad \frac{1}{\mu T(X)}\mu T \neq \mu. \quad (3.138)$$

This can be the result of non-commuting symmetries. For instance for the ASEP on $\Lambda_L = \mathbb{Z}/L\mathbb{Z}$ the $\pi_{L,N}$ are invariant under translations, but not under *CP-symmetry*, since CP-invariance and particle conservation do not commute. CP-invariance is a simple symmetry and corresponds to particle-hole and space inversion, given by

$$\tau\eta(x) = 1 - \eta(L+1-x). \quad (3.139)$$

A similar situation holds for the blocking measures for the ASEP on the infinite lattice $\Lambda = \mathbb{Z}$, which are not invariant under translations. Symmetry breaking is a form of non-uniqueness of stationary measures and is therefore often regarded as a *phase transition* in analogy to the theory of Gibbs measures. However, the use of this analogy is doubtful, because if we take it literally phase transitions are all over the place (e.g. CP-invariance is broken even on finite lattices) and the concept becomes less and less useful.

3.2.3 Currents and conservation laws

Consider the one-dimensional ASEP on $\Lambda = \mathbb{Z}$ or $\Lambda_L = \mathbb{Z}/L\mathbb{Z}$. Remember the forward equation from Theorem 3.1.13:

$$\frac{d}{dt}S(t)f = S(t)\mathcal{L}f \quad \text{which holds for all } f \in C_0(X). \quad (3.140)$$

Integrating w.r.t. the initial distribution μ the equation becomes

$$\frac{d}{dt}\mu\big(S(t)f\big) = \mu\big(S(t)\mathcal{L}f\big) = (\mu S(t))(\mathcal{L}f). \quad (3.141)$$

Using $f(\eta) = \eta(x)$ and writing $\mu_t := \mu S(t)$ for the distribution at time t we have

$$\mu_t(f) = \mathbb{E}^\mu\big(\eta_t(x)\big) =: \rho(x,t) \quad (3.142)$$

for the *particle density* at site x at time t. Note that $\eta(x)$ is a cylinder function and we have

$$(\mathcal{L}f)(\eta) = \sum_{y \in \Lambda} \Big(p\eta(y)\big(1 - \eta(y+1)\big) + q\eta(y+1)\big(1 - \eta(y)\big) \Big)$$
$$\big(f(\eta^{y,y+1}) - f(\eta)\big)$$
$$= -p\eta(x)\big(1 - \eta(x+1)\big) + q\eta(x+1)\big(1 - \eta(x)\big)$$
$$- q\eta(x)\big(1 - \eta(x-1)\big) + p\eta(x-1)\big(1 - \eta(x)\big). \quad (3.143)$$

Taking expectations w.r.t. μ_t and writing

$$\mu_t\big(\eta(x)(1 - \eta(x+1))\big) = \mu_t(1_x 0_{x+1}), \quad (3.144)$$

we get with (3.140)

$$\frac{d}{dt}\rho(x,t) = \underbrace{p\mu_t(1_{x-1}0_x) + q\mu_t(0_x 1_{x+1})}_{\text{gain}}$$
$$\underbrace{-p\mu_t(1_x 0_{x+1}) - q\mu_t(0_{x-1}1_x)}_{\text{loss}}. \quad (3.145)$$

Definition 3.2.9 The *average current* of particles across a directed edge (x, y) on a general lattice (graph) is given by

$$j(x, y, t) := \mu_t\big(c(x, y, \eta) - c(y, x, \eta)\big). \quad (3.146)$$

For the ASEP this is non-zero only across nearest-neighbour bonds and given by

$$j(x, x+1, t) = p\mu_t(1_x 0_{x+1}) - q\mu_t(0_x 1_{x+1}). \quad (3.147)$$

Then we can write, using the lattice derivative $\nabla_x j(x-1, x, t) = j(x, x+1, t) - j(x-1, x, t)$,

$$\frac{d}{dt}\rho(x, t) + \nabla_x j(x - 1, x, t) = 0, \tag{3.148}$$

which is the (lattice) *continuity equation*. It describes the time evolution of the density $\rho(x, t)$ in terms of higher order (two-point) correlation functions. The form of this equation implies that the particle density is conserved, i.e. on the finite lattice $\Lambda_L = \mathbb{Z}/L\mathbb{Z}$ with periodic boundary conditions we have

$$\frac{d}{dt}\sum_{x \in \Lambda_L} \rho(x, t) = -\sum_{x \in \Lambda_L} \nabla_x j(x - 1, x, t) = 0. \tag{3.149}$$

In general, on any finite subset $A \in \Lambda$

$$\frac{d}{dt}\sum_{x \in A} \rho(x, t) = -\sum_{x \in \partial A} \nabla_x j(x - 1, x, t), \tag{3.150}$$

where ∂A is the boundary of A. The other terms in the telescoping sum on the right-hand side cancel, which is a primitive version of Gauss' integration theorem (we have not been very careful with the notation at the boundary here).

In the special case $p = q$ (3.148) simplifies significantly. Let's take $p = q = 1$, then adding and subtracting an auxiliary term we see

$$j(x, x + 1, t) = \mu_t(1_x 0_{x+1}) + \mu_t(1_x 1_{x+1}) - \mu_t(1_x 1_{x+1}) - \mu_t(0_x 1_{x+1})$$
$$= \mu_t(1_x) - \mu_t(1_{x+1}) = \rho(x, t) - \rho(x + 1, t)$$
$$= -\nabla_x \rho(x, t). \tag{3.151}$$

So the current is given by the lattice derivative of the density, and (3.148) turns into a closed equation

$$\frac{d}{dt}\rho(x, t) = \Delta_x \rho(x, t) = \rho(x - 1, t) - 2\rho(x, t) + \rho(x + 1, t). \tag{3.152}$$

Thus the particle density of the SSEP behaves like the probability density of a single simple random walk with jump rates $p = q = 1$.

To describe this behaviour on large scales we scale the lattice constant by a factor of $1/L$ and embed it in the continuum, i.e. $\frac{1}{L}\Lambda \subseteq \mathbb{R}$ and $\frac{1}{L}\Lambda_L \subseteq \mathbb{T} = \mathbb{R}/\mathbb{Z}$ for the torus. Using the macroscopic space variable $y = x/L \in \mathbb{R}, \mathbb{T}$ we define

$$\tilde{\rho}(y, t) := \rho([yL], t) \tag{3.153}$$

for the macroscopic *density field* and use a Taylor expansion

$$\rho(x \pm 1, t) = \tilde{\rho}(y \pm \tfrac{1}{L}, t)$$
$$= \tilde{\rho}(y, t) \pm \tfrac{1}{L}\partial_y \tilde{\rho}(y, t) + \tfrac{1}{2L^2}\partial_y^2 \tilde{\rho}(y, t) + o(\tfrac{1}{L^2}) \quad (3.154)$$

to compute the lattice Laplacian in (3.152). This leads to

$$\Delta_x \rho(x, t) = \frac{1}{L^2}\partial_y^2 \tilde{\rho}(y, t), \qquad (3.155)$$

since first-order terms vanish due to symmetry. In order to get a non-degenerate equation in the limit $L \to \infty$, we have to scale time as $s = t/L^2$. This corresponds to speeding up the process by a factor of L^2, in order to see diffusive motion of the particles on the scaled lattice. Using both in (3.152) we obtain in the limit $L \to \infty$

$$\partial_s \tilde{\rho}(y, s) = \partial_y^2 \tilde{\rho}(y, s), \qquad (3.156)$$

the *heat equation*, describing the diffusion of particles on large scales.

If we use a stationary measure $\mu_t = \mu$ in the continuity equation (3.148) we get

$$0 = \frac{d}{dt}\mu(1_x) = j(x - 1, x) - j(x, x + 1), \qquad (3.157)$$

which implies that the *stationary current* $j(x, x + 1) := p\mu(1_x 0_{x+1}) - q\mu(0_x 1_{x+1})$ is site-independent. Since we know the stationary measures for the ASEP from the previous section we can compute it explicitly. For the homogeneous product measure $\mu = \nu_\rho$ we get

$$j(x, x + 1) := p\nu_\rho(1_x 0_{x+1}) - q\nu_\rho(0_x 1_{x+1})$$
$$= (p - q)\rho(1 - \rho) =: \phi(\rho), \qquad (3.158)$$

which is actually just a function of the total particle density $\rho \in [0, 1]$. We can use this to arrive at a scaling limit of the continuity equation for the asymmetric case $p \neq q$. We use the same space scaling $y = x/L$ as above and write

$$\nabla_x j(x - 1, x, t) = \tfrac{1}{L}\partial_y \tilde{j}(y - \tfrac{1}{L}, y, t) + o(\tfrac{1}{L}), \qquad (3.159)$$

with a similar notation $\tilde{j}$ as for $\tilde{\rho}$ above. In the asymmetric case the first-order terms in the spatial derivative do not vanish and we have to scale time as $s = t/L$, speeding up the process only by a factor L to see ballistic motion. In the limit $L \to \infty$ this leads to the *conservation law* (PDE)

$$\partial_s \tilde{\rho}(y, s) + \partial_y \tilde{j}(y, s) = 0, \qquad (3.160)$$

where we have redefined $\tilde{j}$ as

$$\tilde{j}(y, s) := \lim_{L \to \infty} j([yL] - 1, [yL], sL). \tag{3.161}$$

Since we effectively take microscopic time $t = sL \to \infty$ in that definition, it is plausible to assume that

$$\tilde{j}(y, s) = \phi(\tilde{\rho}(y, s)) \tag{3.162}$$

is in fact the stationary current corresponding to the local density $\tilde{\rho}(y, s)$. This is equivalent to the process becoming locally stationary in the limit $L \to \infty$, the only (slowly) varying quantity remaining on a large scale is the macroscopic density field. Local stationarity (also called local equilibrium) implies for example as $L \to \infty$

$$\mu S(sL)(1_{[yL]} 0_{[yL]+1}) \to \nu_{\tilde{\rho}(y,s)}(1_0 0_1) = \tilde{\rho}(y, s)(1 - \tilde{\rho}(y, s)). \tag{3.163}$$

Definition 3.2.10 The ASEP on $\frac{1}{L}\mathbb{Z}$ or $\frac{1}{L}\mathbb{Z}/L\mathbb{Z}$ with initial distribution μ, such that

$$\tilde{\rho}(y, 0) = \lim_{L \to \infty} \mu(1_{[yL]}) \tag{3.164}$$

exists, is in *local equilibrium* if

$$\mu S(Ls)\tau_{-[yL]} \to \nu_{\tilde{\rho}(y,s)} \quad \text{weakly (locally), as } L \to \infty, \tag{3.165}$$

where $\tilde{\rho}(y, s)$ is a solution of the *Burgers equation*

$$\partial_s \tilde{\rho}(y, s) + \partial_y \phi(\tilde{\rho}(y, s)) = 0 \quad \text{where } \phi(\rho) = (p - q)\rho(1 - \rho), \tag{3.166}$$

with initial condition $\tilde{\rho}(y, 0)$.

By local weak convergence we mean

$$\mu S(Ls)\tau_{-[yL]}(f) \to \nu_{\tilde{\rho}(y,s)}(f) \quad \text{for all } f \in C_0(X). \tag{3.167}$$

Local equilibrium has been established rigorously for the ASEP in a so-called *hydrodynamic limit*, the formulation of this result requires the following definition.

Definition 3.2.11 For each $t \geq 0$ we define the *empirical measure*

$$\pi_t^L := \frac{1}{L} \sum_{x \in \Lambda} \eta_t(x) \delta_{x/L} \in \mathcal{M}(\mathbb{R}) \text{ or } \mathcal{M}(\mathbb{T}), \tag{3.168}$$

and the measure-valued process $(\pi_t^L : t \geq 0)$ is called the *empirical process*.

The π_t^L describe the discrete particle densities on $\mathbb{R}$, $\mathbb{T}$. They are (random) measures depending on the configurations η_t and for $A \subseteq \mathbb{R}, \mathbb{T}$ we have

$$\pi_t^L(A) = \frac{1}{L}\left(\# \text{ of particles in } A \cap \frac{1}{L}\Lambda \text{ at time } t\right). \qquad (3.169)$$

Theorem 3.2.12   *Consider the ASEP ($\eta_t : t \geq 0$) on the lattice $\frac{1}{L}\mathbb{Z}$ or $\frac{1}{L}\mathbb{Z}/L\mathbb{Z}$ with initial distribution μ which has a limiting density $\tilde{\rho}(y, 0)$ analogous to (3.164). Then as $L \to \infty$*

$$\pi_{sL}^L \to \tilde{\rho}(\cdot, s)\, dy \quad \text{weakly, in probability}, \qquad (3.170)$$

where $\tilde{\rho}(y, s)$ is a solution of (3.166) on $\mathbb{R}$ or $\mathbb{T}$ with initial condition $\tilde{\rho}(y, 0)$.

Here weak convergence means that for every $g \in C_0(\mathbb{R})$ continuous with compact support

$$\pi_{sL}^L(g) = \frac{1}{L} \sum_{x \in \Lambda} g(x/L)\, \eta_t(x) \to \int_{\mathbb{R},\mathbb{T}} g(y)\, \tilde{\rho}(y, s)\, dy. \qquad (3.171)$$

The left-hand side is still random, and convergence holds in probability, i.e. for all $\epsilon > 0$

$$\mathbb{P}^\mu\left(\left|\frac{1}{L} \sum_{x \in \Lambda} g(x/L)\, \eta_t(x) - \int_{\mathbb{R},\mathbb{T}} g(y)\, \tilde{\rho}(y, s)\, dy\right| > \epsilon\right) \to 0$$

as $L \to \infty$. \hfill (3.172)

The proof is far beyond the scope of this course. The basic idea consists of two steps:

1. For large L the empirical distribution π_{sL} should be close to the distribution $\mu S(sL)$ at time sL due to a law of large numbers effect resulting from the space scaling.
2. Establish a local equilibrium according to Definition 3.2.10, which should follow from the time scaling and the process reaching local stationarity.

Of course space and time scaling are carried out simultaneously. Both approximations above will give error terms depending on L, which have to be shown to vanish in the limit $L \to \infty$. Hydrodynamic limits are still an area of major research and technically quite involved. Relevant results and references can be found in [16], Chapter 8. The above result was first proved in [22] for the TASEP ($q = 0$), and in [23] for a more general class of models using attractivity, a concept that will be discussed in Section 3.4.

3.2.4 Hydrodynamics and the dynamic phase transition

In the previous section we talked about solutions to the Burgers equation (3.166), not mentioning that it is far from clear whether that equation actually has a unique solution. A useful method to solve a *hyperbolic conservation law* of the form

$$\partial_t \rho(x,t) + \partial_x \phi(\rho(x,t)) = 0, \quad \rho(x,0) = \rho_0(x) \qquad (3.173)$$

with general *flux function* ϕ are characteristics (see [24] for full details). For a more general account of solutions and numerics of PDEs see Chapter 5. In this section we write again ρ for the macroscopic density to avoid notational overload, the notation $\tilde{\rho}$ was only introduced to make the scaling argument clear in the previous section. We consider (3.173) for $x \in \mathbb{R}$ or with periodic boundary conditions $x \in \mathbb{T}$.

Definition 3.2.13 A curve $x : [0, \infty) \to \mathbb{R}, \mathbb{T}$ with $t \mapsto x(t)$ is a *characteristic* for the PDE (3.173) if

$$\frac{d}{dt} \rho(x(t), t) = 0 \quad \text{for all } t \geq 0, \qquad (3.174)$$

i.e. the solution is constant along $x(t)$ and given by the initial conditions, $\rho(x(t), t) = \rho_0(x(0))$.

Using the PDE (3.173) to compute the total derivative we get

$$\begin{aligned}
\frac{d}{dt} \rho(x(t), t) &= \partial_t \rho(x(t), t) + \partial_x \rho(x(t), t) \, \dot{x}(t) \\
&= -\phi'(\rho(x(t), t)) \partial_x \rho(x(t), t) + \partial_x \rho(x(t), t) \, \dot{x}(t) \\
&= 0, \qquad (3.175)
\end{aligned}$$

which implies that

$$\dot{x}(t) = \phi'(\rho(x(t), t)) = \phi'(\rho_0(x(0))) \qquad (3.176)$$

is a constant given by the derivative of the flux function. This is called the *characteristic velocity* $u(\rho)$, and for the ASEP we have

$$u(\rho) = \phi'(\rho) = (p - q)(1 - 2\rho). \qquad (3.177)$$

It turns out (see [23]) that a general solution theory for hyperbolic conservation laws of the form (3.173) can be based on understanding the solutions to the *Riemann problem*, which is given by step initial data

$$\rho_0(x) = \begin{cases} \rho_l, & x \leq 0, \\ \rho_r, & x > 0. \end{cases} \qquad (3.178)$$

Discontinuous solutions of a PDE have to be understood in a weak sense.

Definition 3.2.14 $\rho : \mathbb{R} \times [0, \infty) \to \mathbb{R}$ is a *weak solution* to the conservation law (3.173) if $\rho \in L^1_{loc}(\mathbb{R} \times [0, \infty))$ and for all $\psi \in C^1(\mathbb{R} \times [0, \infty))$ with compact support and $\psi(x, 0) = 0$,

$$\int_{\mathbb{R}} \int_0^\infty \partial_t \psi(x, t) \rho(x, t) \, dx \, dt + \int_{\mathbb{R}} \int_0^\infty f(\rho(x, t)) \partial_x \psi(x, t) \, dx \, dt = 0. \quad (3.179)$$

L^1_{loc} means that for all compact $A \subseteq \mathbb{R} \times [0, \infty)$, $\int_A |\rho(x, t)| \, dx \, dt < \infty$.

The characteristics do not necessarily uniquely determine a solution everywhere, so weak solutions are in general not unique. They can be undetermined or over-determined, and both cases appear already for the simple Riemann problem (3.178) (cf. Figure 3.3). However, for a given initial density profile, the corresponding IPS which lead to the derivation of the PDE shows a unique time evolution on the macroscopic scale.

This unique *admissible solution* can be recovered from the variety of weak solutions to (3.173) by several regularization methods. The *viscosity method* is directly related to the derivation of the continuum equation in a scaling limit. For every $\epsilon > 0$ consider the equation

$$\partial_t \rho^\epsilon(x, t) + \partial_x \phi(\rho^\epsilon(x, t)) = \epsilon \partial_x^2 \phi(\rho^\epsilon(x, t)), \quad \rho^\epsilon(x, 0) = \rho_0(x). \quad (3.180)$$

This is a parabolic equation and has a unique smooth global solution for all $t > 0$, even when starting from non-smooth initial data ρ_0. This is due to the regularizing effect of the diffusive term (consider e.g. the heat equation starting with initial condition $\delta_0(x)$), which captures the fluctuations in large finite IPS. The term can be interpreted as a higher order term of order $1/L^2$ in the expansion (3.159), which disappears in the scaling limit from a particle system. Then one can define the unique admissible weak solution to (3.173) as

$$\rho(\cdot, t) := \lim_{\epsilon \to 0} \rho^\epsilon(\cdot, t) \quad \text{in } L^1_{loc}\text{-sense as above for all } t > 0. \quad (3.181)$$

It can be shown that this limit exists, and further that for one-dimensional conservation laws the precise form of the viscosity is not essential, i.e. one could also add the simpler term $\epsilon \partial_x^2 \rho^\epsilon(x, t)$ leading to the same weak limit solution [24]. There are also other admissibility criteria for hyperbolic conservation laws such as entropy conditions, which can be shown to be equivalent to the viscosity method in one dimension. We do not discuss this further here, for details see [24].

For the Riemann problem with flux function $\phi(\rho) = (p - q)\rho(1 - \rho)$ for the ASEP, there are two basic scenarios for the time evolution of step initial data shown in Figure 3.3. For $\rho_r < \rho_l$ the characteristic speeds are

$u(\rho_r) > u(\rho_l)$, and the characteristics point away from each other and open a cone of points (x, t) where the solution is not determined. The admissibility criteria described above show that the consistent solution in this case is given by the *rarefaction fan*

$$\rho(x, t) = \begin{cases} \rho_l, & x \leq u(\rho_l)t, \\ \rho_r, & x > u(\rho_r)t, \\ \rho_l + (x - tu(\rho_l))\frac{\rho_l - \rho_r}{t(u(\rho_l) - u(\rho_r))}, & u(\rho_l)t < x \leq u(\rho_r)t. \end{cases} \quad (3.182)$$

So the step dissolves and the solution interpolates linearly between the points uniquely determined by the characteristics. An illustrative extreme version of this case is the 'traffic light problem', where $\rho_l = 1$ and $\rho_r = 0$ corresponding to cars piling up behind a red traffic light. When the traffic light turns green not all cars start moving at once, but the density gradually decreases following a continuous linear profile like in real situations.

For $\rho_r > \rho_l$ we have $u(\rho_r) < u(\rho_l)$ and the characteristics point towards each other so that the solution is over-determined in a cone around the origin. Admissibility criteria show that in this case the step is stable, called a *shock solution*,

$$\rho(x, t) = \begin{cases} \rho_l, & x \leq vt, \\ \rho_r, & x > vt. \end{cases} \quad (3.183)$$

In the traffic analogy shocks correspond sharp ends of traffic jams, where density and flow change rather abruptly. The *shock speed* $v = v(\rho_l, \rho_r)$ can be derived by the conservation of mass. The average number of particles $m \geq 0$ transported through the shock in negative direction during a time interval Δt is given by $m = \Delta t(\phi(\rho_r) - \phi(\rho_l))$. If $m > 0$ ($m < 0$) this causes the shock to move with positive (negative) speed v. Therefore m is also given by $m = \Delta t\, v\,(\rho_r - \rho_l)$, leading to

$$v(\rho_l, \rho_r) = \frac{\phi(\rho_r) - \phi(\rho_l)}{\rho_r - \rho_l}. \quad (3.184)$$

As mentioned before, understanding the Riemann problem is sufficient to construct solutions to general initial data by approximations with piecewise constant functions.

In the following we will use our knowledge on solutions to the Riemann problem to understand the time evolution of the ASEP with step initial distribution

$$\mu = \nu_{\rho_l, \rho_r} \quad \text{product with} \quad \nu_{\rho_l, \rho_r}(\eta(x)) = \begin{cases} \rho_l, & x \leq 0, \\ \rho_r, & x > 0. \end{cases} \quad (3.185)$$

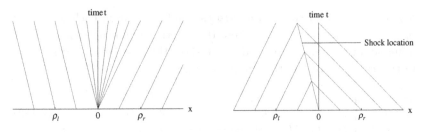

Figure 3.3 Characteristics for the Riemann problem with $\rho_l < \rho_r$ (left) showing a rarefaction fan, and $\rho_l > \rho_r$ (right), showing a shock. The curve's shock location is indicated and the speed is given by (3.184).

Theorem 3.2.15 *For the ASEP on* $\Lambda = \mathbb{Z}$ *with* $p > q$ *we have as* $t \to \infty$,

$$\nu_{\rho_l,\rho_r} S(t) \to \begin{cases} \nu_{\rho_r}, & \rho_r \geq \frac{1}{2}, \rho_l > 1 - \rho_r, & \text{(I)} \\ \nu_{\rho_l}, & \rho_l \leq \frac{1}{2}, \rho_r < 1 - \rho_r, & \text{(II)} \\ \nu_{1/2}, & \rho_l \geq \frac{1}{2}, \rho_r \leq \frac{1}{2}. & \text{(III)} \end{cases} \quad (3.186)$$

Proof By studying shock and rarefaction fan solutions of the conservation law (3.173).

Note that all the limiting distributions are stationary product measures of the ASEP, as required by Theorem 3.1.19. But depending on the initial distribution, the system selects different stationary measures in the limit $t \to \infty$, which do not depend smoothly on ρ_l and ρ_r. Therefore this phenomenon is called a *dynamic phase transition*. The set $\mathcal{I}$ of stationary measures is not changed, but the long-time behaviour of the process depends on the initial conditions in a non-smooth way. This behaviour can be captured in a *phase diagram* (see below), whose axes are given by the (fixed) parameters of our problem, ρ_l and ρ_r. We choose the limiting density

$$\rho_\infty := \lim_{t \to \infty} \nu_{\rho_l,\rho_r} S(t) \big(\eta(0) \big) \quad (3.187)$$

as the *order parameter*, which characterizes the phase transition. The different *phase regions* correspond to areas of qualitatively distinct behaviour of ρ_∞ as a function of ρ_l and ρ_r.

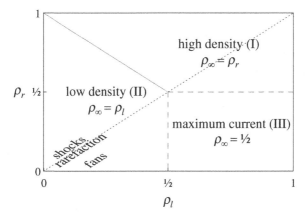

Above the dotted diagonal the solutions of the conservation law (3.173) are given by shocks, and below by rarefaction fans. Analyzing the different cases reveals the following behaviour:

(I) *High density phase:* The limiting density $\rho_\infty = \rho_r \geq 1/2$, since particles drifting to the right are jamming behind the region of high density.

(II) *Low density phase:* The limiting density is $\rho_\infty = \rho_l \leq 1/2$, since particles can drift to the right without jamming.

(III) *Maximum current phase:* The solution to the PDE is a rarefaction fan with negative (positive) characteristic velocity u on the left (right). Thus the limiting density is given by the density $1/2$ with vanishing $u(1/2) = 0$.

The dashed line is a continuous phase transition line, i.e. crossing this line the order parameter $\rho_\infty(\rho_l, \rho_r)$ is continuous. The solid line is a first-order transition line, across which the order parameter jumps from $\rho_l < 1/2$ to $\rho_r > 1/2$. The exact behaviour of the system on that line is given by

$$\nu_{\rho_l, \rho_r} S(t) \to \tfrac{1}{2}\nu_{\rho_l} + \tfrac{1}{2}\nu_{\rho_r}. \tag{3.188}$$

So the limiting distribution is a mixture, and with equal probability all local observables are determined by the left or the right product measure. Formally this leads to $\rho_\infty = 1/2$ as $\rho_l + \rho_r = 1$, but this is misleading. The local density at the origin averaged over space is typically either ρ_l or ρ_r with equal probability, but never $1/2$ as it would be for $\nu_{1/2}$. This difference can be detected by looking at a higher order correlation functions such as $\eta(0)\eta(1)$, which leads to

$$\left(\tfrac{1}{2}\nu_{\rho_l} + \tfrac{1}{2}\nu_{\rho_r}\right)\left(\eta(0)\eta(1)\right) = \tfrac{1}{2}(\rho_l^2 + \rho_r^2), \tag{3.189}$$

as opposed to $\nu_{1/2}\big(\eta(0)\eta(1)\big) = 1/4$.

The characteristics of the hyperbolic conservation law (3.173) provide a powerful tool to describe the transport properties of an IPS on a macroscopic scale. Their counterpart on a microscopic lattice scale are so-called *second class particles*, which move randomly along the characteristics depending on the local density. Since characteristics meet in shocks, second class particles are attracted by shocks, and provide a good microscopic marker for the position of a shock. This is important since a priori shocks do not look sharp on the lattice scale and do not have a well-defined location. Therefore second class particles are an important concept and have been studied in great detail (see e.g. [25], Section III.2 and references therein).

3.3 Zero-range processes

3.3.1 From ASEP to ZRPs

Consider the ASEP on the lattice $\Lambda_L = \mathbb{Z}/L\mathbb{Z}$. For each configuration $\eta \in X_{L,N}$ with $N = \sum_{x \in \Lambda_L} \eta(x)$ label the particles $j = 1, \ldots, N$ and let $x_j \in \Lambda_L$ be the position of the jth particle. We attach the labels such that the positions are ordered $x_1 < \cdots < x_N$. We map the configuration η to a configuration $\xi \in \mathbb{N}^{\Lambda_N}$ on the lattice $\Lambda_N = \{1, \ldots, N\}$ by

$$\xi(j) = x_{j+1} - x_j - 1. \qquad (3.190)$$

Here the lattice site $j \in \Lambda_N$ corresponds to particle j in the ASEP and $\xi_j \in \mathbb{N}$ to the distance to the next particle $j + 1$. Note that η and ξ are equivalent descriptions of an ASEP configuration up to the position x_1 of the first particle.

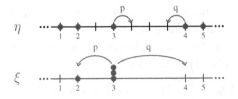

As can be seen from the construction, the dynamics of the ASEP $(\eta_t : t \geq 0)$ induce a process $(\xi_t : t \geq 0)$ on the space $\mathbb{N}^{\Lambda_N}$ with rates

$$c(\xi, \xi^{j \to j+1}) = q(1 - \delta_{0,\xi(j)}) \text{ and } c(\xi, \xi^{j \to j-1}) = p(1 - \delta_{0,\xi(j)}), \quad (3.191)$$

where we write $\xi^{x \to y} = \begin{cases} \xi(x) - 1, & z = x, \\ \xi(y) + 1, & z = y, \\ \xi(z), & z \neq x, y. \end{cases}$

Since the order of particles in the ASEP is conserved, we have $\xi_t(j) \geq 0$ and therefore $\xi_t \in \mathbb{N}^{\Lambda_N}$ for all $t \geq 0$. Note also that the number of ξ-particles is

$$\sum_{j \in \Lambda_N} \xi(j) = L - N = \text{ number of holes in ASEP}, \quad (3.192)$$

which is conserved in time, and therefore $(\xi_t : t \geq 0)$ is a lattice gas. There is no exclusion interaction for this process, i.e. the number of particles per site is not restricted. With analogy to quantum mechanics this process is sometimes called a *bosonic* lattice gas, whereas the ASEP is a *fermionic* system.

The ξ-process defined above is an example of a more general class of bosonic lattice gases, zero-range processes, which we introduce in the following. From now on we will switch back to our usual notation, denoting configurations by η and lattice sizes by L.

Definition 3.3.1 Consider a lattice Λ (any discrete set) and the state space $X = \mathbb{N}^\Lambda$. Let $p(x, y)$ be the irreducible, finite range transition probabilities of a single random walker on Λ with $p(x, x) = 0$, called the *jump probabilities*. For each $x \in \Lambda$ define the *jump rates* $g_x : \mathbb{N} \to [0, \infty)$ as a non-negative function of the number of particles $\eta(x)$ at site x, where

$$g_x(n) = 0 \quad \Leftrightarrow \quad n = 0 \quad \text{for all } x \in \Lambda. \quad (3.193)$$

Then the process $(\eta_t : t \geq 0)$ on X defined by the generator

$$\mathcal{L}f(\eta) = \sum_{x,y \in \Lambda} g_x(\eta(x)) p(x, y)(f(\eta^{x \to y}) - f(\eta)) \quad (3.194)$$

is called a *zero-range process* (ZRP).

Remarks

- ZRPs are interacting random walks with *zero-range* interaction, since the jump rate of a particle at site $x \in \Lambda$ depends only on the number of particles $\eta(x)$ at that site. The interpretation of the generator is

that each site x loses a particle with rate $g(\eta(x))$, which then jumps
to a site y with probability $p(x, y)$.

- The above ξ-process is a simple example of a ZRP with $\Lambda = \mathbb{Z}/N\mathbb{Z}$
  and

$$g_x(n) \equiv p + q, \; p(x, x + 1) = \frac{q}{p + q}, \; p(x, x - 1) = \frac{p}{p + q}. \quad (3.195)$$

- On finite lattices Λ_L of size L, irreducibility of $p(x, y)$ and (3.193)
  imply that ZRPs are irreducible finite state Markov chains on

$$X_{L,N} = \left\{ \eta \in \mathbb{N}^{\Lambda_L} \,\middle|\, \Sigma_L(\eta) = N \right\} \quad (3.196)$$

for all fixed particle numbers $N \in \mathbb{N}$ (remember the shorthand
$\Sigma_L(\eta) = \sum_{x \in \Lambda_L} \eta(x)$). Therefore they have a unique stationary distribution $\pi_{L,N}$ on $X_{L,N}$.

Examples

- For the rates $g_x(n) = g_x > 0$ for all $n \geq 0$ and $x \in \Lambda$ the ZRP can
  be interpreted as a network of $M/M/1$ server queues,[12] where at each
  site x a single server completes jobs with rate g_x and passes them on
  to another server y according to $p(x, y)$.
- For the rates $g_x(n) = g_x n$ for all $x \in \Lambda$, we have a network of $M/M/\infty$
  server queues, i.e. each queue can serve all the particles present at the
  same time. That means that each particle individually exits the queue
  at rate g_x independently of all others, leading to a total exit rate $g_x n$.
  (Remember from Section 3.1.1 that the sum of n independent $PP(g_x)$
  processes is a $PP(g_x n)$ process.) Thus this corresponds to a system
  of independent random walkers moving with rates $g_x p(x, y)$.

On infinite lattices the number of particles is in general also infinite,
but as opposed to exclusion processes the local state space of a ZRP is $\mathbb{N}$.
This is not compact, and therefore in general also X is not compact and
the construction of the process with semigroups and generators given in
Chapter 1 does not apply directly and has to be modified.

In addition to non-degeneracy (3.193) we assume a sub-linear growth
of the jump rates, i.e.

$$\bar{g} := \sup_{x \in \Lambda} \sup_{n \in \mathbb{N}} \left| g_x(n + 1) - g_x(n) \right| < \infty, \quad (3.197)$$

---

[12] $M/M/1$ means that a single server (1) receives input and generates output via
continuous-time Markov processes (M), i.e. with exponential waiting time
distributions. There are more general queueing systems with applications in
traffic routing or process optimization (see e.g. [6], Chapter 11).

and restrict to the state space

$$X_\alpha = \{\eta \in \mathbb{N}^\Lambda \mid \|\eta\|_\alpha < \infty\} \quad \text{with} \quad \|\eta\|_\alpha = \sum_{x \in \Lambda} |\eta(x)| \alpha^{|x|} \quad (3.198)$$

for some $\alpha \in (0,1)$. Let $L(X) \subseteq C(X)$ be the set of *Lipshitz-continuous* test functions $f : X_\alpha \to \mathbb{R}$, i.e.

$$|f(\eta) - f(\zeta)| \le l(f) \|\eta - \zeta\|_\alpha \quad \text{for all } \eta, \zeta \in X_\alpha. \quad (3.199)$$

Theorem 3.3.2 *Under the above conditions (3.197) to (3.199) the generator $\mathcal{L}$ given in (3.194) is well defined for $f \in L(X) \cap C_0(X)$ and generates a Markov semigroup $(S(t) : t \ge 0)$ on $L(X)$ which uniquely specifies a ZRP $(\eta_t : t \ge 0)$.*

Proof [31]. The proof includes in particular the statement that $\eta_0 \in X_\alpha$ implies $\eta_t \in X_\alpha$ for all $t \ge 0$, which follows from showing that the semigroup is contractive, i.e.

$$|S(t)f(\eta) - S(t)f(\zeta)| \le l(f) e^{3\bar{g} t/(1-\alpha)} \|\eta - \zeta\|_\alpha.$$

Remarks

- Let μ be a measure on $\mathbb{N}^\Lambda$ with density

$$\mu(\eta(x)) \le C_1 C_2^{|x|} \quad \text{for some } C_1, C_2 > 0 \quad (3.200)$$

  (this includes in particular uniformly bounded densities). Then for all $\alpha < 1/C_1$ we have $\mu(X_\alpha) = 1$, so the restricted state space is very large and contains most cases of interest.
- The conditions (3.197) to (3.199) are sufficient but not necessary, in particular (3.197) can be relaxed when looking on regular lattices and imposing a finite range condition on $p(x, y)$.

3.3.2 Stationary measures

Let $(\eta_t : t \ge 0)$ be a (non-degenerate, well-defined) ZRP on a lattice Λ with jump probabilities $p(x, y)$ and jump rates g_x.

Lemma 3.3.3 *There exists a positive harmonic function $\lambda = (\lambda_x : x \in \Lambda)$ such that*

$$\sum_{y \in \Lambda} p(y, x) \lambda_y = \lambda_x, \quad (3.201)$$

which is unique up to multiples for finite Λ.

Proof Existence of non-negative λ_x follows directly from $p(x,y)$ being the transition probabilities of a random walk on Λ, irreducibility of $p(x,y)$ implies uniqueness up to multiples and strict positivity. $\quad\square$

Note that we do not assume λ to be normalizable, which is only the case if the corresponding random walk is positive recurrent. Since (3.201) is homogeneous, every multiple of λ is again a solution. In the following we fix $\lambda_0 = 1$ (for some lattice site $0 \in \Lambda$, say the origin) and denote the one-parameter family of solutions to (3.201) by

$$\{\phi\lambda : \phi \geq 0\}, \tag{3.202}$$

where the parameter ϕ is called the *fugacity*.

Theorem 3.3.4 *For each $\phi \geq 0$, the product measure ν_ϕ with marginals*

$$\nu_\phi^x(\eta(x) = n) = \frac{w_x(n)(\phi\lambda_x)^n}{z_x(\phi)} \quad and \quad w_x(n) = \prod_{k=1}^{n} \frac{1}{g_x(k)} \tag{3.203}$$

*is stationary, provided that the local normalization (also called **partition function**)*

$$z_x(\phi) = \sum_{n=0}^{\infty} w_x(n)(\phi\lambda_x)^n < \infty \quad for\ all\ x \in \Lambda. \tag{3.204}$$

Proof To simplify notation in the proof we write

$$\nu_\phi^x(n) := \nu_\phi^x(\eta(x) = n), \tag{3.205}$$

and assume that Λ is finite. Our argument can be immediately extended to infinite lattices.

First note that using $w_x(n) = 1/\prod_{k=1}^{n} g_x(k)$ we have for all $n \geq 0$

$$\nu_\phi^x(n+1) = \frac{1}{z_x(\phi)} w_x(n+1)(\phi\lambda_x)^{n+1} = \frac{\phi\lambda_x}{g_x(n+1)} \nu_\phi^x(n). \tag{3.206}$$

We have to show that for all cylinder test functions f

$$\nu_\phi(\mathcal{L}f) = \sum_{\eta \in X} \sum_{x,y \in \Lambda} g_x(\eta(x))p(x,y)(f(\eta^{x \to y}) - f(\eta))\nu_\phi(\eta), \tag{3.207}$$

vanish, which will be done by two changes of variables:

1. For all $x, y \in \Lambda$ we change variables in the sum over η

$$\sum_{\eta \in X} g_x(\eta(x))p(x,y) f(\eta^{x \to y})\nu(\eta)$$

$$= \sum_{\eta \in X} g_x(\eta(x)+1)p(x,y) f(\eta)\nu(\eta^{y \to x}). \tag{3.208}$$

Using (3.206) we have

$$\nu_\phi(\eta^{y\to x}) = \nu_\phi^x\big(\eta(x)+1\big)\,\nu_\phi^y\big(\eta(y)-1\big)\,\prod_{z\neq x,y}\nu_\phi^z\big(\eta(z)\big)$$

$$= \frac{\phi\lambda_x}{g_x\big(\eta(x)+1\big)}\,\nu_\phi^x\big(\eta(x)\big)\,\frac{g_y\big(\eta(y)\big)}{\phi\lambda_y}\,\nu_\phi^y\big(\eta(y)\big)\,\prod_{z\neq x,y}\nu_\phi^z\big(\eta(z)\big)$$

$$= \nu_\phi(\eta)\,\frac{\lambda_x}{\lambda_y}\frac{g_y\big(\eta(y)\big)}{g_x\big(\eta(x)\big)}. \tag{3.209}$$

Plugging this into (3.207) we get

$$\nu_\phi(\mathcal{L}f)$$
$$= \sum_{\eta\in X}f(\eta)\nu_\phi(\eta)\sum_{x,y\in\Lambda}\Big(g_y\big(\eta(y)\big)p(x,y)\frac{\lambda_x}{\lambda_y} - g_x\big(\eta(x)\big)p(x,y)\Big). \tag{3.210}$$

2. Exchanging summation variables $x \leftrightarrow y$ in the first part of the above sum we get

$$\nu_\phi(\mathcal{L}f) = \sum_{\eta\in X}f(\eta)\nu_\phi(\eta)\sum_{x\in\Lambda}\frac{g_x\big(\eta(x)\big)}{\lambda_x}\sum_{y\in\Lambda}\big(p(y,x)\lambda_y - p(x,y)\lambda_x\big)$$
$$= 0, \tag{3.211}$$

since

$$\sum_{y\in\Lambda}\big(p(y,x)\lambda_y - p(x,y)\lambda_x\big) = \sum_{y\in\Lambda}\big(p(y,x)\lambda_y\big) - \lambda_x = 0. \tag{3.212}$$

Note that terms of the form $\nu_\phi^y(-1)$ do not appear in the above sums, since $g_y(0) = 0$. $\qquad\square$

Examples Take $\Lambda = \Lambda_L = \mathbb{Z}/L\mathbb{Z}$, $p(x,y) = p\,\delta_{y,x+1} + q\,\delta_{y,x-1}$ corresponding to nearest-neighbour jumps on a one-dimensional lattice with periodic boundary conditions. Then we simply have $\lambda_x = 1$ for all $x \in \Lambda_L$ as the solution to (3.201).

For the constant jump rates $g_x(n) = 1$ for all $n \geq 1$, $x \in \Lambda_L$ the stationary weights are just $w_x(n) = 1$ for all $n \geq 0$.[13] So the stationary product measures ν_ϕ have geometric marginals

$$\nu_\phi^x(\eta(x) = n) = (1-\phi)\phi^n \quad\text{since}\quad z_x(\phi) = \sum_{k=0}^\infty \phi^n = \frac{1}{1-\phi}, \tag{3.213}$$

which are well defined for all $\phi \in [0,1)$.

---

[13] We always use the convention that the empty product $\prod_{k=1}^0 1/g_x(k) = 1$.

For independent particles with jump rates $g_x(n) = n$ for all $x \in \Lambda_L$ we have $w_x(n) = 1/n!$ and the ν_ϕ have Poisson marginals

$$\nu_\phi^x(\eta(x) = n) = \frac{\phi^n}{n!} e^{-\phi} \quad \text{since} \quad z_x(\phi) = \sum_{k=0}^{\infty} \frac{\phi^k}{k!} = e^\phi, \quad (3.214)$$

which are well defined for all $\phi \geq 0$.

Remarks

- The partition function $z_x(\phi) = \sum_{n=0}^{\infty} w_x(n)(\phi\lambda_x)^n$ is a power series with radius of convergence

$$r_x = \left(\limsup_{n \to \infty} w_x(n)^{1/n} \right)^{-1}$$

$$\text{and so} \quad z_x(\phi) < \infty \quad \text{if} \quad \phi < r_x/\lambda_x. \quad (3.215)$$

If $g_x^\infty = \lim_{k \to \infty} g_x(k) \in [0, \infty]$ exists, we have

$$w_x(n)^{1/n} = \left(\prod_{k=1}^{n} g_x(k)^{-1} \right)^{1/n}$$

$$= \exp\left(-\frac{1}{n} \sum_{k=1}^{n} \log g_x(k) \right) \to 1/g_x^\infty \quad (3.216)$$

as $n \to \infty$, so that $r_x = g_x^\infty$.
- The density at site $x \in \Lambda$ is given by

$$\rho_x(\phi) = \nu_\phi^x(\eta(x)) = \frac{1}{z_x(\phi)} \sum_{k=1}^{\infty} k \, w_x(k)(\phi\lambda_x)^k. \quad (3.217)$$

Multiplying the coefficients $w_x(k)$ by k (or any other polynomial) does not change the radius of convergence of the power series and therefore $\rho_x(\phi) < \infty$ for all $\phi < r_x/\lambda_x$.

Furthermore $\rho_x(0) = 0$ and it can be shown that $\rho_x(\phi)$ is a monotone increasing function of ϕ. Note that for $\phi > r_x/\lambda_x$ the partition function and $\rho_x(\phi)$ diverge, but for $\phi = r_x/\lambda_x$ both, convergence or divergence, are possible.
- With Definition 3.2.9 the expected stationary current across a bond (x, y) is given by

$$j(x, y) = \nu_\phi^x(g_x) \, p(x, y) - \nu_\phi^y(g_y) \, p(y, x), \quad (3.218)$$

and using the form $w_x(n) = 1/\prod_{k=1}^{n} g_x(k)$ of the stationary weight we have

$$\nu_\phi^x(g_x) = \frac{1}{z_x(\phi)} \sum_{n=1}^{\infty} g_x(n)\, w_x(n) (\phi\lambda_x)^n$$

$$= \frac{\phi\lambda_x}{z_x(\phi)} \sum_{n=1}^{\infty} w_x(n-1)(\phi\lambda_x)^{n-1} = \phi\lambda_x. \qquad (3.219)$$

So the current is given by

$$j(x,y) = \phi\big(\lambda_x p(x,y) - \lambda_y p(y,x)\big), \qquad (3.220)$$

which is proportional to the fugacity ϕ and the stationary probability current of a single random walker (as long as λ can be normalized).

Examples For the above example with $\Lambda_L = \mathbb{Z}/L\mathbb{Z}$, $p(x,y) = p\,\delta_{y,x+1} + q\,\delta_{y,x-1}$ and $g_x(n) = 1$ for $n \geq 1$, $x \in \Lambda$ the density is of course x-independent and given by

$$\rho_x(\phi) = \rho(\phi) = (1-\phi)\sum_{k=1}^{\infty} k\phi^k = \frac{\phi}{1-\phi}. \qquad (3.221)$$

The stationary current $j(x, x+1) = \phi(p-q)$ for all $x \in \Lambda_L$, and as we have seen before in one-dimensional systems it is bond-independent. Using the invertible relation (3.221) we can write the stationary current as a function of the density ρ analogous to the ASEP in Section 3.2,

$$j(\rho) = (p-q)\frac{\rho}{1+\rho}, \qquad (3.222)$$

where we use the same letter j to avoid notational overload.

For independent particles with $g_x(n) = n$ for all $x \in \Lambda$, we get the very simple relation

$$\rho(\phi) = e^{-\phi}\sum_{k=1}^{\infty} k\frac{\phi^k}{k!} = \phi e^{-\phi}\sum_{k=0}^{\infty} \frac{\phi^k}{k!} = \phi. \qquad (3.223)$$

For the current this implies the linear relation

$$j(\rho) = (p-q)\rho, \qquad (3.224)$$

which is to be expected for independent particles.

3.3.3 Equivalence of ensembles and relative entropy

In this section let $(\eta_t : t \geq 0)$ be a homogeneous ZRP on the lattice $\Lambda_L = \mathbb{Z}/L\mathbb{Z}$ with state space $X_L = \mathbb{N}^{\Lambda_L}$, jump rates $g_x(n) \equiv g(n)$ and translation invariant jump probabilities $p(x, y) = q(y - x)$. This implies that the stationary product measures ν_ϕ given in Theorem 3.3.4 are translation invariant with marginals

$$\nu_\phi^x\big(\eta(x) = n\big) = \frac{w(n)\,\phi^n}{z(\phi)}. \tag{3.225}$$

Analogous to Section 3.2.1 for exclusion processes, the family of measures

$$\{\nu_\phi^L : \phi \in [0, \phi_c)\} \text{ is called the \textbf{grand-canonical ensemble},} \tag{3.226}$$

where ϕ_c is the radius of convergence of the partition function $z(\phi)$ (called r_x in the previous section for more general processes). We further assume that the jump rates are bounded away from 0, i.e. $g(k) \geq C > 0$ for all $k > 0$, which implies that $w(k) \leq C^{-k}$ and thus $\phi_c \geq C > 0$ using (3.216). The particle density $\rho(\phi)$ is characterized uniquely by the fugacity ϕ as given in (3.217).

As noted before the ZRP is irreducible on

$$X_{L,N} = \big\{\eta \in \mathbb{N}^{\Lambda_L} \,\big|\, \Sigma_L(\eta) = N\big\} \tag{3.227}$$

for all fixed particle numbers $N \in \mathbb{N}$. It has a unique stationary measure $\pi_{L,N}$ on $X_{L,N}$, and analogous to the ASEP in Section 3.2.2 it can be written as a conditional product measure

$$
\begin{aligned}
\pi_{L,N}(\eta) &= \nu_\phi^L(\eta | X_{L,N}) \\
&= \mathbb{1}_{X_{L,N}}(\eta) \frac{\phi^N \prod_x w(\eta(x))}{z(\phi)^L} \frac{z(\phi)^L}{\phi^N \sum_{\eta \in X_{L,N}} \prod_x w(\eta(x))} \\
&= \frac{\mathbb{1}_{X_{L,N}}(\eta)}{Z_{L,N}} \prod_{x \in \Lambda_L} w(\eta(x)),
\end{aligned}
\tag{3.228}
$$

where we write $Z_{L,N} = \sum_{\eta \in X_{L,N}} \prod_x w(\eta(x))$ for the *canonical partition function*.

The family of measures

$$\{\pi_{L,N} : N \in \mathbb{N}\} \text{ is called the \textbf{canonical ensemble}.} \tag{3.229}$$

In general these two ensembles are expected to be 'equivalent' as $L \to \infty$, in vague analogy to the law of large numbers for iid random variables. We will make this precise in the following. To do this we need to quantify the 'distance' of two probability measures.

Definition 3.3.5 Let $\mu_1, \mu_2 \in \mathcal{M}_1(\Omega)$ be two probability measures on a countable space Ω. Then the *relative entropy* of μ_1 w.r.t. μ_2 is defined as

$$H(\mu_1; \mu_2) = \begin{cases} \mu_1\left(\log \frac{\mu_1}{\mu_2}\right) = \sum_{\omega \in \Omega} \mu_1(\omega) \log \frac{\mu_1(\omega)}{\mu_2(\omega)}, & \mu_1 \ll \mu_2, \\ \infty, & \mu_1 \not\ll \mu_2, \end{cases} \tag{3.230}$$

where $\mu_1 \ll \mu_2$ is a shorthand for $\mu_2(\omega) = 0 \Rightarrow \mu_1(\omega) = 0$ (called *absolute continuity*).

Lemma 3.3.6 (Properties of relative entropy) *Let $\mu_1, \mu_2 \in \mathcal{M}_1(\Omega)$ be two probability measures on a countable space Ω.*

(i) **Non-negativity:** $H(\mu_1; \mu_2) \geq 0$ *and* $H(\mu_1; \mu_2) = 0 \Leftrightarrow \mu_1(\omega) = \mu_2(\omega)$ *for all $\omega \in \Omega$.*

(ii) **Sub-additivity:** *Suppose $\Omega = S^\Lambda$ with some local state space $S \subseteq \mathbb{N}$ and a lattice Λ. Then for $\Delta \subseteq \Lambda$ and marginals μ_i^Δ, $H\left(\mu_1^\Delta; \mu_2^\Delta\right)$ is increasing in Δ and*

$$H(\mu_1; \mu_2) \geq H\left(\mu_1^\Delta; \mu_2^\Delta\right) + H\left(\mu_1^{\Lambda \backslash \Delta}; \mu_2^{\Lambda \backslash \Delta}\right). \tag{3.231}$$

If μ_1 and μ_2 are product measures, then equality holds.

(iii) **Entropy inequality:** *For all bounded $f \in C_b(\Omega)$ and all $\epsilon > 0$ we have*

$$\mu_1(f) \leq \frac{1}{\epsilon}\left(\log \mu_2\left(e^{\epsilon f}\right) + H(\mu_1; \mu_2)\right). \tag{3.232}$$

Proof In the following let $\mu_1 \ll \mu_2$ and $h(\omega) = \mu_1(\omega)/\mu_2(\omega) \geq 0$.
  (i) Then

$$H(\mu_1; \mu_2) = \mu_2(h \log h) = \mu_2\left(\phi(h)\right) \quad \text{with}$$
$$\phi(u) := u \log u + 1 - u, \tag{3.233}$$

since $\mu_2(1 - h) = 1 - \mu_1(1) = 1 - 1 = 0$. Elementary properties of ϕ are

$$\phi(u) \geq 0 \text{ for } u \geq 0 \quad \text{and} \quad \phi(u) = 0 \Leftrightarrow u = 1, \tag{3.234}$$

which implies that $H(\mu_1; \mu_2) \geq 0$. If $\mu_1 = \mu_2$ the relative entropy obviously vanishes.

On the other hand, if $H(\mu_1; \mu_2) = 0$ then $\phi\left(h(\omega)\right) = 0$ whenever $\mu_2(\omega) > 0$, which implies $h(\omega) = 1$ and thus $\mu_1(\omega) = \mu_2(\omega)$. Since $\mu_1 \ll \mu_2$ equality also holds when $\mu_2(\omega) = 0$.

(ii) For $\Omega = S^\Lambda$ we fix some $\Delta \subsetneq \Lambda$ and write $h(\eta) = \mu_1(\eta)/\mu_2(\eta)$ and $h^\Delta(\eta(\Delta)) = \mu_1^\Delta(\eta(\Delta))/\mu_2^\Delta(\eta(\Delta))$ for marginal distributions with

$\Delta \subseteq \Lambda_L$. Then h^Δ is given by an expectation conditioned on the sub-configuration $\eta(\Delta)$ on Δ,

$$h^\Delta(\eta(\Delta)) = \frac{\mu_1^\Delta}{\mu_2^\Delta}(\eta(\Delta)) = \mu_2\Big(\frac{\mu_1}{\mu_2}\Big|\eta(\Delta)\Big) = \mu_2\big(h\big|\eta(\Delta)\big). \quad (3.235)$$

Since ϕ is convex we can apply Jensen's inequality to get

$$\phi(h^\Delta(\eta(\Delta))) = \phi\Big(\mu_2\big(h\big|\eta(\Delta)\big)\Big) \leq \mu_2\big(\phi(h)\big|\eta(\Delta)\big). \quad (3.236)$$

Therefore with $\mu_2\Big(\mu_2\big(\phi(h)\big|\eta(\Delta)\big)\Big) = \mu_2\big(\phi(h)\big)$ we have

$$H\big(\mu_1^\Delta;\mu_2^\Delta\big) = \mu_2\big(\phi(h^\Delta)\big) \leq \mu_2\big(\phi(h)\big) = H\big(\mu_1;\mu_2\big), \quad (3.237)$$

which implies that in general $H\big(\mu_1^\Delta;\mu_2^\Delta\big)$ is increasing in Δ.

Using the auxiliary measure $\nu = \frac{\mu_1^\Delta}{\mu_2^\Delta}\mu_2$ monotonicity in Δ implies

$$H(\mu_1;\mu_2) - H\big(\mu_1^\Delta;\mu_2^\Delta\big) = \mu_1\Big(\log\frac{\mu_1\,\mu_2^\Delta}{\mu_2\,\mu_1^\Delta}\Big) = \mu_1\Big(\log\frac{\mu_1}{\nu}\Big)$$

$$= H(\mu_1;\nu) \geq H\big(\mu_1^{\Lambda\backslash\Delta};\nu^{\Lambda\backslash\Delta}\big) = H\big(\mu_1^{\Lambda\backslash\Delta};\mu_2^{\Lambda\backslash\Delta}\big), \quad (3.238)$$

since $\nu^{\Lambda\backslash\Delta} = \mu_2^{\Lambda\backslash\Delta}$ by definition ($\mu_1^\Delta/\mu_2^\Delta$ does not change μ_2 on $\Lambda\backslash\Delta$).

If μ_1 and μ_2 are product measures, then $h = \mu_1/\mu_2$ factorizes, leading to equality.

Proof of (iii) is harder, see e.g. [16], Appendix 1. $\qquad\qquad\square$

Remarks

- $H(\mu_1;\mu_2)$ is not symmetric and therefore not a metric on $\mathcal{M}_1(X)$.
- (i) only holds if $\mu_1, \mu_2 \in \mathcal{M}_1(X)$ are normalized probability measures, for general distributions in $\mathcal{M}(X)$ the relative entropy can also be negative.
- $H(\mu_1;\mu_2)$ is a well studied concept from information theory, often also called *Kullback–Leibler divergence* or *information gain*.

Theorem 3.3.7 *Consider the canonical and grand-canonical ensembles for a homogeneous ZRP as defined above. Then the **specific relative entropy***

$$h_L(\phi) := \frac{1}{L}H(\pi_{L,N};\nu_\phi^L) \to 0 \qquad\qquad (3.239)$$

*in the **thermodynamic limit** $L \to \infty$ and $N/L \to \bar{\rho} \geq 0$, provided that $\phi \in [0, \phi_c)$ solves $\rho(\phi) = \bar{\rho}$.*

Proof First we fix some $L \geq 0$. Note that for all $\eta \in X_L$ and $\phi > 0$, $\nu_\phi(\eta) > 0$, so in particular $\pi_{L,N} \ll \nu_\phi$ and we have

$$h_L(\phi) = \frac{1}{L} \sum_{\eta \in X_{L,N}} \pi_{L,N}(\eta) \log \frac{\pi_{L,N}(\eta)}{\nu_\phi^L(\eta)}. \tag{3.240}$$

Using the form (3.225) and (3.228) of the two measures we get for $\eta \in X_{L,N}$

$$\frac{\pi_{L,N}(\eta)}{\nu_\phi^L(\eta)} = \frac{\prod_x w(\eta(x))}{Z_{L,N}} \frac{z(\phi)^L}{\prod_x w(\eta(x))\phi^{\eta(x)}} = \frac{z(\phi)^L}{Z_{L,N}\phi^N}. \tag{3.241}$$

So due to the special form of the ensembles we get the simple expression

$$h_L(\phi) = \frac{1}{L} \sum_{\eta \in X_{L,N}} \pi_{L,N}(\eta) \log \frac{z(\phi)^L}{Z_{L,N}\phi^N} = -\frac{1}{L} \log \frac{Z_{L,N}\phi^N}{z(\phi)^L}. \tag{3.242}$$

Further note that

$$Z_{L,N} = \sum_{\eta \in X_{L,N}} \prod_{x \in \Lambda_L} w(\eta(x)) = \nu_\phi^L(\Sigma_L(\eta) = N)\phi^{-N}z(\phi)^L, \tag{3.243}$$

and thus

$$h_L(\phi) = -\frac{1}{L} \log \nu_\phi^L(\Sigma_L(\eta) = N). \tag{3.244}$$

Since $\phi < \phi_c$ we have $\sum_n n^2 w(n)\phi^n < \infty$. So under ν_ϕ the $\eta(x)$ are iid random variables with finite variance and mean $\nu_\phi^x(\eta(x)) = \rho(\phi) = \bar{\rho}$. Now taking $L \to \infty$ with $N/L \to \bar{\rho}$ by the local central limit theorem (see e.g. [26], Chapter 9)

$$\nu_\phi^L(\Sigma_L(\eta) = N) = \nu_\phi^L\left(\sum_{x \in \Lambda_L} \eta(x) = N \right) = O(L^{-1/2}), \tag{3.245}$$

which corresponds to the width $\sqrt{L}$ of the distribution of a sum of iid random variables. This implies that

$$h_L(\phi) = O\left(\frac{1}{L} \log L \right) \to 0 \quad \text{as } L \to \infty. \tag{3.246}$$

$\square$

Note that this convergence result only holds if $\bar{\rho}$ is in the range of the function $\rho(\phi)$ for $\phi \in [0, \phi_c)$. Whenever this is not the case the system exhibits an interesting phase transition which is discussed in detail in the next section.

Corollary 3.3.8 *Let $f \in C_0(X)$ be a cylinder test function with $\nu_\phi(e^{\epsilon f}) < \infty$ for some $\epsilon > 0$. Then with ν_ϕ being the product measure on the whole lattice,*

$$\mu_{L,N}(f) \to \nu_\phi(f) \quad as \quad L \to \infty, \tag{3.247}$$

provided that $\phi \in [0, \phi_c)$ solves $\rho(\phi) = \bar\rho = \lim_{L\to\infty} N/L$.

Proof Let $\Delta \subseteq \Lambda_L$ be the finite range of dependence of the cylinder function $f \in C_0(X)$. Then we can plug $f - \nu_\phi^\Delta(f)$ and $\nu_\phi^\Delta(f) - f$ in the entropy inequality (3.232) to show that

$$\left| \pi_{L,N}(f) - \nu_\phi(f) \right| \leq H(\pi_{L,N}^\Delta; \nu_\phi^\Delta). \tag{3.248}$$

This involves extending the inequality to unbounded functions f with finite exponential moments and a standard $\epsilon-\delta$ argument. It is rather lengthy and we do not present this here, for a reference see e.g. [27], Lemma 3.1.

Then sub-additivity (Lemma 3.3.6(ii)) gives

$$H(\pi_{L,N}^\Delta; \nu_\phi^\Delta) \leq \frac{|\Delta|}{L} H(\pi_{L,N}; \nu_\phi^L) = |\Delta| h_L(\phi) \to 0 \tag{3.249}$$

as $L \to \infty$ which implies the statement. $\square$

Remarks

- The above corollary implies e.g. convergence of the test function $f(\eta) = \eta(x)$, since for all $\phi < \phi_c$

$$\sum_{n=0}^{\infty} e^{\epsilon n} w(n) \phi^n < \infty \quad for \quad e^\epsilon \phi < \phi_c \text{, i.e. } \epsilon < \log \frac{\phi_c}{\phi}. \tag{3.250}$$

So $\pi_{L,N}(\eta(x)) = N/L \to \nu_\phi(\eta(x)) = \rho(\phi)$, which is not very surprising since ϕ is chosen to match the limiting density $\bar\rho$.

- The function $f(\eta) = \eta(x)^2$ corresponding to the second moment is not covered by the above result, since $e^{\epsilon n^2}$ grows too fast with n for all $\epsilon > 0$. However, convergence can be extended to functions $f \in L^2(\nu_\phi)$ (with considerable technical effort, see e.g. the appendix of [16]). Since $\phi < \phi_c$ leads to an exponential decay of $w(n)\phi^n$, this extension includes all polynomial correlation functions.

3.3.4 Phase separation and condensation

Since ZRPs are bosonic lattice gases, they exhibit a condensation transition under certain conditions which is similar to Bose–Einstein condensation for bosons. For more details and other applications and related results to this section see [28] and references therein. As in the previous section we consider a homogeneous ZRP on the lattice $\Lambda_L = \mathbb{Z}/L\mathbb{Z}$ with jump rates $g(n)$ bounded away from 0 for $n > 0$ and translation invariant jump probabilities $p(x, y) = q(y - x)$.

Definition 3.3.9 Let $\rho(\phi) = \nu_\phi(\eta(x))$ be the density of the grand-canonical product measure ν_ϕ and $\phi_c \in [0, \infty]$ be the radius of convergence of the partition function $z(\phi)$. Then we define the *critical density*

$$\rho_c = \lim_{\phi \nearrow \phi_c} \rho(\phi) \in [0, \infty]. \qquad (3.251)$$

ρ_c can take the value ∞, as we have seen above for the example

$$g(k) = 1 - \delta_{k,0} \quad \Rightarrow \quad \rho(\phi) = \frac{\phi}{1 - \phi} \nearrow \infty \quad \text{as } \phi \nearrow \phi_c = 1. \qquad (3.252)$$

In fact, this is the 'usual' situation since it implies that the function $\rho(\phi)$ is a bijection and there exists a grand-canonical stationary measure for all densities $\rho \geq 0$.

To have an example with $\rho_c < \infty$ we need

$$\sum_{n=0}^{\infty} n\, w(n)\, \phi_c^n < \infty, \qquad (3.253)$$

i.e. the power series has to converge **at** the radius of convergence ϕ_c.

Therefore $w(n)\phi_c^n$ has to decay sub-exponentially (by definition of ϕ_c), but fast enough for the sum to converge. A generic example is a power law decay

$$w(n)\, \phi_c^n \simeq n^{-b} \quad \text{as} \quad n \to \infty \quad \text{with} \quad b > 2. \qquad (3.254)$$

Since we have the explicit formula $w(n) = \prod_{k=1}^n g(k)^{-1}$ this implies for the jump rates

$$g(n) = \frac{w(n-1)}{w(n)} \simeq \frac{(n-1)^{-b}\phi_c^{-(n-1)}}{n^{-b}\phi_c^{-n}}$$
$$= \phi_c(1 - 1/n)^{-b} \simeq \phi_c(1 + b/n) \qquad (3.255)$$

to leading order. Such a ZRP with rates

$$g(n) = 1 + b/n \quad \text{with} \quad \phi_c = 1 \quad \text{and} \quad w(n) \simeq \Gamma(1 + b)\, n^{-b} \qquad (3.256)$$

was introduced [29]. For this model ρ_c can be computed explicitly,

$$\rho_c = \frac{1}{b-2} < \infty \quad \text{for} \quad b > 2. \qquad (3.257)$$

The interesting question is now, what happens to the equivalence of ensembles in the limit $L \to \infty$ with $N/L \to \bar{\rho} > \rho_c$?

Theorem 3.3.10 *Consider the canonical $\pi_{L,N}$ and the grand-canonical measures ν_ϕ^L of a homogeneous ZRP, for which we assume that*

$$\lim_{n \to \infty} \frac{1}{n} \sum_{k=1}^{n} \log g(k) \in \mathbb{R} \quad \text{exists.} \qquad (3.258)$$

Then

$$h_L(\phi) := \frac{1}{L} H(\pi_{L,N}; \nu_\phi^L) \to 0 \text{ as } L \to \infty \text{ and } N/L \to \bar{\rho} \geq 0, \quad (3.259)$$

*provided that for $\bar{\rho} \leq \rho_c$, $\phi \in [0, \phi_c]$ solves $\rho(\phi) = \rho$ (**sub-critical case**) and for $\bar{\rho} > \rho_c$, $\phi = \phi_c$ (**super-critical case**).*

Proof Analogous to the proof of Theorem 3.3.7 we have

$$h_L(\phi) = -\frac{1}{L} \log \nu_\phi^L \big(\Sigma_L(\eta) = N\big), \qquad (3.260)$$

and for $\bar{\rho} \leq \rho_c$ or $\rho_c = \infty$ this implies the result as before.

For $\bar{\rho} > \rho_c$, $\sum_{x \in \Lambda_L} \eta(x) = N$ is a large deviation event, and to get an upper bound on (3.260) we need a lower bound on its probability under the critical measure $\nu_{\phi_c}^L$.

$$\nu_{\phi_c}^L \Big(\sum_{x \in \Lambda_L} \eta(x) = N \Big) \geq \nu_{\phi_c}^1 \big(\eta(1) = N - [\rho_c(L-1)]\big)$$

$$\nu_{\phi_c}^{\Lambda_L \setminus \{1\}} \Big(\sum_{x \in \Lambda_L \setminus \{1\}} \eta(x) = [\rho_c(L-1)]\big), \qquad (3.261)$$

which corresponds to putting an extensive amount of particles on a single lattice site (we arbitrarily chose 1), and distributing an amount which is typical under ν_{ϕ_c} on the remaining sites.

The second term can be treated by local limit theorems analogous to the previous result* (see remark below). Since ϕ_c is the radius of convergence of the partition function, $\nu_{\phi_c}^1$ has a subexponential tail, i.e.

$$\frac{1}{L} \log \nu_{\phi_c}^1 \big(\eta(1) = N - [\rho_c(L-1)]\big) \to 0 \quad \text{as } L \to \infty, \quad (3.262)$$

since $N - [\rho_c(L-1)] \simeq (\bar{\rho} - \rho_c)L \to \infty$ for $\bar{\rho} > \rho_c$. The fact that this holds not only along a subsequence but the limit really exists, is guaranteed by assumption (3.258) using

$$\log \nu^1_{\phi_c}\big(\eta(1) = n\big) = n \log \big(\phi_c \, w(n)^{1/n}\big) - \log z(\phi_c) \qquad (3.263)$$

and (3.216). Plugging these results for (3.261) into (3.260) we get $h_L(\phi_c) \to 0$ for $\bar{\rho} > \rho_c$. $\qquad \square$

Remarks

- Existence of the (*Cesàro*) limit in (3.258) is a very weak assumption, it is certainly fulfilled if $g(k)$ has a limit as $k \to \infty$ as in our example above. It only excludes pathological cases where $g(k)$ has an exponentially diverging subsequence.
- *For $b > 3$ the $\eta(x)$ are iid random variables with finite variance and the second term in (3.261) is of order $1/\sqrt{L}$. For $2 < b \le 3$ the variance is infinite and the sum of $\eta(x)$ has a non-normal limit distribution. Using adapted local limit theorems (see also [26], Chapter 9), the second term can still be bounded below by terms of order $1/L$ for all $b > 2$.
- Corollary 3.3.8 still applies, but note that in the super-critical case $\nu_{\phi_c}(e^{\epsilon \eta(x)}) = \infty$ for all $\epsilon > 0$ due to sub-exponential tails. So the test function $f(\eta) = \eta(x)$ is not included in the result, which is to be expected, since for $\rho > \rho_c$

$$\pi_{L,N}(\eta(x)) = N/L \to \rho > \rho_c = \nu_{\phi_c}(\eta(x)). \qquad (3.264)$$

Interpretation

- Elements ν_ϕ of the grand-canonical ensemble are also called *fluid phases*. For $\rho > \rho_c$ the ensemble

$$\{\nu_\phi : \phi \in [0, \phi_c]\} \quad \text{has density range} \quad [0, \rho_c], \qquad (3.265)$$

and there are no fluid phases with density $\rho > \rho_c$.
- The limiting distribution in any finite fixed volume Δ is given by the fluid phase $\nu^\Delta_{\phi_c}$ with density is ρ_c. Therefore for large systems the *excess mass* $(\rho - \rho_c)L$ concentrates in a region with vanishing volume fraction (volume $o(L)$), the so-called *condensed phase*. This phenomenon is called *phase separation* in general, and since one of the phases covers only a vanishing fraction of the system this particular form of phase separation is called *condensation*.

- It can be shown (see [30]) that in fact the condensed phase concentrates on a *single* lattice site, i.e. for $\rho > \rho_c$ we have a law of large numbers for the maximal occupation number in the canonical ensemble, as $L \to \infty$

$$\pi_{L,N}\left(\left|\frac{1}{L}\max_{x \in \Lambda_L}\eta(x) - (\rho - \rho_c)\right| > \epsilon\right) \to 0 \quad \text{for all } \epsilon > 0. \quad (3.266)$$

For the above example with $g(k) = 1 + b/k$, $k > 0$ and $\rho_c(b) = 1/(b-2)$ these results can be summarized in the following phase diagram:

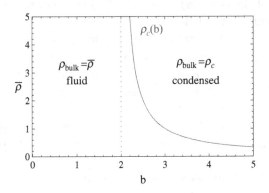

The axes are given by the system parameters b and the density $\bar{\rho} = \lim_{L\to\infty} N/L$. As the order parameter we took the limiting *bulk density* $\rho_{bulk} := \nu_\phi(\eta(x))$, where ν_ϕ is the limit measure of Theorem 3.3.10. This leads to

$$\rho_{bulk} = \begin{cases} \bar{\rho}, & \bar{\rho} \leq \rho_c, \\ \rho_c, & \bar{\rho} > \rho_c, \end{cases} \quad (3.267)$$

corresponding to two phase regions which we call *fluid* and *condensed*. ρ_{bulk} is continuous across the phase transition line (the curve), and therefore condensation is a continuous phase transition with respect to the order parameter ρ_{bulk}.

3.4 The contact process

The lattice Λ, an arbitrary countable set, is endowed with a graph structure by a directed edge set $E \subseteq \Lambda \times \Lambda$. We assume that (Λ, E) is strongly connected, i.e. for all $x, y \in \Lambda$ there exists a directed path of edges connecting x to y. The state space of the contact process (CP) is $X = \{0, 1\}^\Lambda$

and the generator is

$$\mathcal{L}f(\eta) = \sum_{z \in \Lambda} \Big(\eta(z) + \lambda \big(1 - \eta(z) \big) \sum_{y \sim z} \eta(y) \Big) \big(f(\eta^z) - f(\eta) \big), \quad (3.268)$$

where $y \sim x$ if $(y, x) \in E$. Infected sites ($\eta(x) = 1$) recover independently with rate 1, and infect neighbouring sites independently with rate $\lambda > 0$.

3.4.1 Mean-field rate equations

Choosing $f(\eta) = \eta(x)$, denoting by $\mu_t = \mu_0 S(t)$ the distribution at time t and writing $\rho(x, t) = \mu_t(\eta(x)) \in [0, 1]$ for the density, we get from the forward equation (3.44)

$$\frac{d}{dt} \rho(x, t) = \mu_t(\mathcal{L}f) = -\rho(x, t) + \lambda \sum_{y \sim x} \mu_t \Big(\eta(y) \big(1 - \eta(x) \big) \Big). \quad (3.269)$$

This follows from plugging $f(\eta) = \eta(x)$ and $f(\eta^x) = 1 - \eta(x)$ into (3.268), which leads to

$$\mathcal{L}f(\eta) = \eta(x) \big(1 - \eta(x) \big) + \lambda \big(1 - \eta(x) \big)^2 \sum_{y \sim x} \eta(y)$$

$$- \eta(x)^2 - \lambda \eta(x) \big(1 - \eta(x) \big) \sum_{y \sim x} \eta(y)$$

$$= -\eta(x) + \lambda \big(1 - \eta(x) \big) \sum_{y \sim x} \eta(y), \quad (3.270)$$

since $\eta(x) \in \{0, 1\}$ leads to simplifications $\eta(x) \big(1 - \eta(x) \big) = 0$ and $\eta(x)^2 = \eta(x)$. Note that only the term $z = x$ in the sum in (3.268) contributes.

So the time evolution of the first moment $\rho(t)$ involves second moments and is not a closed equation, similar to what we have seen for the ASEP in Section 3.2. The simplest way to close these equations is called the *mean-field assumption*:

$$\mu_t \big(\eta(y)(1 - \eta(x)) \big) = \mu_t \big(\eta(y) \big) \mu_t \big(1 - \eta(x) \big)$$

$$= \rho(y, t) \big(1 - \rho(x, t) \big), \quad (3.271)$$

i.e. μ_t is assumed to be a product measure and the $\eta(x)$ to be independent. If the graph (Λ, E) is translation invariant, e.g. a regular lattice such as $\mathbb{Z}^d$ or $(\mathbb{Z}/L\mathbb{Z})^d$ or homogeneous trees, and the initial distribution μ_0 is as well, the system is homogeneous and we have the additional identity $\rho(x,t) \equiv \rho(t)$ for all $x \in \Lambda$. Using this and the mean-field assumption in (3.269) we get the *mean-field rate equation* for the CP

$$\frac{d}{dt}\rho(t) = -\rho(t) + m\lambda\rho(t)\big(1 - \rho(t)\big), \qquad (3.272)$$

where m is the coordination number or vertex degree of the lattice Λ, i.e. the number of neighbours of a lattice site, such as $m = 2d$ for d-dimensional cubic lattices.

Remarks

- Of course there is no reason why the mean-field assumption should be correct, in fact it is known to be false (see later sections). However, it turns out that for high coordination number the replacement

$$\mu_t\Big(\sum_{y \sim x} \eta(y)(1 - \eta(x))\Big) \approx \sum_{y \sim x} \rho_t(y)\big(1 - \rho_t(x)\big) \qquad (3.273)$$

leads to quantitatively good predictions. Due to a 'law of large numbers'-effect $\sum_{y \sim x} \eta(y)$ can be replaced by its expected value when the number m of terms is large. For example, this is the case for high dimensional cubic lattices with $m = 2d$, and it can even be shown that mean-field results are 'exact' as long as $d > 4$. The highest dimension for which the mean-field assumption is not exact is often referred to as the *upper critical dimension* in the physics literature.
- Another situation with high coordination number is when the lattice Λ is actually a complete graph, i.e. $E = \Lambda_L \times \Lambda_L$. Here it can be shown that (3.272) holds rigorously with $m = L$ for $\rho(t) = \frac{1}{L}\sum_{x \in \Lambda_L} \rho(x,t)$.
- For low dimensions/coordination numbers the mean-field assumption still is useful to get a first idea of the critical behaviour of the system, since it is typically easy to derive and analyze. In most cases quantitative predictions are wrong (such as location of phase boundaries and critical exponents), but qualitative features are often predicted correctly (such as the number of phase regions or existence of critical points).

Analysis of the rate equation

The long-time behaviour of solutions to an equation of the form $\frac{d}{dt}\rho(t) = f\big((\rho(t))\big)$ is given by stationary points of the right-hand side $f(\rho) = 0$. In our case for (3.272) these are given by

$$0 = -\rho + m\lambda\rho(1 - \rho) = -m\lambda\rho^2 + (m\lambda - 1)\rho, \qquad (3.274)$$

which are the roots of a downward parabola, given by $\rho_1 = 0$ and

$\rho_2 = 1 - 1/(m\lambda)$. $\rho \equiv \rho_1 = 0$ is always a stationary solution to the equation, corresponding to the absorbing state $\eta = 0$ of the CP, called the *inactive phase*. For $m\lambda > 1$ there is a second stationary density $\rho_2 = 1 - 1/(m\lambda) \in (0,1)$ called the *active phase*. The domains of attraction of these stationary points are determined by the sign of $f(\rho)$, and ρ_i is locally stable if $f'(\rho_i) < 0$. In summary we have

$$f'(0) = m\lambda - 1 \quad \Rightarrow \quad \rho = 0 \quad \begin{array}{ll} \text{stable for} & m\lambda \leq 1, \\ \text{unstable for} & m\lambda > 1, \end{array}$$

$$f'(\rho_2) = 1 - m\lambda \quad \Rightarrow \quad \rho = \rho_2 \quad \begin{array}{ll} \notin (0,1] \text{ for} & m\lambda \leq 1, \\ \text{stable for} & m\lambda > 1, \end{array} \quad (3.275)$$

which leads to the following mean-field prediction of the phase diagram of the CP with the *critical value* $\lambda_c = 1/m$.

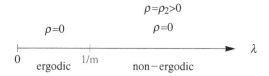

As opposed to previous sections the diagram is one-dimensional, since the number of particles in the CP is not conserved and λ is the only system parameter. The two phase regions can be characterized by ergodicity of the infinite system, as explained below.

Remarks

- The mean-field rate equation does not take into account fluctuations. Since the CP is irreducible on $X \setminus \{0\}$, on a finite lattice the states in the active phase are transient and the CP is ergodic with unique stationary measure $\mu = \delta_0$.

  However, if the infection rate λ is large enough and we start the system in the active phase (e.g. $\eta_0(x) = 1$ for all x), it remains active for a (random) time with mean of the order $\exp(CL)$, where L is the size of the lattice. If L is large it takes the system very long to reach its stationary distribution and the active phase is said to be *metastable* (see e.g. [25], Chapter I.3).

- The lifetime of the active phase diverges for infinite lattice size. Therefore infinite systems exhibit a truly stationary active phase if λ is large enough. The system is no longer ergodic since it has two stationary distributions, δ_0 corresponding to the absorbing state (inactive phase) and μ corresponding to the active phase (more details on μ follow later).

- On $\mathbb{Z}$ ($d = 1$) precise numerical estimates (and rigorous bounds) show that $\lambda_c = 1.64893$, which is quite far from the mean-field value $1/m = 1/2$ predicted by (3.272). Nevertheless, the qualitative prediction of a phase transition turns out to be true. Comparing to the first remark it is actually not surprising that mean-field underestimates the critical value, since even for $\lambda > 1/2$ the system can still die out due to fluctuations. Clearly λ_c should decrease with m since the total infection rate of one infected site is actually $m\lambda$, and in fact the numerical estimate for $\mathbb{Z}^2$ is 0.4119 (MF prediction $1/m = 0.25$).

3.4.2 Stochastic monotonicity and coupling

In this section we introduce a powerful technique which can be used to get rigorous results on the contact process. Let $X = S^\Lambda$ be the state space of a particle system with $S \subseteq \mathbb{N}$ and Λ some arbitrary discrete lattice. X is a partially ordered set, given by

$$\eta \leq \zeta \quad \text{if} \quad \eta(x) \leq \zeta(x) \quad \text{for all } x \in \Lambda. \tag{3.276}$$

Definition 3.4.1 A function $f \in C(X)$ is *increasing* if

$$\eta \leq \zeta \quad \text{implies} \quad f(\eta) \leq f(\zeta). \tag{3.277}$$

This leads to the concept of *stochastic monotonicity* for probability measures μ_1, μ_2 on X:

$$\mu_1 \leq \mu_2 \quad \text{if} \quad \mu_1(f) \leq \mu_2(f) \quad \text{for all increasing } f \in C(X). \tag{3.278}$$

This definition is quite hard to work with, and the best way to understand and use stochastic monotonicity is in terms of *couplings*.

Definition 3.4.2 A coupling of two measures $\mu_1, \mu_2 \in \mathcal{M}_1(X)$ is a measure μ on the product state space $X \times X$ of pair configurations $\eta = (\eta^1, \eta^2)$, such that the marginals for $i = 1, 2$ are

$$\mu^i = \mu_i \quad \text{i.e.} \quad \mu(\{\eta : \eta^i \in A\}) = \mu_i(A), \tag{3.279}$$

for all measurable $A \subseteq X$.

Remark In other words, a coupling means constructing the random variables $\eta^1(\omega)$ and $\eta^2(\omega)$ on the same probability space $(\Omega, \mathcal{A}, \mathbb{P})$, such that

$$\mathbb{P}(\{\omega : \eta^i(\omega) \in A\}) = \mu_i(A) \quad \text{for all measurable } A \subseteq X. \quad (3.280)$$

Theorem 3.4.3 **(Strassen)** *Suppose $\mu_1, \mu_2 \in \mathcal{M}_1(X)$. Then $\mu_1 \leq \mu_2$ if and only if there exists a coupling $\mu \in \mathcal{M}_1(X \times X)$ such that*

$$\mu(\{\eta : \eta^1 \leq \eta^2\}) = 1 \quad (\eta^1 \leq \eta^2 \ \mu - a.s.). \quad (3.281)$$

Proof Suppose such a coupling μ exists. If $f \in C(X)$ is increasing then $f(\eta^1) \leq f(\eta^2) \ \mu - a.s.$ and writing $\pi^i : X \times X \to X$ for the projection on the i-th coordinate $\pi^i(\eta) = \eta^i$, we have

$$\mu_1(f) = \mu(f \circ \pi^1) \leq \mu(f \circ \pi^2) = \mu_2(f), \quad (3.282)$$

so that $\mu_1 \leq \mu_2$.

Inferring the existence of the coupling from monotonicity involves a construction of the coupling on a probability space, see e.g. [25], Theorem 2.4, p. 72. $\qquad \square$

Example Let $\nu_{\rho_1}, \nu_{\rho_2}$ be product measures on $X = \{0,1\}^\Lambda$ with $\rho_1 \leq \rho_2$. Then for each $i = 1, 2$ the $\eta^i(x)$ are iid $Be(\rho_i)$ random variables. We construct a (so-called *maximal*) coupling μ on $X \times X$ that concentrates on configurations $\eta^1 \leq \eta^2$. Let $\Omega_x = (0,1)$ and $\mathbb{P}_x = U(0,1)$ be the uniform measure independently for each $x \in \Lambda$. Then define

$$\eta^i(x)(\omega) := \begin{cases} 1, & \omega_x \leq \rho_i, \\ 0, & \omega_x > \rho_i, \end{cases} \quad (3.283)$$

which implies that $\eta^1(x)(\omega) \leq \eta^2(x)(\omega)$ for all $\omega \in \Omega$ and $x \in \Lambda$. Taking the product over all lattice sites with $\mathbb{P} = \prod_x \mathbb{P}_x$, we can define a coupling measure on $X \times X$ by

$$\mu := \mathbb{P} \circ \eta^{-1} \quad \text{i.e.} \quad \mu(A) = \mathbb{P}(\{\omega : \eta(\omega) \in A\}), \quad (3.284)$$

for all $A \in X \times X$, and we have $\eta^1 \leq \eta^2 \ \mu - a.s.$ Therefore the theorem implies $\nu_{\rho_1} \leq \nu_{\rho_2}$.

In practice, to sample from μ (i.e. choose a coupled pair of configurations $\eta^1 \leq \eta^2$), first fix η^1 by choosing iid $Be(\rho_1)$ variables. Then under the coupling measure $\eta^1(x) = 1$ implies $\eta^2(x) = 1$, which fixes η_2 on those sites. On the remaining empty sites, choose iid $Be\left(\frac{\rho_2 - \rho_1}{1 - \rho_1}\right)$ variables. Then the $\eta^2(x)$ are independent and since $\mu(\eta^1(x) = 1) = \nu_{\rho_1}(\eta^1(x) = 1) = \rho_1$ we have

$$\mu(\eta^2(x) = 1) = \rho_1 + (1 - \rho_1)\frac{\rho_2 - \rho_1}{1 - \rho_1} = \rho_2, \quad (3.285)$$

so $\eta^2 \sim \nu_{\rho_2}$ has the right marginal.

The idea of monotinicity and coupling can be extended to processes.

Definition 3.4.4 Consider an IPS on X with generator $(S(t) : t \geq 0)$. The process is *attractive* or *monotone* if

$$f \text{ increasing} \quad \Rightarrow \quad S(t)f \text{ increasing for all } t \geq 0, \qquad (3.286)$$

or equivalently

$$\mu_1 \leq \mu_2 \quad \Rightarrow \quad \mu_1 S(t) \leq \mu_2 S(t) \quad \text{for all } t \geq 0. \qquad (3.287)$$

Let $\mathbb{P}_1, \mathbb{P}_2 \in \mathcal{M}_1\big(D[0,\infty)\big)$ be the path space measures of two IPS $(\eta_t^1 : t \geq 0)$ and $(\eta_t^2 ; t \geq 0)$. Then a *coupling* of the processes is given by a Markov process $\big((\eta_t^1, \eta_t^2) : t \geq 0\big)$ on $X \times X$ with measure $\mathbb{P} \sim \mathcal{M}_1\big(D[0,\infty) \times D[0,\infty)\big)$, having marginal processes $(\eta_t^i : t \geq 0) \sim \mathbb{P}_i$, i.e. $\mathbb{P}^i = \mathbb{P}_i$.

Lemma 3.4.5 *The contact process is attractive.*

Proof We couple two contact processes $(\eta_t^1 : t \geq 0)$ (full line) and $(\eta_t^2 ; t \geq 0)$ (dashed line) using a graphical construction.

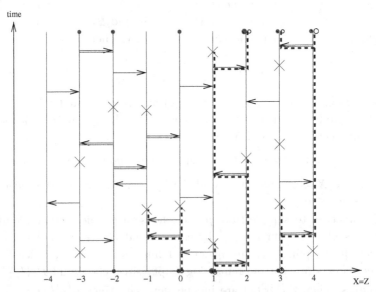

Both processes use the same realization of infection and recovery processes $\rightarrow$, $\leftarrow$ and $\times$, and the initial conditions fulfill $\eta_0^2 \leq \eta_0^1$. Then by inspection of the coupling construction this immediately implies that

$\eta_t^2 \leq \eta_t^1$ for all $t \geq 0$ (example shown above). Therefore we have for all increasing $f \in C(X)$,

$$S(t)f(\eta_0^2) = \mathbb{E}^{\eta_0^2}\left(f(\eta_t^2)\right) \leq \mathbb{E}^{\eta_0^1}\left(f(\eta_t^1)\right) = S(t)f(\eta_0^1), \quad (3.288)$$

and since this holds for all ordered initial conditions the CP is attractive as given in Definition 3.4.4. □

More generally it can be shown that:

Proposition 3.4.6 *A general spin system on $\{0,1\}^\Lambda$ with generator*

$$\mathcal{L}f(\eta) = \sum_{x \in \Lambda} c(x, \eta)\left(f(\eta^x) - f(\eta)\right) \quad (3.289)$$

is attractive if and only if the jump rates (spin flip rates) fulfill

$$\eta \leq \zeta \quad implies \quad \begin{cases} c(x, \eta) \leq c(x, \zeta), & if \ \eta(x) = \zeta(x) = 0, \\ c(x, \eta) \geq c(x, \zeta), & if \ \eta(x) = \zeta(x) = 1. \end{cases} \quad (3.290)$$

Proof Suppose the spin system is attractive, i.e. f increasing implies $S(t)f$ increasing for all $t \geq 0$. Since $f(\eta) = \eta(x)$ is increasing and in $C_0(X)$ we have

$$\mathcal{L}f(\eta) = \lim_{t \searrow 0} \frac{S(t)f(\eta) - f(\eta)}{t}, \quad (3.291)$$

and for all $\eta \leq \zeta$ with $\eta(x) = \zeta(x)$

$$\mathcal{L}f(\eta) - \mathcal{L}f(\zeta) = \lim_{t \searrow 0} \frac{S(t)f(\eta) - S(t)f(\zeta) + \eta(x) - \zeta(x)}{t} \leq 0. \quad (3.292)$$

Therefore $\mathcal{L}f(\eta) \leq \mathcal{L}f(\zeta)$ and since

$$\mathcal{L}f(\eta) = c(x, \eta)\left(1 - 2\eta(x)\right) \quad (3.293)$$

this implies (3.290).

The other direction involves a more general version of the coupling given in the proof of Lemma 3.4.5 above, see e.g. [25], Theorem 2.2, p. 134. □

Remark Property (3.290) asserts that 0 is more likely to flip to 1 in an environment of more 1s ($\zeta \geq \eta$), and vice versa. That means that local occupation numbers 'attract' one another, explaining the term 'attractive' for such particle systems.

Lemma 3.4.7   (Monotonicity in λ) *Let $(\eta_t^\lambda : t \geq 0)$ and $(\eta_t^{\lambda'} : t \geq 0)$ be two CPs with infection rates $\lambda \leq \lambda'$. Then*

$$\mu^\lambda \leq \mu^{\lambda'} \quad implies \quad \mu^\lambda S_\lambda(t) \leq \mu^{\lambda'} S_{\lambda'}(t) \quad for\ all\ t > 0, \quad (3.294)$$

i.e. there exists a coupling such that

$$\eta_0^\lambda \leq \eta_0^{\lambda'} \quad and \quad \eta_t^\lambda \leq \eta_t^{\lambda'} \quad for\ all\ t > 0. \quad (3.295)$$

Proof By Strassen's theorem, $\mu^\lambda \leq \mu^{\lambda'}$ implies existence of a coupling $\mu \in \mathcal{M}_1(X \times X)$ such that $\eta_0^\lambda \leq \eta_0^{\lambda'} \ \mu - a.s.$ Suppose first that $\eta_0^\lambda = \eta_0^{\lambda'}$ and couple the processes $(\eta_t^\lambda : t \geq 0)$ and $(\eta_t^{\lambda'} : t \geq 0)$ by using coupled infection processes $PP(\lambda)$ and $PP(\lambda) + PP(\lambda' - \lambda) \sim PP(\lambda')$ in the graphical construction. Then clearly $\eta_t^\lambda \leq \eta_t^{\lambda'}$ for all $t > 0$. Now by attractivity of the process $(\eta_t^\lambda : t \geq 0)$ this also holds for initial conditions $\eta_0^\lambda \leq \eta_0^{\lambda'}$. $\qquad \square$

3.4.3 Invariant measures and critical values

Consider a CP with infection rate λ on some connected graph (Λ, E) and let δ_0 be the point mass on the empty configuration and δ_1 on the full configuration $\eta(x) = 1$, $x \in \Lambda$. Since $\eta \equiv 0$ is absorbing, δ_0 is stationary.

Proposition 3.4.8   *For all $0 \leq s \leq t$ we have*

$$\delta_1 S(t) \leq \delta_1 S(s), \quad \bar\nu_\lambda = \lim_{t\to\infty} \delta_1 S(t) \quad exists\ and \quad \bar\nu_\lambda \in \mathcal{I}_e. \quad (3.296)$$

$\bar\nu_\lambda$ *is called the **upper invariant measure**, and we have $\delta_0 \leq \mu \leq \bar\nu_\lambda$ for all $\mu \in \mathcal{I}$.*

Furthermore, $\lambda < \lambda'$ implies $\bar\nu_\lambda \leq \bar\nu_{\lambda'}$, and for each $x \in \Lambda$

$$\rho_x(\lambda) := \bar\nu_\lambda\big(\eta(x)\big) \quad is\ monotone\ increasing\ in\ \lambda. \quad (3.297)$$

Proof Since δ_1 is maximal on X we have

$$\delta_1 \geq \delta_1 S(t - s) \quad for\ all\ 0 \leq s \leq t. \quad (3.298)$$

By attractivity of the CP and the Markov property this implies

$$\delta_1 S(s) \geq \delta_1 S(t - s) S(s) = \delta_1 S(t). \quad (3.299)$$

Therefore $\delta_1 S(t)$ is a monotone sequence, and by compactness of $\mathcal{M}_1(X)$ (in the topology of weak convergence) the limit exists and is stationary by Theorem 3.1.19(b). Furthermore, $\delta_0 \leq \pi \leq \delta_1$ for all $\pi \in \mathcal{M}_1$. Every stationary measure $\mu \in \mathcal{I}$ can be written as $\mu = \lim_{t\to\infty} \pi S(t)$ for some $\pi \in \mathcal{M}_1$, so by attractivity

$$\delta_0 S(t) \le \pi S(t) \le \delta_1 S(t) \quad \text{and after } t \to \infty, \quad \delta_0 \le \mu \le \bar{\nu}_\lambda. \quad (3.300)$$

Suppose that $\bar{\nu}_\lambda \in \mathcal{I}$ is not extremal, i.e. $\bar{\nu}_\lambda = \alpha\mu_1 + (1-\alpha)\mu_2$ for $\mu_1, \mu_2 \in \mathcal{I}$ and $\alpha \in (0,1)$. Then $\mu_1, \mu_2 \le \bar{\nu}_\lambda$, so for all increasing $f \in C(X)$ we have $\mu_1(f), \mu_2(f) \le \bar{\nu}_\lambda(f)$. Suppose now that $\mu_1(f) < \bar{\nu}_\lambda(f)$, then

$$\alpha\mu_1(f) + (1-\alpha)\mu_2(f) < \alpha\bar{\nu}_\lambda(f) + (1-\alpha)\bar{\nu}_\lambda(f) = \bar{\nu}_\lambda(f) \ (3.301)$$

in contradiction to the assumption. So $\mu_1(f) = \mu_2(f) = \bar{\nu}_\lambda(f)$ for all increasing $f \in C(X)$, and thus $\mu_1 = \mu_2 = \bar{\nu}_\lambda$ and $\bar{\nu}_\lambda \in \mathcal{I}_e$.

By monotonicity in λ we have for all $t \ge 0$

$$\delta_1 S^\lambda(t) \le \delta_1 S^{\lambda'}(t), \quad (3.302)$$

provided that $\lambda \le \lambda'$, which implies $\bar{\nu}_\lambda \le \bar{\nu}_{\lambda'}$. Since $\eta(x)$ is increasing this also holds for the corresponding densities. $\qquad\square$

On a finite lattice $\eta \equiv 0$ can be reached in finite time from any other configuration, and since $\eta \equiv 0$ is absorbing this implies

$$\mu S(t) \to \delta_0 \quad \text{as } t \to \infty \quad \text{for all } \mu \in \mathcal{M}_1(X). \quad (3.303)$$

This holds in particular for $\mu = \delta_1$, and thus the upper invariant measure is $\bar{\nu}_\lambda = \delta_0$ and the CP is ergodic for all $\lambda \ge 0$. On the other hand, on an infinite lattice it might be possible that $\bar{\nu}_\lambda \ne \delta_0$ and the mean-field prediction of an active phase is correct. It turns out that this is indeed the case for high enough infection rate λ as we will see below.

Definition 3.4.9 Denote by

$$\alpha_\eta := \mathbb{P}^\eta(\eta_t \ne 0 \text{ for all } t \ge 0) \quad (3.304)$$

the *survival probability* with initial configuration $\eta \in X$. For each $x \in \Lambda$ denote by $\xi_x \in X$ the configuration with $\xi_x(y) = \delta_{y,x}$ having a single infection at x. The CP $(\eta_t : t \ge 0)$ is said to *die out* if $\alpha_{\xi_x} = 0$ for some $x \in \Lambda$, otherwise it is said to *survive*.

Note that the condition in Definition 3.4.9 actually does not depend on the lattice site x, since Λ is connected and therefore the CP is irreducible on $X \setminus \{0\}$.

Proposition 3.4.10 *If the CP dies out for infection rate $\lambda' > 0$, then it dies out for all $\lambda \in [0, \lambda']$. The **critical value** $\lambda_c \in [0, \infty]$ is then given by*

$$\lambda_c := \sup\{\lambda \geq 0 : \ CP \ with \ infection \ rate \ \lambda \ dies \ out\}. \quad (3.305)$$

Proof Monotonicity in λ of the CP (Lemma 3.4.7) and $\eta_0^\lambda = \eta_0^{\lambda'}$ imply that if $(\eta_t^{\lambda'} : t \geq 0)$ dies out so does $(\eta_t^\lambda : t \geq 0)$.

Since the CP with $\lambda = 0$ certainly dies out, the supremum λ_c is well defined in $[0, \infty]$. $\qquad\qquad\qquad\qquad\qquad\qquad\qquad\qquad\qquad\qquad \square$

Proposition 3.4.11 *Analogous to above for any $A \subseteq \Lambda$ write $\xi_A \in X$ for $\xi_A(y) = \mathbb{1}_A(y)$. Then the survival probability is*

$$\alpha_{\xi_A} = \mathbb{P}^{\xi_A}(\eta_t \neq 0 \ for \ all \ t \geq 0) = \bar{\nu}_\lambda(\{\xi_B : B \cap A \neq \emptyset\}), \quad (3.306)$$

and for $\lambda < \lambda_c$ we have $\bar{\nu}_\lambda = \delta_0$ for $\lambda > \lambda_c$, $\bar{\nu}_\lambda \neq \delta_0$.

Proof The result is based on the following *duality property* of the CP. For all $A, B \subseteq \Lambda$ where one of them, say A is finite, we have

$$\mathbb{P}^{\xi_A}(\eta_t(x) = 1 \ \text{for a } x \in B) = \mathbb{P}^{\xi_B}(\eta_t(x) = 1 \ \text{for a } x \in A). \quad (3.307)$$

For a proof of this see e.g. [15], Theorem VI.1.7. Now choosing $B = \Lambda$ we have $\xi_B(x) = 1$ for all $x \in \Lambda$ and

$$\mathbb{P}^{\xi_A}(\eta_t \neq 0) = \mathbb{P}^{\delta_1}(\eta_t(x) = 1 \ \text{for some } x \in A). \quad (3.308)$$

Taking the limit $t \to \infty$ implies the first statement. For $\lambda < \lambda_c$ the process dies out with probability 1 for all initial configurations ξ_x and thus with $A = \{x\}$ in (3.306) we have

$$\bar{\nu}_\lambda(\eta(x) = 1) = \bar{\nu}_\lambda(\eta(x)) = \rho_x(\lambda) = 0 \quad \text{for all } x \in \Lambda, \quad (3.309)$$

which implies that $\bar{\nu}_\lambda = \delta_0$. For $\lambda > \lambda_c$ the process survives, and thus (3.309) has non-zero value and $\bar{\nu}_\lambda \neq \delta_0$. $\qquad\qquad\qquad\qquad\qquad \square$

Remark Note that Proposition 3.4.11 implies in particular that the density

$$\rho_x(\lambda) = \bar{\nu}_\lambda(\eta(x)) = \mathbb{P}^{\xi_x}(\eta_t \neq 0 \ \text{for all } t \geq 0) \quad (3.310)$$

is equal to the survival probability.

Our results so far imply that there is a well-defined critical value $\lambda_c \in [0, \infty]$ such that the CP dies out and $\bar{\nu}_\lambda = \delta_0$ for $\lambda < \lambda_c$, and the CP survives and $\bar{\nu}_\lambda \neq \delta_0$ for $\lambda > \lambda_c$. On a finite lattice we have discussed above that $\lambda_c = \infty$. The crucial question on infinite lattices is now whether λ_c is *non-trivial*, i.e. $\lambda_c \in (0, \infty)$. Certainly the value of λ_c

will depend on the lattice Λ but at least one can derive a quite general lower bound.

Let $(\eta_t : t \geq 0)$ be the CP with infection rate λ on a connected graph (Λ, E). Consider the auxiliary process $(\zeta_t : t \geq 0)$ on the same graph with state space $X = \mathbb{N}^\Lambda$ and generator

$$\mathcal{L}f(\zeta) = \sum_{x \in \Lambda} \Big(\eta(x)\big(f(\zeta^{-x}) - f(\zeta)\big) + \lambda \sum_{y \sim x} \zeta(y)\big(f(\zeta^{+x}) - f(\zeta)\big) \Big), \quad (3.311)$$

where we write $\zeta^{\pm x}(y) = \begin{cases} \zeta(y) \pm 1, & y = x, \\ \zeta(y), & y \neq x. \end{cases}$

In this process particles independently create new particles at connected sites with rate λ and die independently with rate 1, so the number of particles per site can be larger than 1. We couple this process to a CP $(\eta_t : t \geq 0)$ by using the same Poisson processes $PP(\lambda)$ and $PP(1)$ for infection/creation and death/recovery in the graphical construction. If for the auxiliary process $\zeta_t > 1$, we use independent creation and death processes for the extra particles. This construction implies that the CP is *dominated* by the ζ-process, i.e.

$$\eta_0 \leq \zeta_0 \quad \Rightarrow \quad \eta_t \leq \zeta_t \quad \text{for all } t \geq 0. \qquad (3.312)$$

Therefore if $(\zeta_t : t \geq 0)$ dies out then the CP dies out as well. Now let m be the maximal vertex degree of the graph (Λ, E). Then the number of particles in the ζ-process is dominated by a Markov chain $N(t)$ on the state space $\mathbb{N}$ with transition rates

$$c(n, n+1) = n\, m\, \lambda \quad \text{for } n \geq 0, \quad c(n, n-1) = n \quad \text{for } n \geq 1. \quad (3.313)$$

All the particles independently create new particles at rate $m\lambda$ and die at rate 1. Again there exists an obvious coupling such that

$$\sum_{x \in \Lambda} \zeta_t(x) \leq N(t) \quad \text{for all } t \geq 0. \qquad (3.314)$$

$N(t)$ is a well-known birth–death chain with absorbing state $n = 0$, and dies out with probability 1 if and only if $m\lambda \leq 1$. For $m\lambda > 1$ the average $\mathbb{E}(N(t))$ is monotone increasing and the process can survive with positive probability.

Proposition 3.4.12 **(Lower bound for λ_c)** *Consider a CP on a connected graph (Λ, E) with maximal vertex degree m. Then $\lambda_c \geq 1/m$.*

Proof With initial condition ξ_x as in Definition 3.4.9 and using the above coupling the number of active sites in the CP is dominated by the birth–death chain

$$\sum_{x \in \Lambda} \eta_t(x) \leq N(t) \quad \text{with} \quad N(0) = 1. \tag{3.315}$$

Therefore $\lambda \leq 1/m$ implies that the CP dies out and thus $\lambda_c \geq 1/m$. $\square$

Note that the lower bound coincides with the mean-field prediction $\lambda_c = 1/m = 1/(2d)$ of Section 3.4.1. To get an upper bound on λ_c is in general harder. In the following we will concentrate on $\Lambda = \mathbb{Z}^d$ and only give a small part of the proof.

3.4.4 Results for $\Lambda = \mathbb{Z}^d$

Consider the CP on the regular lattice $\Lambda = \mathbb{Z}^d$.

Theorem 3.4.13 *For the critical value $\lambda_c(d)$ of a CP on the lattice $\Lambda = \mathbb{Z}^d$ we have*

$$\frac{1}{2d} \leq \lambda_c(d) \leq \frac{2}{d} \quad \text{for all } d \geq 1. \tag{3.316}$$

Proof The lower bound is given by Proposition 3.4.12, for the proof of $\lambda_c(1) \leq 2$ see Theorem VI.1.33 in [15]. For higher dimensions the required inequality $\lambda_c(d) \leq \lambda_c(1)/d$ follows from

$$\mathbb{P}^{\xi_x}(\eta_t^d \neq 0) \geq \mathbb{P}^{\xi_x}(\eta_t^1 \neq 0), \quad t \geq 0, \tag{3.317}$$

where $(\eta_t^d : t \geq 0)$ is the d-dimensional CP with rate λ, and $(\eta_t^1 : t \geq 0)$ is a 1-dimensional CP with rate $d\lambda$. We show this by coupling the two processes such that for each $y \in \mathbb{Z}$

$$\eta_t^1(y) = 1 \quad \text{implies} \quad \eta_t^d(x) = 1 \text{ for some } x \text{ with } \pi_d(x) = y, \tag{3.318}$$

where for all $x \in \mathbb{Z}^d$ we denote

$$\pi_d(x) = \pi_d(x_1, \ldots, x_d) = x_1 + \ldots + x_d \in \mathbb{Z}. \tag{3.319}$$

Suppose that $A \subseteq \mathbb{Z}^d$ and $B \subseteq \mathbb{Z}$ are finite and such that

$$B \subseteq \pi_d(A) = \{\pi_d(x) : x \in A\}, \tag{3.320}$$

i.e. for each $y \in B$ there is (at least) one $x \in A$ such that $y = \pi_d(x)$.

Choose one of these $\bar{x}$, and associate its $PP(1)$ death process with site y. Also, for all of the $2d$ neighbours of $\bar{x}$ we have

$$x \sim \bar{x} \quad \text{implies} \quad \pi_d(x) = y \pm 1 \sim y. \tag{3.321}$$

Now associate the infection processes $PP(\lambda)$ pointing towards $\bar{x}$ from all its neighbours with infections at y, which leads to a net infection rate of $d\lambda$ from each of the two neighbours $y \pm 1$. Note that all other deaths and infections in the d-dimensional CP that would correspond to y are not used in the coupling. With this construction both marginal processes $(\eta_t^1 : t \geq 0)$ and $(\eta_t^d : t \geq 0)$ have the right law, and clearly (3.318) is fulfilled, which finishes the proof. $\qquad\square$

Using more involved techniques than we do here, lower and upper bound can be improved significantly, depending on the dimension d. Further it can be shown that

$$d\,\lambda_c(d) \to \frac{1}{2} \quad \text{as } d \to \infty, \tag{3.322}$$

supporting the physics wisdom that 'mean-field theory is exact in high dimensions'.

Theorem 3.4.14 **(Complete convergence)** *Consider the CP on* $\Lambda = \mathbb{Z}^d$. *For every* $\eta \in X$ *as* $t \to \infty$ *we have*

$$\delta_\eta S(t) \to \alpha_\eta \bar{\nu}_\lambda + (1 - \alpha_\eta)\delta_0 \quad \text{weakly (locally)}, \tag{3.323}$$

where $\alpha_\eta = \mathbb{P}^\eta(\eta_t \neq 0 \text{ for all } t \geq 0)$ *is the survival probability.*

Proof See e.g. [25], Theorem I.2.27.

Remark Taking the expected value w.r.t. an initial distribution μ in (3.323) we get weak convergence of

$$\mu S(t) \to \mu(\alpha_\eta)\bar{\nu}_\lambda + \big(1 - \mu(\alpha_\eta)\big)\delta_0. \tag{3.324}$$

This holds in particular for all stationary $\mu \in \mathcal{M}_1(X)$, and therefore every stationary distribution is a convex combination of δ_0 and $\bar{\nu}_\lambda$ and we have

$$\mathcal{I}_e = \{\delta_0, \bar{\nu}_\lambda\}. \tag{3.325}$$

Theorem 3.4.15 **(Extinction time)** *Suppose* $\lambda > \lambda_c$ *and for the CP* $(\eta_t : t \geq 0)$ *let*

$$\tau := \inf\{t \geq 0 : \eta_t = 0\} \tag{3.326}$$

be the **extinction time** of the process. Then there exists $\epsilon > 0$ such that for every initial condition $\eta_0 = \eta \in X$

$$\mathbb{P}^\eta(\tau < \infty) \le e^{-\epsilon|\eta|} \quad \text{where} \quad |\eta| = \sum_{x \in \Lambda} \eta(x). \qquad (3.327)$$

Proof See [25], Theorem I.2.30

Note that this implies that the supercritical CP can only die out with positive probability if the initial condition is finite $|\eta| < \infty$. If, however, $\mu \in \mathcal{M}_1(X)$ is translation invariant and $\mu(\eta(x)) > 0$, then we have $\mu(|\eta| = \infty) = 1$, and therefore

$$\mathbb{P}^\eta(\tau = \infty) = \alpha_\eta = 1 \qquad (3.328)$$

and the process survives with probability 1. With Theorem 3.4.14 this implies

$$\mu S(t) \to \bar{\nu}_\lambda \quad \text{as } t \to \infty. \qquad (3.329)$$

Theorem 3.4.16 *The critical contact process dies out.*

Proof See [25], Theorem I.2.25

This implies that the density

$$\rho(\lambda) = \bar{\nu}_\lambda(\eta(x)) = \mathbb{P}^{\xi_x}(\eta_t \ne 0 \text{ for all } t \ge 0) \qquad (3.330)$$

which is independent of x due to translation invariance, is a continuous function of λ. By Proposition 3.4.8 it is also monotone increasing, for $\lambda > \lambda_c$ and vanishes for $\lambda < \lambda_c$ by Proposition 3.4.11.

In particular, to leading order the behaviour at the critical point is given by

$$\rho(\lambda) \sim C(\lambda - \lambda_c)^\beta \qquad (3.331)$$

for some exponent $\beta > 0$. The only rigorous bound is $\beta \le 1$, and our mean-field result from section 4.1 predicts $\lambda_c = 1/(2d)$ and for $\lambda \ge \lambda_c$ similar we have to leading order

$$\rho(\lambda) = 1 - \frac{1}{2d\lambda} = 1 - \frac{1}{2d\lambda_c}\left(1 + \frac{\lambda - \lambda_c}{\lambda_c}\right)^{-1} \simeq \frac{\lambda - \lambda_c}{\lambda_c}, \qquad (3.332)$$

which implies $\beta = 1$. In fact numerical estimates give values $\beta \approx 0.28$ ($d = 1$), 0.58 ($d = 2$), 0.81 ($d = 3$), and for $d \ge 4$ the mean-field value $\beta = 1$ should be 'exact'.

The CP has also been analyzed on other regular lattices, in particular homogeneous trees T^d (see e.g. [25], Chapter I.4). In this case the critical behaviour turns out to be more complicated, there exists a second critical value $\lambda_2 > \lambda_c$ and complete convergence in the sense of Theorem 3.4.14 only holds outside the interval $[\lambda_c, \lambda_2]$. Inside this interval there exist infinitely many extremal invariant measures and the infection survives globally but dies out locally.

3.4.5 Duality

Definition 3.4.17   Consider two independent Markov processes $(\eta_t : t \geq 0)$ on X and $(\xi_t : t \geq 0)$ on $\tilde{X}$ with corresponding path measures $\mathbb{P}^\eta$ and $\tilde{\mathbb{P}}^\xi$. $(\xi_t : t \geq 0)$ is the *dual* of $(\eta_t : t \geq 0)$ with *duality function* $D : X \times \tilde{X} \to \mathbb{R}$ if

$$\mathbb{E}^\eta D(\eta_t, \xi) = \tilde{\mathbb{E}}^\xi D(\eta, \xi_t) \quad \text{for all} \quad \eta \in X \text{and } \xi \in \tilde{X}. \quad (3.333)$$

An equivalent formulation using semigroups $\big(S(t) : t \geq 0\big)$ and $\big(\tilde{S}(t) : t \geq 0\big)$ is

$$S(t)\, D(\cdot, \xi)(\eta) = \tilde{S}(t)\, D(\eta, \cdot)(\xi) \quad \text{for all} \quad \eta \in X \text{and } \xi \in \tilde{X}. \quad (3.334)$$

If $X = \tilde{X}$ and $\mathbb{P}^\eta = \tilde{\mathbb{P}}^\eta$ for all $\eta \in X$, $(\eta_t : t \geq 0)$ is called *self-dual*.

Proposition 3.4.18   *Consider the processes* $(\eta_t : t \geq 0)$ *on* X *with generator* $\mathcal{L}$ *and* $(\xi_t : t \geq 0)$ *on* $\tilde{X}$ *with generator* $\tilde{\mathcal{L}}$. *The processes are dual with duality function* $D : X \times \tilde{X} \to \mathbb{R}$ *if and only if*

$$\mathcal{L}D(\cdot, \xi)(\eta) = \tilde{\mathcal{L}}D(\eta, \cdot)(\xi) \quad \text{for all} \quad \eta \in X \text{and } \xi \in \tilde{X}. \quad (3.335)$$

This holds provided that $\mathcal{L}D(\cdot, \xi)$ *and* $\tilde{\mathcal{L}}D(\eta, \cdot)$ *are well defined for all* $\eta \in X$ *and* $\xi \in \tilde{X}$.

Proof Assume duality of $(\eta_t : t \geq 0)$ and $(\xi_t : t \geq 0)$. Then

$$\frac{1}{t}\big(S(t)D(\cdot, \xi)(\eta) - D(\eta, \xi)\big) = \frac{1}{t}\big(\tilde{S}(t)D(\eta, \cdot)(\xi) - D(\eta, \xi)\big) \quad (3.336)$$

for all $t > 0$. Taking the limit $t \searrow 0$ implies (3.335) using the definition (3.42) of the generator. By the Hille–Yosida Theorem 3.1.13 the reverse follows from taking the limit $n \to \infty$ in the identity

$$\Big(Id + \frac{t}{n}\mathcal{L}\Big)^n D(\cdot, \xi)(\eta) = \Big(Id + \frac{t}{n}\tilde{\mathcal{L}}\Big)^n D(\eta, \cdot)(\xi), \quad (3.337)$$

which holds for all $n \in \mathbb{N}$ by induction over n.   □

Remarks

- $\mathcal{L}D$ and $\tilde{\mathcal{L}}D$ are well defined e.g. if $(\eta_t : t \geq 0)$ and $(\xi_t : t \geq 0)$ are Markov chains with countable state space. If they are IPS with state spaces X and $\tilde{X}$ then $D(.,\xi)$ and $D(\eta,.)$ should be cylinder functions for all $\eta \in X$ and $\xi \in \tilde{X}$.
- Duality is a symmetric relation, i.e. if $(\eta_t : t \geq 0)$ is dual to $(\xi_t : t \geq 0)$ then $(\xi_t : t \geq 0)$ is dual to $(\eta_t : t \geq 0)$ with the same duality function modulo coordinate permutation.

Proposition 3.4.19   *The CP with $X = \{0,1\}^\Lambda$, Λ connected, is self-dual.*

Proof For $\eta \in X$ and $A \subseteq \Lambda$ finite define

$$D(\eta, A) := \prod_{x \in A} \big(1 - \eta(x)\big) = \begin{cases} 1, & \text{if } \eta \equiv 0 \text{ on } A, \\ 0, & \text{otherwise.} \end{cases} \qquad (3.338)$$

Then, using $D(\eta, A) = \big(1 - \eta(x)\big)D\big(\eta, A \setminus \{x\}\big)$ for $x \in A$, we have

$$D(\eta^x, A) - D(\eta, A) = \begin{cases} D\big(\eta, A \setminus \{x\}\big), & x \in A, \eta(x) = 1, \\ -D(\eta, A), & x \in A, \eta(x) = 0, \\ 0, & x \notin A. \end{cases} \qquad (3.339)$$

This implies for the generator of the contact process $(\eta_t : t \geq 0)$

$$\mathcal{L}D(\cdot, A)(\eta) = \sum_{x \in \Lambda} \Big(\eta(x) + \lambda\big(1 - \eta(x)\big) \sum_{y \sim x} \eta(y)\Big)\big(D(\eta^x, A) - D(\eta, A)\big)$$

$$= \sum_{x \in A} \Big(\eta(x) D\big(\eta, A \setminus \{x\}\big) - \lambda \sum_{y \sim x} \eta(y)\big(1 - \eta(x)\big) D(\eta, A)\Big). \qquad (3.340)$$

Using $\big(1 - \eta(x)\big)D(\eta, A) = \big(1 - \eta(x)\big)D\big(\eta, A \setminus \{x\}\big) = D(\eta, A)$ for $x \in A$ and writing $\eta(x) = \eta(x) - 1 + 1$ we get

$$\mathcal{L}D(\cdot, A)(\eta) = \sum_{x \in A} \Big(D\big(\eta, A \setminus \{x\}\big) - D(\eta, A)$$

$$+ \lambda \sum_{y \sim x} \Big(D\big(\eta, A \cup \{y\}\big) - D(\eta, A)\Big)$$

$$=: \tilde{\mathcal{L}}D(\eta, \cdot)(A). \qquad (3.341)$$

Now $\tilde{\mathcal{L}}$ is a generator on $\tilde{X} = \{A \subseteq \Lambda \text{ finite}\}$ with transitions

$$\begin{aligned} A &\to A \setminus \{x\} \quad \text{at rate } 1, \text{ if } x \in A, \\ A &\to A \cup \{y\} \quad \text{at rate } \lambda \big| \{x \in A : x \sim y\} \big|, \text{ if } y \notin A. \quad (3.342) \end{aligned}$$

If we identify $A = \{x : \tilde{\eta}(x) = 1\}$ to be the set of infected sites of a process $(\tilde{\eta}_t : t \geq 0)$, then this is again a CP on X with infection rate λ. $\square$

Remark It is often convenient to describe a CP $(\eta_t : t \geq 0)$ also in terms of the set of infections $(A_t : t \geq 0)$. We use the same notation for the path measures $\mathbb{P}$ to indicate that we really have the same process only in different notation. In that sense (3.338) is a duality function for the CP and we have

$$\mathbb{E}^\eta D(\cdot, A) = \mathbb{P}^\eta(\eta_t \equiv 0 \text{ on } A) = \mathbb{P}^A(\eta \equiv 0 \text{ on } A_t) = \mathbb{E}^A D(\eta, \cdot). \quad (3.343)$$

Note that this is the relation we used in the proof of Proposition 3.4.11 in slightly different notation.

Proposition 3.4.20 *Let $(\eta_t : t \geq 0)$ on X be dual to $(\xi_t : t \geq 0)$ on $\tilde{X}$ w.r.t. $D : X \times \tilde{X} \to \mathbb{R}$. If $T : C(X) \to C(X)$ is a simple symmetry or a conservation law for $(\eta_t : t \geq 0)$ according to Propositions 3.2.5 and 3.2.7, then*

$$D'(\eta, \xi) = \big(T\, D(\cdot, \xi)\big)(\eta) \quad (3.344)$$

is also a duality function.

Proof For a symmetry T we have $S(t)T = TS(t)$ for all $t \geq 0$, so

$$\begin{aligned} S(t)\, D'(\cdot, \xi)(\eta) = S(t)\, T\, D(\cdot, \xi)(\eta) &= T\, S(t)\, D(\cdot, \xi)(\eta) \\ &= T\big(\tilde{S}(t)\, D(\cdot, \cdot)(\xi)\big)(\eta). \quad (3.345) \end{aligned}$$

Now, if T is a simple symmetry with $Tf = f \circ \tau$, $\tau : X \to X$ for all $f \in C(X)$, we have

$$\begin{aligned} T\big(\tilde{S}(t)D(\cdot, \cdot)(\xi)\big)(\eta) = \big(\tilde{S}(t)D(\cdot, \cdot)(\xi)\big)(\tau\eta) &= \tilde{S}(t)D(\tau\eta, \cdot)(\xi) \\ &= \tilde{S}(t)D'(\eta, \cdot)(\xi). \quad (3.346) \end{aligned}$$

If T is a conserved quantity with $Tf = gf$ for all $f \in C(x)$ and some $g \in C(X)$,

$$\begin{aligned} T\big(\tilde{S}(t)D(\cdot, \cdot)(\xi)\big)(\eta) = g(\eta)\tilde{S}(t)D(\eta, \cdot)(\xi) &= \tilde{S}(t)g(\eta)D(\eta, \cdot)(\xi) \\ &= \tilde{S}(t)D'(\eta, \cdot)(\xi), \quad (3.347) \end{aligned}$$

since $g(\eta)$ is a constant under $\tilde{S}(t)$, and the latter is a linear operator. □

Remarks

- Of course it is possible that $D' = TD = D$. For example, translation invariance is a symmetry of the CP on $\Lambda = \mathbb{Z}^d$, and the duality function D in (3.338) is translation invariant. But the linear voter model (see Definition 3.1.6) has a particle-hole symmetry, which can generate two different duality functions.
- The result of Proposition 3.4.20 holds for all symmetries for which the commutation relation $T\tilde{S}(t)D = \tilde{S}(t)TD$ holds. As seen in the proof this holds for general duality functions D as long as we restrict to simple symmetries or conservation laws. For more general symmetries regularity assumptions on D are necessary. Even though T and $\tilde{S}(t)$ act on different arguments of $D(\eta, \xi)$, they do not necessarily commute in general, like e.g. partial derivatives of a function $f : \mathbb{R}^2 \to \mathbb{R}$ only commute if f is differentiable.

References

[1] F. Spitzer: Interaction of Markov processes. *Adv. Math.* **5**, 246–290 (1970).
[2] R. E. Wilson: Mechanisms for spatio-temporal pattern formation in highway traffic models. *Philos. Transact A Math. Phys. Eng. Sci.* **366**(1872), 2017–2032 (2008).
[3] A. Ali, E. Somfai, S. Grosskinsky: Reproduction time statistics and segregation patterns in growing populations. *Phys. Rev. E* **85**(2), 021923 (2012).
[4] O. Hallatschek et al.: Genetic drift at expanding frontiers promotes gene segregation. *PNAS* **104**(50), 19926–19930 (2007).
[5] D. Helbing: Traffic and related self-driven many-particle systems. *Rev. Mod. Phys.* **73**, 1067–1141 (2001).
[6] G. Grimmett, D. Stirzaker: *Probability and Random Processes*, 3rd edition, Oxford (2001).
[7] J. R. Norris: *Markov Chains*, Cambridge (1997).
[8] O. Kallenberg: *Foundations of Modern Probability*, 2nd edition, Springer (2002).
[9] O. Golinelli, K. Mallick: The asymmetric simple exclusion process: an integrable model for non-equilibrium statistical mechanics. *J. Phys. A: Math. Gen.* **39**, 12679–12705 (2006).
[10] D. Chowdhury, L. Santen and A. Schadschneider: Statistical physics of vehicular traffic and some related systems. *Phys. Rep.* **329**(4), 199–329 (2000).

[11] A. Franc: Metapopulation dynamics as a contact process on a graph. *Ecological Complexity* **1**, 49–63 (2004).

[12] J. Marro, R. Dickman: *Nonequilibrium Phase Transitions in Lattice Models.* Cambridge (2005).

[13] W. Rudin: *Real and Complex Analysis*, McGraw-Hill (1987).

[14] K. Yosida: *Functional Analysis*, 6th edition, Springer (1980).

[15] T. M. Liggett: *Interacting Particle Systems*, Springer (1985).

[16] C. Kipnis, C. Landim: *Scaling limits of Interacting Particle Systems*, Springer (1999).

[17] W. Rudin: *Functional Analysis*, 2nd edition, McGraw-Hill (1991).

[18] P. Billingsley: *Convergence of Probability Measures*, 2nd edition, Wiley (1999).

[19] H.-O. Georgii: *Gibbs Measures and Phase Transitions*, de Gruyter (1988).

[20] D.-Q. Jiang, F.-X. Zhang: The Green-Kubo formula and power spectrum of reversible Markov processes; *J. Math. Phys.* **44**(10), 4681–4689 (2003).

[21] T. M. Liggett: Coupling the simple exclusion process. *Ann. Probab.* **4**, 339–356 (1976).

[22] H. Rost: Non-equilibrium behaviour of a many particle process: density profile and local equilibria. *Z. Wahrscheinlichkeitstheorie verw. Gebiete* **58**, 41–53 (1981).

[23] F. Rezakhanlou: Hydrodynamic limit for attractive particle systems on $\mathbb{Z}^d$. *Commun. Math. Phys.* **140**, 417–448 (1991).

[24] C. M. Dafermos: *Hyperbolic Conservation Laws in Continuum Physics.* Springer (2000).

[25] T. M. Liggett: *Stochastic Interacting Systems*, Springer (1999).

[26] B. W. Gnedenko, A. N. Kolmogorov: *Limit Distributions for Sums of Indepenent Random Variables.* Addison Wesley (1954).

[27] Csiszár: *I*-Divergence geometry of probability distributions and minimization problems. I; *Ann. Prob.* **3**, 146–158 (1975).

[28] M. R. Evans, T. Hanney: Nonequilibrium statistical mechanics of the zero-range process and related models; *J. Phys. A: Math. Theor.* **38**, R195–R240 (2005).

[29] M. R. Evans: Phase transitions in one-dimensional nonequilibrium systems; *Braz. J. Phys.* **30**(1), 42–57 (2000).

[30] I. Jeon, P. March, B. Pittel: Size of the largest cluster under zero-range invariant measures. *Ann. Probab.* **28**(3), 1162–1194 (2000).

[31] E. D. Andjel: Invarient measures for the zero range process. *Ann. Probab.* **10**(3), 525–547 (1982).

4

Statistical mechanics of complex systems

Ellák Somfai

Abstract

In this chapter we introduce statistical mechanics in a very general form, and explore how the tools of statistical mechanics can be used to describe complex systems.

To illustrate what *statistical mechanics* is, let us consider a physical system made of a number of interacting particles. When it is just a single particle in a given potential, it is an easy problem: one can write down the solution (even if one could not calculate everything in closed form). Having two particles is equally easy, as this so-called "two-body problem" can be reduced to two modified one-body problems (one for the centre of mass, other for the relative position). However, a dramatic change occurs when the number of particles is increased to three. The study of the three-body problem started with Newton, Lagrange, Laplace and many others, but the general form of the solution is still unknown. Even relatively recently, in 1993 a new type of periodic solution has been found, where three equal mass particles interacting gravitationally chase each other in a figure-of-eight shaped orbit. This and other systems where the degrees of freedom is low belongs to the subject of dynamical systems, and is discussed in detail in Chapter 2 of this volume. When the number of interacting particles increases to very large numbers, like 10^{23}, which is typical for the number of atoms in a macroscopic object, surprisingly it gets simpler again, as long as we are interested only in aggregate quantities. This is the subject of statistical mechanics.

Statistical mechanics is also the microscopic foundation of thermodynamics. It developed a number of powerful tools, which can be used outside of the conventional physics domain, like biology, finance (see

Chapter 6), traffic, and more. It can also be considered as *the science of ignorance*: how to handle a system where we do not know (even prefer not to know) everything. A particular example is renormalisation theory, developed in the second half of the 20th century, which gives a systematic framework to dispense successively with the non-interesting degrees of freedom.

Our approach is that of a physicist: (i) based on models, (ii) fundamentally quantitative, (iii) but not rigorous. Probably the most important concept in science is that it is possible to construct an abstraction ("model") of the world around us which is admittedly much simpler than the real thing, but nevertheless captures important characteristics, and which enables to make predictions that are not "explicitly put in" the model. The models are fundamentally quantitative, and we use mathematical language to describe them. We stop here, however, and leave rigorous treatment to mathematicians – this approach enables one to progress much quicker, at the expense of losing that what we do is absolutely unshakable.

We start with elements of information theory, which in turn are used to derive the foundations of statistical mechanics based on the maximum entropy principle. Unlike the typical treatment in physics textbooks, this approach has the advantage that the abstract formalism developed can be used in a more straightforward way for systems outside the typical range of thermal applications. We will follow by considering the effects of fluctuations to provide a link to thermodynamics. One of the most characteristic collective phenomena of complex systems is phase transition, which we approach from the direction of statistical physics. The rest of the chapter will deal with dynamics in some form: we will consider interface growth and collective biological motion like flocking.

4.1 Introduction to information theory

Random variables

While probability theory can (and should) be founded rigorously, in these notes we take a relaxed approach and attempt to define everything without mentioning the probability space. We call a *random variable* an object which can take values when observed, say random variable X can take any values from $x_1, x_2, \ldots, x_n$. When observed many times, x_1 is taken n_1 times, x_2 is taken n_2 times, etc. The *probabilities* of the outcomes can be defined as relative frequencies in the limit of a large number of observations, for example:

$$\frac{n_1}{\sum n_i} \to p_1 \,.$$

It immediately follows that the probabilities add up to 1:

$$\sum_{i=1}^{n} p_i = 1 \,.$$

We then say the the probability of X taking a given value x_i is p_i:

$$P(X = x_i) = p_i \,.$$

A *function of a random variable* is another random variable: if X takes x_i with probability p_i, then $f(X)$ takes the value $f(x_i)$ with the same probability p_i.

The *expectation* or *average* of a random variable can be considered as observing it many times and taking the average of the observed values. In the statistical mechanics literature the standard notation is angular brackets:

$$\langle X \rangle = \sum_i x_i p_i \,, \qquad \text{or} \qquad \langle f(X) \rangle = \sum_i f(x_i) p_i \,.$$

The above concepts can be easily extended to random variables which can take infinitely many discrete values, and even to ones which take values from a continuum. In the latter case e.g. if X can take real values, the probability that X takes any value in $[x, x+dx]$ is $p(x)dx$, where dx is small, and $p(x)$ is called the probability density function. The sums above are replaced with integrals, e.g. $\int p(x)dx = 1$. While this naive approach to continuous random variables is sufficient for these notes, in general, especially when dealing with continuous random variables, one needs a rigorous foundation of probability theory.

The above *frequentist* approach to probabilities is not the only one. In a sentence like "Tomorrow we will have a 10% chance of rain", probabilities are interpreted as a degree of belief or confidence.

The information entropy

Suppose we want to describe the outcome of a sequence of coin tosses (heads or tails): HTHHTHTTTH... This sequence looks very different from that of a lottery play: LLLLL...LLLL...LLWLLLL... The second sequence is much more boring, one can describe it as e.g. "trial no. 857,923 was a win, all others lost". In the first sequence, however, we cannot get away much better than quoting the whole sequence verbatim.

To quantify this difference, we introduce the *information entropy H* of a random variable. Its intuitive meaning is the amount of uncertainty in an observation of a random variable, or in other words the amount of information we gain when observing a random variable. One can think about it as the "amount of answers" needed on average to learn the outcome of an observation as a response to an optimally crafted question-tree. It is a function of the probabilities only: $H(p_1, p_2, \dots)$.

We require certain regularity properties:

(i) Continuity: $H(p_1, p_2, \dots)$ is a continuous function of its arguments.

(ii) "Sense of direction": of the random variables that take all outcomes with equal probability, the ones with more outcomes carry more information: the function

$$h(n) := H\left(\frac{1}{n}, \frac{1}{n}, \dots, \frac{1}{n}\right) \tag{4.1}$$

is monotonically increasing with n.

(iii) "Consistency": if H is calculated in two different ways, they should agree. For example, to calculate the information entropy of a three-state random variable, we can group the last two states and first obtain the information entropy for the obtained two-state random variable, and then with probability $p_2 + p_3$ need to resolve the grouped states:

$$H_3(p_1, p_2, p_3) = H_2(p_1, q) + qH_2\left(\frac{p_2}{q}, \frac{p_3}{q}\right),$$

where $q = p_2 + p_3$.

It can be shown that these three requirements restrict the functional form of $H(\cdot)$, see [1]:

$$H(p_1, p_2, \dots, p_r) = -K \sum_{i=1}^{r} p_i \ln(p_i) \tag{4.2}$$

This is the information entropy of a random variable with probabilities $p_1, p_2, \dots, p_r$. The constant K sets the units, which can be fused into the logarithm as setting its base. In many of the following formulae we will use the notation

$$H(p_1, p_2, \ldots, p_r) = -\sum_{i=1}^{r} p_i \log(p_i) \qquad (4.3)$$

without explicitly specifying the base of the logarithm. Then setting $K = 1/\ln(2)$ in (4.2), or equivalently using $\log_2$ in (4.3) the information entropy is measured in *bits*. When setting $K = 1$ or using $\log_e$ the units are *nats*, and finally the decimal case is $K = 1/\ln(10)$ or using $\log_{10}$, when the units are called *bans*.

Multiple random variables

When X and Y are random variables, we can look at the probabilities of (X, Y) pairs. These are called *joint probabilities*:

$$p_{ij} = P(X = x_i, Y = y_j).$$

The probabilities of one of the random variables (the *marginal probabilities*) are obtained by summing up the joint probabilities on all states of the other random variable:

$$p_i^{(X)} = P(X = x_i) = \sum_j P(X = x_i, Y = y_j) = \sum_j p_{ij},$$

and similarly $p_j^{(Y)} = \sum_i p_{ij}$.

Two random variables are called *independent*, if the joint probabilities factorise into marginals for all (i, j) pairs:

$$\text{if} \quad P(X = x_i, Y = y_j) = P(X = x_i)P(Y = x_j) \qquad \text{for all } i, j,$$

or equivalently

$$\text{if} \quad p_{ij} = p_i^{(X)} p_j^{(Y)} \qquad \text{for all } i, j.$$

The *conditional probabilities* tell the probability of one random variable when we know the value of another:

$$p_{i|j} := P(X = x_i \mid Y = y_j) = \frac{P(X = x_i, Y = y_j)}{P(Y = y_j)} = \frac{p_{ij}}{p_j^{(Y)}}.$$

The *joint information entropy* is the uncertainty of the (X, Y) pair:

$$H(X, Y) = -\sum_{i,j} p_{ij} \log p_{ij}.$$

The *conditional information entropy* gives the uncertainty of X when Y is known:

$$H(X \mid Y) := \langle H(X \mid Y = y_j) \rangle_Y = \sum_j p_j^{(Y)}(-1) \sum_i p_{i|j} \log p_{i|j}$$

$$= -\sum_{ij} p_{ij} \log \frac{p_{ij}}{p_j^{(Y)}} = H(X,Y) - H(Y),$$

where in the last step we used $\sum_i p_{ij} = p_j^{(Y)}$.

Finally the *mutual information* is defined as

$$I(X;Y) := \sum_{i,j} p_{ij} \log \frac{p_{ij}}{p_i^{(X)} p_j^{(Y)}} = H(X) + H(Y) - H(X,Y)$$

$$= H(X) - H(X \mid Y),$$

so its meaning is the reduction in uncertainty of X due to the knowledge of Y, or in other words how much Y tells about X.

4.2 The maximum entropy framework[1]

The maximum entropy principle – an example

Suppose we have a random variable X with known states (values of the observations, $x_1, \ldots, x_n$) but unknown probabilities $p_1, \ldots, p_n$; plus some extra constrains, e.g. $\langle X \rangle$ is known. We are given the task to attempt to have a good guess for the probabilities.

Let's start with one of the simplest examples: X can take 1, 2 or 3 with unknown probabilities, and $\langle X \rangle = \bar{x}$ is known. Fixing $\langle X \rangle$ does not determine the probabilities, for example for $\bar{x} = 2$ any $(p_1, p_2, p_3) = (\frac{1-p_2}{2}, p_2, \frac{1-p_2}{2})$ satisfies the constraint, including e.g. $(0, 1, 0)$ or $(\frac{1}{2}, 0, \frac{1}{2})$ or $(\frac{1}{3}, \frac{1}{3}, \frac{1}{3})$. Which one is the "best"? According to the *maximum entropy principle*, the best guess is the one which maximises the information entropy under the given constraints.

To calculate this solution, we need to find the maximum of $H(p_1, p_2, p_3)$ as a function of p_1, p_2, p_3, under two constraints: $\langle X \rangle = 1p_1 + 2p_2 + 3p_3 = \bar{x}$ and $p_1 + p_2 + p_3 = 1$. We use the method of Lagrange multipliers: first calculate the unconditional maximum of the original function plus the constraints added with some multiplying factors (the Lagrange multipliers), which give the probabilities in a functional form with the Lagrange

---

[1] In this section we follow the treatment of [1].

multipliers as parameters.

$$
\begin{aligned}
0 &= d\left[H(p_1, p_2, p_3) - \lambda\left(\sum_{i=1}^{3} ip_i - \overline{x}\right) - \mu\left(\sum_{i=1}^{3} p_i - 1\right)\right] \\
&= d\left[-\sum_{i=1}^{3} p_i \log p_i - \lambda\sum_{i=1}^{3} ip_i - \mu\sum_{i=1}^{3} p_i\right] \\
&= \sum_{i=1}^{3} \left\{ -\log p_i - 1 - \lambda i - \mu\right\} dp_i = 0.
\end{aligned}
$$

Since this has to hold for any dp_i, the curly brackets need to be zero:

$$
-\log(p_i) - 1 - \lambda i - \mu = 0, \qquad i = 1, 2, 3,
$$

which with the notation $\lambda_0 = \mu + 1$ gives

$$
p_i = e^{-\lambda_0 - \lambda i}.
$$

Now we set the Lagrange multipliers by requiring the constraints to be satisfied. The constraint on the sum of probabilities gives

$$
1 = \sum_{i=1}^{3} p_i = e^{-\lambda_0}\sum_{i=1}^{3} e^{-\lambda i} \quad\Rightarrow\quad e^{-\lambda_0} = \frac{1}{e^{-\lambda} + e^{-2\lambda} + e^{-3\lambda}},
$$

so

$$
p_i = \frac{e^{-\lambda i}}{e^{-\lambda} + e^{-2\lambda} + e^{-3\lambda}} = \frac{e^{\lambda(1-i)}}{1 + e^{-\lambda} + e^{-2\lambda}}.
$$

The other constraint, $\langle X\rangle = \overline{x}$ gives

$$
\overline{x} = \sum_{i=1}^{3} ip_i = \frac{1 + 2e^{-\lambda} + 3e^{-2\lambda}}{1 + e^{-\lambda} + e^{-2\lambda}}. \tag{4.4}
$$

Multiplying the equation with the denominator gives a second degree equation for $e^{-\lambda}$, which has the solution

$$
e^{-\lambda} = \frac{2 - \overline{x} \pm \sqrt{4 - 3(\overline{x} - 2)^2}}{2(\overline{x} - 3)}.
$$

Now if we rewrite (4.4) as

$$
\overline{x} = \frac{e^\lambda + 2 + 3e^{-\lambda}}{e^\lambda + 1 + e^{-\lambda}} = 1 + \frac{1 + 2e^{-\lambda}}{e^\lambda + 1 + e^{-\lambda}},
$$

then p_2 becomes

$$
p_2 = \frac{-1 + \sqrt{4 - 3(\overline{x} - 2)^2}}{3}.
$$

Note that one of the roots have been dropped to keep p_2 non-negative. Finally the other probabilities become

$$p_1 = \frac{3 - \bar{x} - p_2}{2}, \qquad p_3 = \frac{\bar{x} - 1 - p_2}{2}.$$

This solution has the right behaviour in the limiting cases: when $\bar{x} = 1$, the probabilities $(p_1, p_2, p_3) = (1, 0, 0)$; and when $\bar{x} = 3$, they are $(0, 0, 1)$. For $\bar{x} = 2$, the solution is $(\frac{1}{3}, \frac{1}{3}, \frac{1}{3})$. The maximum entropy solution assigns zero probabilities only when no other possibilities are allowed. This is a very desirable property: it would be a sure failure to propose that a certain state has zero probability, and then find out that a given observation happened to yield that state. The maximum entropy solution is guaranteed not to fail there.

Maximum entropy principle – general form

After having this worked out example, we state the maximum entropy principle in a more general form. Suppose we have a random variable X taking known values $x_1, \ldots, x_n$ with unknown probabilities $p_1, \ldots, p_n$. In addition, we have m constraint functions $f_k(x)$ with $1 \leq k \leq m < n$, where

$$\langle f_k(X) \rangle = F_k ,$$

the F_ks are fixed. Then the maximum entropy principle assigns probabilities in such a way that maximises the information entropy of X under the above constraints. This is the "best guess" in the absence of any further knowledge about the random variable. Since any extra assumption would bring a reduction in uncertainty (see mutual information), we explicitly deny those extra assumptions by maximising the uncertainty.

In the following we calculate various properties of the maximum entropy solution. This may sound dry, but has the advantage that these abstract results can be very easily applied later for concrete examples.

To obtain a formal solution we proceed in a similar way as in the example, maximise the information entropy using Lagrange multipliers:

$$0 = d \left[H(p_1, \ldots, p_n) - \sum_{k=1}^{m} \lambda_k \left(\sum_{i=1}^{n} f_k(x_i) p_i - F_k \right) - \underbrace{\mu}_{\lambda_0 - 1} \left(\sum_{i=1}^{n} p_i - 1 \right) \right]$$

$$= \sum_{i=1}^{n} \left\{ -\log(p_i) - 1 - \sum_{k=1}^{m} \lambda_k f_k(x_i) - (\lambda_0 - 1) \right\} dp_i.$$

Since this is zero for any dp_i, all n braces have to be zero, giving

$$p_i = \exp\left(-\lambda_0 - \sum_{k=1}^{m} \lambda_k f_k(x_i)\right). \qquad (4.5)$$

Then all the Lagrange multipliers $(\lambda_0, \lambda_1, \ldots \lambda_m)$ are fixed by substituting back into the constraints. The sum of probabilities give

$$1 = \sum_{i=1}^{n} p_i = e^{-\lambda_0} \sum_{i=1}^{n} \exp\left(-\sum_{k=1}^{m} \lambda_k f_k(x_i)\right).$$

The sum after $e^{-\lambda_0}$ appears frequently, so it is useful to consider it separately: we will call it *partition function*

$$Z(\lambda_1, \ldots, \lambda_m) := \sum_{i=1}^{n} \exp\left(-\sum_{k=1}^{m} \lambda_k f_k(x_i)\right). \qquad (4.6)$$

With this notation

$$e^{-\lambda_0} = \frac{1}{Z(\lambda_1, \ldots, \lambda_m)}. \qquad (4.7)$$

The other constraints are

$$
\begin{aligned}
F_k &= \sum_{i=1}^{n} f_k(x_i) p_i = e^{-\lambda_0} \sum_{i=1}^{n} f_k(x_i) \exp\left(-\sum_{k=1}^{m} \lambda_k f_k(x_i)\right) \\
&= -\frac{1}{Z} \frac{\partial Z(\lambda_1, \ldots, \lambda_m)}{\partial \lambda_k} = -\frac{\partial \log Z(\lambda_1, \ldots, \lambda_m)}{\partial \lambda_k},
\end{aligned}
\qquad (4.8)
$$

which is m implicit equations, just enough to determine in principle the m unknowns λ_k. Using (4.7) then the probabilities (4.5) are then fully determined:

$$p_i = \frac{1}{Z(\lambda_1, \ldots, \lambda_m)} \exp\left(-\sum_{k=1}^{m} \lambda_k f_k(x_i)\right). \qquad (4.9)$$

Unlike the simple example we had with three states, in practice it is usually not possible to calculate the λ_ks explicitly as a function of F_ks, but as we see later this does not prevent us obtaining lots of useful results.

Consider now the value of the maximised information entropy. It is no longer function of the probabilities, but instead of the constraint values

F_k, and to reflect this we change notation to S:

$$S(F_1, \ldots, F_m) := H(\underbrace{p_1, \ldots, p_n}_{\text{from (4.9)}}) = -\sum_{i=1}^{n} p_i \left(-\lambda_0 - \sum_{k=1}^{m} \lambda_k f_k(x_i) \right)$$

$$= \lambda_0 + \sum_{k=1}^{m} \lambda_k \sum_{i=1}^{n} f_k(x_i) p_i = \log Z(\lambda_1, \ldots, \lambda_m) + \sum_{k=1}^{m} \lambda_k F_k. \quad (4.10)$$

Now calculate the partial derivatives of S w.r.t. the F_ks, being careful about what is kept constant in the partial derivatives:[2]

$$\left. \frac{\partial S}{\partial F_k} \right|_{\{F\}} = \sum_{\ell=1}^{m} \underbrace{\left. \frac{\partial \log Z}{\partial \lambda_\ell} \right|_{\{\lambda\}}}_{-F_\ell} \left. \frac{\partial \lambda_\ell}{\partial F_k} \right|_{\{F\}} + \sum_{\ell=1}^{m} \left. \frac{\partial \lambda_\ell}{\partial F_k} \right|_{\{F\}} F_\ell + \lambda_k = \lambda_k.$$

$$(4.11)$$

Here either $S(F_1, \ldots, F_m)$ or $\log Z(\lambda_1, \ldots, \lambda_m)$ give a full description of the system, as the other can be calculated using (4.10), and there is a symmetric relation between their partial derivatives: (4.8) and (4.11). We look at this kind of relation between two functions more closely below.

Legendre transform

Consider a convex function $f(x)$, and define the following function:

$$f^*(p) := \max_x \left(px - f(x) \right). \quad (4.12)$$

We call this[3] the *Legendre transform* of $f(x)$. If f is differentiable as well, then the maximum can be calculated as

$$0 = \frac{d}{dx} (px - f(x)) = p - \frac{df(x)}{dx}.$$

Its solution for x depends on p, which we call $x(p)$:

$$\left. \frac{df(x)}{dx} \right|_{x=x(p)} = p,$$

---

[2] In thermodynamics and statistical physics functions of many variables are used extensively, and the notation is not always clear on what the free variables are. When taking partial derivatives, it is essential to be clear on what is kept constant; therefore it is often shown at the bottom of the vertical bar after the partial differential. For example, the notation $\{\lambda\}$ means all λ_js are kept fixed except the one we differentiate with.

[3] The Legendre transform is sometimes defined with a sign difference: $f^*(p) = \max(f(x) - px)$. The advantage of our notation is that the inverse, as we will soon see, is completely symmetric.

which plugged into (4.12) gives

$$f^*(p) = px(p) - f(x(p)).$$

Now let's calculate the Legendre transform of f^*:

$$(f^*)^*(y) = \max_p \left(yp - f^*(p)\right).$$

Again, if f^* is differentiable then

$$\frac{df^*(p)}{dp}\bigg|_{p=p(y)} = y.$$

However,

$$\frac{df^*(p)}{dp} = \frac{px(p) - f(x(p))}{dp} = x(p) + p\frac{dx(p)}{dp} - \underbrace{\frac{df(x)}{dx}\bigg|_{x(p)}}_{p}\frac{dx(p)}{dp} = x(p),$$

so

$$y = \frac{df^*(p)}{dp}\bigg|_{p=p(y)} = x(p(y)),$$

thus

$$f^{**}(y) = yp(y) - f^*(p(y)) = yp(y) - p(y)x(p(y)) + f(x(p(y))) = f(y).$$

We just obtained that the functions $f^{**}(\cdot)$ and $f(\cdot)$ are equal, or equivalently the Legendre transform is its own inverse.

The Legendre transform can be easily generalised to concave functions: in the definition max needs to be replaced by min.

The other generalisation applies to functions of multiple variables: the Legendre transform of $f(x_1, \ldots, x_m)$ is

$$f^*(p_1, \ldots, p_k) = \sum_{k=1}^m x_k p_k - f(x_1, \ldots, x_m), \qquad \text{where } p_k = \frac{\partial f}{\partial x_k}.$$

Now looking back to the maximum entropy solution, (4.10) and (4.8) establish that $S(F_1, \ldots, F_m)$ and $-\log Z(\lambda_1, \ldots, \lambda_m)$ are Legendre transforms of each other. Having seen the symmetric structure of the Legendre transform, (4.11) is no longer surprising. The only remaining bit is to show that $-\log Z$ is indeed either convex or concave so that the Legendre transform is defined, to which we come back soon.

Reciprocity laws and covariances

We can easily derive relationships between partial derivatives of the constraints F_k and Lagrange multipliers λ_k. By changing the order of partial differentiations we obtain

$$\left.\frac{\partial F_k}{\partial \lambda_j}\right|_{\{\lambda\}} = \left.\frac{\partial^2 - \log Z}{\partial \lambda_j \partial \lambda_k}\right|_{\{\lambda\}} = \left.\frac{\partial^2 - \log Z}{\partial \lambda_k \partial \lambda_j}\right|_{\{\lambda\}} = \left.\frac{\partial F_j}{\partial \lambda_k}\right|_{\{\lambda\}}. \qquad (4.13)$$

Similarly

$$\left.\frac{\partial \lambda_k}{\partial F_j}\right|_{\{F\}} = \left.\frac{\partial^2 S}{\partial F_j \partial F_k}\right|_{\{F\}} = \left.\frac{\partial^2 S}{\partial F_k \partial F_j}\right|_{\{F\}} = \left.\frac{\partial \lambda_j}{\partial F_k}\right|_{\{F\}}.$$

By cursory observation one might say the second equation is just the reciprocal of the first one, so it is not telling anything new. This is wrong, as the quantities that are kept fixed at differentiation are not the same. However, the naive notion of inverse holds in a more intricate way: the matrices with elements $A_{jk} = \partial F_j/\partial \lambda_k$ and $B_{jk} = \partial \lambda_j/\partial F_k$ are inverses of each other: $A = B^{-1}$.

When we set $\langle f_k(X) \rangle = F_k$, we required that *the expectation of $f_k(X)$ is what is prescribed*, but still it varies from observation to observation. Now we look at how large these fluctuations are.

The *covariance* of two random variables is defined as

$$\mathrm{Cov}(X, Y) := \langle [X - \langle X \rangle][Y - \langle Y \rangle] \rangle = \langle XY \rangle - \langle X \rangle \langle Y \rangle,$$

which is a measure of "how much Y is above its average at the same time when X is above its average". A covariance of a random variable with itself is called *variance*:

$$\mathrm{Var}(X) := \mathrm{Cov}(X, X) = \langle (X - \langle X \rangle)^2 \rangle = \langle X^2 \rangle - \langle X \rangle^2,$$

with the convenient meaning that its square root (the *standard deviation* σ) measures how much a random variable differs from its average, suitably weighted. (The variance is always non-negative, as it is the average of a non-negative quantity: a square.)

So we can calculate the covariance of $f_k(X)$ and $f_j(X)$:

$$\mathrm{Cov}(f_j(X), f_k(X)) = \langle f_j(X) f_k(X) \rangle - \langle f_j(X) \rangle \langle f_k(X) \rangle.$$

The first term using (4.9) is

$$\langle f_j(X)f_k(X)\rangle = \frac{1}{Z}\sum_{i=1}^{n}f_j(x_i)f_k(x_i)\exp\left(-\sum_{\ell=1}^{m}\lambda_\ell f_\ell(x_i)\right)$$

$$= \frac{1}{Z}\frac{\partial^2 Z(\lambda_1,\ldots,\lambda_m)}{\partial\lambda_j\partial\lambda_k}.$$

As a side remark, the above calculation easily generalises to averages of arbitrary products of f_ks:

$$\langle f_j^{m_j}(X)f_k^{m_k}(X)\cdots\rangle = \frac{1}{Z}\left(\frac{\partial^{m_j}}{\partial\lambda_j^{m_j}}\frac{\partial^{m_k}}{\partial\lambda_k^{m_k}}\cdots\right)Z.$$

Coming back to the covariance

$$\mathrm{Cov}(f_j(X),f_k(X)) = \frac{1}{Z}\frac{\partial^2 Z}{\partial\lambda_j\partial\lambda_k} - \frac{1}{Z^2}\frac{\partial Z}{\partial\lambda_j}\frac{\partial Z}{\partial\lambda_k} = \frac{\partial^2\log Z}{\partial\lambda_j\partial\lambda_k}$$

$$= -\frac{\partial F_k}{\partial\lambda_j} = -\frac{\partial F_j}{\partial\lambda_k},$$

where we have seen the last steps already in (4.13). Similarly for variance

$$0 \le \mathrm{Var}(f_k(X)) = \frac{\partial^2\log Z}{\partial\lambda_k^2} = -\frac{\partial F_k}{\partial\lambda_k}. \tag{4.14}$$

This confirms that the second derivative of $\log Z$ is non-negative, ie. $\log Z$ is a convex function, which we implicitly assumed when mentioned that $-\log Z$ and S are Legendre transforms of each other.

Suppose now that the constraint functions f_k depend on an external parameter: $f_k(X;\alpha)$. Everything, including Z and S become dependent on α. To see its effect we calculate partial derivatives:

$$-\frac{\partial\log Z}{\partial\alpha}\bigg|_{\{\lambda\}} = -\frac{1}{Z}\sum_{i=1}^{n}\exp\left(-\sum_{k=1}^{m}\lambda_k f_k(x_i;\alpha)\right)\sum_{k=1}^{m}-\lambda_k\frac{\partial f_k(x_i;\alpha)}{\partial\alpha}$$

$$= \sum_{k=1}^{m}\lambda_k\left\langle\frac{\partial f_k}{\partial\alpha}\right\rangle. \tag{4.15}$$

Similarly, using $S = \log Z + \sum_k\lambda_k F_k$:

$$\frac{\partial S(F_1,\ldots,F_n;\alpha)}{\partial\alpha}\bigg|_{\{F\}} = \sum_{k=1}^{m}\underbrace{\frac{\partial\log Z}{\partial\lambda_k}\bigg|_{\{\lambda\}}}_{-F_k}\frac{\partial\lambda_k}{\partial\alpha}\bigg|_{\{F\}} + \frac{\partial\log Z}{\partial\alpha}\bigg|_{\{\lambda\}}$$

$$+ \sum_{k=1}^{m}\frac{\partial\lambda_k}{\partial\alpha}\bigg|_{\{F\}}F_k = \frac{\partial\log Z}{\partial\alpha}\bigg|_{\{\lambda\}}.$$

So the partial derivatives of $\log Z$ and S with respect to α are equal, though one should note that the variables kept fixed are the natural variables in each case.

4.3 Applications of the maximum entropy framework

The microcanonical ensemble

The simplest system to consider is the isolated one, with no interaction with its environment. A physical example can be a thermally and mechanically isolated box containing some gas, conventionally these are called *microcanonical ensembles*. With no way to communicate, we have no information about the current state of the system. To put it in the maximum entropy framework, we do not have any constraint to apply.

The maximum entropy solution for such a system is

$$Z = \sum_{i=1}^{n} 1, \qquad p_i = \frac{1}{Z}, \qquad S = \log Z.$$

Using the conventions of statistical physics the number of states is denoted by Ω, and the unit of entropy is k_B: recall this sets the prefactor and/or the base of the logarithm in (4.2)–(4.3). Using this notation (the MC subscript denotes microcanonical):

$$Z = \Omega, \qquad p_i = \frac{1}{Z} = \frac{1}{\Omega}, \qquad S_{\mathrm{MC}} = k_B \ln \Omega.$$

In this most simple system all internal states have equal probability.

The canonical ensemble

In the next level of increasing complexity, we allow the exchange of one conserved quantity with the external environment. The physical example is a system which is thermally coupled (allowing energy exchange) with its environment; conventionally these are called *canonical ensembles*. Using this terminology we label the internal states with their energy. By having the ability to interact with the system, we can control e.g. the average energy of the system by changing the condition of the environment, corresponding to having one constraint in the maximum entropy formalism.

The maximum entropy solution for one constraint reads

$$Z(\lambda) = \sum_{i=1}^{n} e^{-\lambda f(x_i)}, \qquad p_i = \frac{1}{Z} e^{-\lambda f(x_i)}, \qquad S(F) = \log Z(\lambda) + \lambda F.$$

The conventional units for entropy is k_B for canonical ensembles as well, and as we mentioned the states are labelled with energy: $f(x_i) = E_i$ with average energy (the value of the constraint) $F = E$. Finally, the Lagrange multiplier λ is called $\beta = 1/(k_B T)$ in statistical physics, where T is temperature (measured in Kelvins), and k_B is the Boltzmann constant. Thus we have

$$Z(\beta) = \sum_{i=1}^{n} e^{-\beta E_i}, \qquad p_i = \frac{e^{-\beta E_i}}{Z} = \frac{e^{-\frac{E_i}{k_B T}}}{Z}, \qquad S_C(\langle E \rangle) = k_B \ln Z + \frac{\langle E \rangle}{T}.$$

In p_i the exponential factor $e^{-\beta E_i}$ is called Boltzmann factor, while Z provides the normalisation.

Having established this connection, we can easily translate the results of the maximum entropy formalism. Equations (4.8) and (4.11) become

$$\langle E \rangle = -\frac{\partial \ln Z}{\partial \beta} \qquad \text{and} \qquad \frac{1}{T} = \frac{\partial S_C}{\partial \langle E \rangle}.$$

Equation (4.14) gives the energy fluctuation:

$$\sigma_E^2 = \text{Var}(E) = \frac{\partial^2 \ln Z}{\partial \beta^2} = -\frac{\partial \langle E \rangle}{\partial \beta} = \underbrace{\frac{\partial \langle E \rangle}{\partial T}}_{C_V} k_B T^2,$$

where C_V is the *heat capacity* of the system. This is an interesting relation, connecting microscopic fluctuations with macroscopic thermodynamic quantities.

In practice it is useful to define the following quantity, called the *Helmholtz free energy*:

$$A := -k_B T \ln Z = \langle E \rangle - T S_C,$$

If we consider it as a function of temperature, $A(T)$, its derivative is

$$\frac{\partial A}{\partial T} = -k_B \ln Z - k_B T \frac{\partial \ln Z}{\partial \beta} \frac{1}{k_B T^2} = -S_C.$$

This leads to a relation with the energy. Our approach so far determined the entropy S_C as a function of average energy $\langle E \rangle$. Considering its inverse function $\langle E \rangle(S_C)$, we see that its Legendre transform is $-A(T)$.

It is interesting to note that

$$\sum_i \exp\left(-\frac{E_i}{k_B T}\right) = Z = \exp\left(-\frac{A}{k_B T}\right),$$

so the sum of Boltzmann factors equals to a single Boltzmann factor with energy replaced with the Helmholtz free energy. We will see its implications later in the grand canonical ensemble.

Next we consider a system made of two subsystems, which are sufficiently uncoupled. The joint partition function can be written as (labeling the left and right subsystem with (L) and (R)):

$$Z = \sum_i \sum_j e^{-\beta\left(E_i^{(L)}+E_j^{(R)}\right)} = \left(\sum_i e^{-\beta E_i^{(L)}}\right)\left(\sum_j e^{-\beta E_j^{(R)}}\right) = Z^{(L)} Z^{(R)}.$$

This means that $\ln Z$ is additive: $A = -k_B T \ln Z = A^{(L)} + A^{(R)}$. Other quantities, like the entropy or the energy have the similar additive property, and we call these *extensive* quantities.

Physical examples for canonical ensembles

We have seen that to calculate any statistical mechanics quantity for a given system, the partition function is calculated first, and then any other quantity is easily expressed. We will consider physical systems, like a particle at position x, momentum $p = mv$, and energy $E = p^2/(2m) + U(x)$, where U is the potential. In systems made of discrete states the formula involves a sum over the states. For continuous systems, however, the sum needs to be replaced by integration:

$$\sum_i (\cdot) \quad \leftrightarrow \quad \frac{1}{h}\int_{-\infty}^{\infty} dx \int_{-\infty}^{\infty} dp\,(\cdot), \qquad (4.16)$$

This is a *semiclassical* formula: not quantum mechanical, as x and p are independent variables and not non-commuting operators; but not purely classical either as the Planck constant h is involved. Instead of fully understanding, we just rationalise this formula as (i) a constant needs to appear in front of the integrals to make the full expression dimensionless, as Z should be, and (ii) in quantities involving $\log Z$ the prefactor $1/h$ becomes an additive constant, and in particular for the entropy it sets its zero level.

The simplest example is a one-dimensional box of length L. The potential can be taken as zero within the box and infinity outside, giving

$$Z = \frac{1}{h} \int_{-\infty}^{\infty} dx \int_{-\infty}^{\infty} dp \, \exp\left(-\beta \left[\frac{p^2}{2m} + U(x)\right]\right)$$

$$= \frac{1}{h} \int_{0}^{L} dx \int_{-\infty}^{\infty} dp \, \exp\left(-\beta \frac{p^2}{2m}\right) = \frac{L}{h} \sqrt{\frac{2\pi m}{\beta}} = \frac{L}{\lambda}, \quad (4.17)$$

where we used the Gaussian integral $1 = \int_{-\infty}^{\infty} \exp(-x^2/2\sigma^2)/\sqrt{2\pi\sigma^2}$. The factors other than L are collected into a quantity of dimension length: $\lambda = h/\sqrt{2\pi m k_B T}$, called the *thermal de Broglie wavelength*. When it is small compared to characteristic length scales, in our case $\lambda \ll L$, the system can be considered as classical; while if $\lambda \gtrsim L$, proper quantum mechanics needs to be used. Interestingly, this does not only involve size, but also mass and temperature. This is the reason why typically electrons are always quantum mechanical, but full atoms can be considered as classical (as is done in molecular dynamics simulations). The exception is very light atoms at very low temperature, when inherently quantum effects like superfluidity of helium can be observed.

The ideal gas is a model of gases where gas atoms or molecules are point particles which do not interact. Since in the energy the x, y, and z components are decoupled, the coordinates of all N particles can be considered as independent, which using (4.16) and (4.17) leads to

$$Z = \frac{1}{N!} \left(\frac{L}{h} \sqrt{\frac{2\pi m}{\beta}}\right)^{3N} = \frac{1}{N!} \left(\frac{V}{\lambda^3}\right)^N. \quad (4.18)$$

The $1/N!$ comes from the fact that the particles are indistinguishable: states where e.g. particle 1 has position and momentum $\mathbf{r}_a, \mathbf{p}_a$ and particle 2 has $\mathbf{r}_b, \mathbf{p}_b$ is identical to the state where particle 1 of $\mathbf{r}_b, \mathbf{p}_b$ and particle 2 of $\mathbf{r}_a, \mathbf{p}_a$; the factor corrects the double counting in the integrals. Having Z, it is easy to show that the average energy $\langle E \rangle = (3/2)N k_B T$, the Helmholtz free energy $A = N k_B T(log(\rho\lambda^3) - 1)$, and the entropy $S = N k_B(5/2 - \log(\rho\lambda^3))$, where $\rho = N/V$ is the number density.

It is interesting to see that considering V as a parameter of the system, we can apply (4.15) to obtain a new relation. Plugging in $\alpha = V$ and $\partial f/\partial \alpha = \partial E/\partial V = -p$ (the latter can be considered as a definition of pressure):

$$-\frac{\partial \log Z}{\partial V} = -\beta \left\langle \frac{\partial E}{\partial V} \right\rangle,$$

which using (4.18) gives

$$N k_B T = \langle p \rangle V.$$

This is called the *equation of state*, as it provides a relation between state variables like pressure, volume and temperature.

Next we consider another fundamental system, the *harmonic oscillator*. One can think about it as a point mass m moving in one dimension, connected to a spring of stiffness k, of which the other end is kept fixed. If the position x is measured from the equilibrium position (unstretched spring), then the force acting on the the point mass is $-kx$, yielding Newton's equation $md^2x/dt^2 = -kx$. This has a solution $x = A\sin(\omega t + \phi)$, where the amplitude A and phase ϕ are parameters set by the initial condition, and the frequency is $\omega = \sqrt{k/m}$. The energy stored in the spring can be written as $kx^2/2 = m\omega^2 x^2/2$, so the total energy is

$$E = \frac{p^2}{2m} + \frac{m\omega^2}{2}x^2. \tag{4.19}$$

Using the standard recipe we first calculate the partition function:

$$Z = \frac{1}{h}\int_{-\infty}^{\infty} dx \int_{-\infty}^{\infty} dp\, e^{-\beta\frac{p^2}{2m}} e^{-\beta\frac{m\omega^2}{2}x^2} = \frac{1}{h}\sqrt{\frac{2\pi m}{\beta}}\sqrt{\frac{2\pi}{m\omega^2\beta}} = \frac{1}{\hbar\omega\beta},$$

where we introduced $\hbar = h/(2\pi)$. Then

$$\langle E \rangle = -\frac{\partial \ln Z}{\partial \beta} = -\frac{\partial}{\partial \beta}\ln\frac{1}{\beta} = k_B T \qquad \text{and} \qquad C_V = \frac{\partial\langle E\rangle}{\partial T} = k_B.$$

This last result is a realisation of the principle of *equipartition*: each quadratic half-degree of freedom (like x and p in (4.19)) contributes $k_B T/2$ to the average energy, and consequently $k_B/2$ to the heat capacity.

We will now apply these results to calculate the heat capacity of solids. Far away from the melting temperature the many-body potential of the atoms in a crystal can be considered quadratic. Collecting all $3N$ coordinates of the N atoms into a vector $\mathbf{x} = (x_1, x_2, \ldots, x_{3N})$, the potential is

$$U(\mathbf{x}) = U_0 + \sum_{i=1}^{3N} \frac{\partial U}{\partial x_i}(x_i - x_i^0) + \frac{1}{2}\sum_{i,j=1}^{3N} \frac{\partial^2 U}{\partial x_i \partial x_j}(x_i - x_i^0)(x_j - x_j^0) + \cdots,$$

where the series expansion is truncated at the quadratic term. The equation of motion involves the $3N \times 3N$ dynamical matrix $\partial^2 U/\partial x_i \partial x_j$, which separates into $3N$ independent one-dimensional harmonic oscillators corresponding to the normal modes and eigenfrequencies. This leads to $C = 3Nk_B$, known as the Dulong–Petit law, which turns out to be correct at high temperatures.

At low temperatures quantum mechanical effects have to be taken into account, which we do simply by replacing the classical harmonic oscillators with quantum harmonic oscillators. For our purposes the quantum harmonic oscillator is a system with discrete energy levels: in the ith state $E_i = (i + \frac{1}{2})\hbar\omega$, where $i = 0, 1, \ldots$. Being a discrete system the partition function involves just a sum, which here is a geometric sum:

$$Z = \sum_{i=0}^{\infty} e^{-\beta\left(i+\frac{1}{2}\right)\hbar\omega} = \frac{e^{-\frac{1}{2}\beta\hbar\omega}}{1 - e^{-\beta\hbar\omega}} = \frac{1}{2\sinh\left(\frac{\beta\hbar\omega}{2}\right)}.$$

The average energy and heat capacity are

$$\langle E \rangle = -\frac{\partial \ln Z}{\partial \beta} = \frac{\hbar\omega}{2} \coth \frac{\hbar\omega}{2k_B T},$$

$$C = \frac{\partial \langle E \rangle}{\partial T} = k_B \left(\frac{\hbar\omega}{2k_B T}\right)^2 \frac{1}{\sinh^2 \frac{\hbar\omega}{2k_B T}}.$$

At high temperature (small β), the argument of sinh is small, which expands to $\sinh x \sim x$. This leads to $C \to k_B$, which is the classical result.

At low temperature (large β), however, the argument of sinh is large, expanding to $\sinh x \sim \frac{1}{2}e^x$. This gives $C \approx k_B \left(\frac{\hbar\omega}{k_B T}\right)^2 e^{-\frac{\hbar\omega}{k_B T}}$, resulting in exponential suppression at low temperatures. Naively applying this result to crystals leads to the *Einstein model* of solids, which at low temperatures simply gives $C = 3Nk_B \left(\frac{\hbar\omega}{2k_B T}\right)^2 / \sinh^2 \frac{\hbar\omega}{2k_B T}$.

This is still incorrect, however, since all quantum harmonic oscillators are assumed to have the same frequency. In the *Debye model* of solids the proper spectrum of frequencies is used, which indeed reproduces experimental measurements at low temperatures as well. The reader is referred to standard solid state physics textbooks for details.

The grand canonical ensemble

We now allow the exchange of two conserved quantities with the external environment: to follow the physical example of *grand canonical ensembles*, these are the energy and the particle number. In the maximum entropy formalism this corresponds to constraining the average energy and the average particle number. As before, the units of entropy is k_B, and the ith state has energy E_i and particle number N_i. The Lagrange multiplier conjugate to energy is $\beta = 1/(k_B T)$ as in the canonical ensemble. The other one, however, is conventionally denoted by $-\mu\beta = -\frac{\mu}{k_B T}$.

Accordingly the grand canonical partition function (denoted by Ξ) and the probabilities of the states are

$$\Xi(\beta, \mu) = \sum_i e^{-\beta(E_i - \mu N_i)},$$

$$p_i = \frac{1}{\Xi} e^{-\frac{1}{k_B T}(E_i - \mu N_i)},$$

while the entropy, now function of the average energy and average particle number, using (4.10) becomes

$$S_{GC}(\langle E \rangle, \langle N \rangle) = k_B \ln \Xi + \frac{\langle E \rangle}{T} - \frac{\mu \langle N \rangle}{T}.$$

The simple relations (4.8) and (4.11) become more complicated due to the fact that the physical variables, especially μ, are not simply the Lagrange multipliers but functions of them:

$$\langle E \rangle = -\left.\frac{\partial \ln \Xi}{\partial \beta}\right|_{-\mu \cdot \beta} = -\left.\frac{\partial \ln \Xi}{\partial \beta}\right|_\mu + \left.\frac{\partial \ln \Xi}{\partial \mu}\right|_\beta \mu k_B T,$$

$$\langle N \rangle = -\left.\frac{\partial \ln \Xi}{\partial -\mu \beta}\right|_\beta = k_B T \left.\frac{\partial \ln \Xi}{\partial \mu}\right|_\beta,$$

$$\frac{1}{T} = \left.\frac{\partial S_{GC}}{\partial \langle E \rangle}\right|_{\langle N \rangle},$$

$$-\frac{\mu}{T} = \left.\frac{\partial S_{GC}}{\partial \langle N \rangle}\right|_{\langle E \rangle}.$$

In the grand canonical ensemble not only the energy fluctuates, but also the particle number:

$$\sigma_N^2 = \text{Var}(N) = \left.\frac{\partial^2 \ln \Xi}{\partial(-\mu \beta)^2}\right|_\beta = k_B T \left.\frac{\partial \langle N \rangle}{\partial \mu}\right|_\beta.$$

The reciprocity relations also become more complicated, for example

$$\left.\frac{\partial \langle E \rangle}{\partial -\mu \beta}\right|_\beta = \left.\frac{\partial \langle N \rangle}{\partial \beta}\right|_{-\mu \cdot \beta}$$

becomes

$$-k_B T \left.\frac{\partial \langle E \rangle}{\partial \mu}\right|_\beta = \left.\frac{\partial \langle N \rangle}{\partial \beta}\right|_\mu - \mu k_B T \left.\frac{\partial \langle N \rangle}{\partial \mu}\right|_\beta.$$

An important quantity is the grand free energy (we will see soon the relevance of the free energies), which is defined as

$$\Phi(T, \mu) := -k_B T \ln \Xi = \langle E \rangle - \mu \langle N \rangle - T S_{GC}.$$

It is interesting to note that the partition function can be written as

$$e^{-\beta\Phi} = \Xi = \sum_i e^{-\beta E_i} e^{\beta\mu N_i} = \sum_{N=0}^{\infty} e^{\beta\mu N} \sum_j e^{-\beta E_{j,N}}$$
$$= \sum_N e^{-\beta\left(A(T;N) - \mu N\right)}.$$

In this expression microscopic states with the same particle number N are lumped together into a macroscopic state, and the sum of their Boltzmann factors is replaced by a single Boltzmann factor where the role of the energy is played by an appropriate free energy. This manipulation is called *partial trace*, a terminology borrowed from the quantum formalism of statistical mechanics.

4.4 Fluctuations and thermodynamics

In the previous section we calculated the energy and particle number fluctuations in the canonical and grand canonical ensembles. Considering how the *relative* fluctuations depend on the system size, we obtain

$$\frac{\sigma_E}{\langle E \rangle}\bigg|_{\text{canonical ensemble}} = \frac{\sqrt{k_B T^2 \frac{\partial \langle E \rangle}{\partial T}}}{\langle E \rangle} \sim \frac{1}{\sqrt{\langle E \rangle}},$$

$$\frac{\sigma_N}{\langle N \rangle}\bigg|_{\text{grand canonical ens.}} = \frac{\sqrt{k_B T \frac{\partial \langle N \rangle}{\partial \mu}}}{\langle N \rangle} \sim \frac{1}{\sqrt{\langle N \rangle}}.$$

In both cases the relative fluctuations decay as the $-\frac{1}{2}$ power of the system size. In the $N \to \infty$ limit, called *thermodynamic limit*, the fluctuating quantities (when rescaling with the system size) become definite, not random. Thus we can replace $\langle E \rangle$ with E, etc. This is why statistical mechanics is the microscopic foundation of thermodynamics.

In many cases fluctuations are the aggregate effect of many independent contributions. To consider this case more rigorously, suppose X_i are *iid* (independent, identically distributed) random variables, with $\langle X_i \rangle = \mu$ and $\text{Var}(X_i) = \sigma^2$. Then the *central limit theorem* states that

$$Z_n := \frac{\overbrace{X_1 + \cdots + X_n}^{S_n} - n\mu}{\sqrt{n}\sigma} \xrightarrow{D} \mathcal{N}(0,1).$$

Here $\mathcal{N}(0,1)$ is the distribution of standard normal (Gaussian) random variables, ie. with zero mean and unit variance. The notation $\xrightarrow{D}$ means convergence in distribution:

$$\lim_{n \to \infty} P(Z_n < z) = P(\zeta < z),$$

where ζ is a standard normal random variable. Note that this is pointwise convergence of the cumulative distribution function, which is weaker than the convergence of the probability density function.

The central limit theorem is behind the fact that the normal distribution is so prevalent: for macroscopic fluctuations often the microscopic contributions are sufficiently independent. As we have seen before the relative fluctuations of the sum decrease as $1/\sqrt{n}$:

$$\frac{\sigma_{S_n}}{\langle S_n \rangle} \to \frac{\sqrt{n}\sigma}{n\mu} \sim \frac{1}{\sqrt{n}}.$$

A simple application is the one-dimensional random walk: X_i takes values ± 1 each with probability $1/2$. The resulting trajectory, $S_n = X_1 + \cdots + X_n$ is *like* a Gaussian variable with mean zero and standard deviation $\sqrt{n}$, when sufficiently coarse grained to remove the discreteness.

Certain important cases fall outside the applicability of the central limit theorem, like distributions where the variance (or the mean as well) is undefined. One such example is the Cauchy (or Lorentz) distribution, defined by the probability density function

$$f(x) = \frac{1}{\pi(1 + x^2)} \qquad \text{or} \qquad f(x) = \frac{1}{\pi\gamma \left(1 + \left(\frac{x - x_0}{\gamma} \right)^2 \right)}.$$

Surprisingly, the average of n iid Cauchy random variables has the same distribution as just one, which means that if one deals with such quantities, taking averages is useless.

When generalising this phenomena one arrives at the concept of *stable distributions*: these are families of distributions where the sum of such random variables is from the same family. More formally, let $\text{Fam}(\Theta)$ represent a family of distributions where Θ denotes all the parameters. Suppose X_1 and X_2 are from this family. If their linear combination is also from this family:

$$X_1 \sim \text{Fam}(\Theta_1), \quad X_2 \sim \text{Fam}(\Theta_2) \quad \Rightarrow \quad aX_1 + bX_2 \sim \text{Fam}(\Theta_3) + c,$$

then we call Fam a stable distribution.

We have seen that both the normal and the Cauchy are stable distributions. One more where the probability density function can be given in closed form is the Levy distribution:

$$f(x) = \sqrt{\frac{c}{2\pi}} \frac{e^{-c/(2x)}}{x^{3/2}},$$

which can be generalised to the four-parameter Levy-skew-α-stable family.

This distribution underpins the *Levy flight*, which is similar to a random walk, but the increments are taken from a heavy tailed distribution,

$$f(x) \sim 1/|x|^{\alpha+1}, \qquad \text{where } 0 < \alpha < 2.$$

Thermodynamics

As we mentioned, thermodynamics is founded by statistical mechanics. However, it can also be considered as a self-standing axiomatic theory, based on the following axioms:

(0) There exists a relation between thermodynamic systems. This relation is called thermodynamic equilibrium, and it is transitive (equivalence relation):

$$\text{if } A \sim B \text{ and } B \sim C, \text{ then } A \sim C.$$

Here A, B and C label different systems. For example, in thermal equilibrium this means a transitive relation between the temperatures of the three systems.

(1) Energy conservation: the total energy of an isolated system is fixed. Thus if during some process a system absorbs heat ΔQ, at the same time that work $\Delta W = -p\,\Delta V + \cdots$ is made on it, then its energy changes by $\Delta E = \Delta Q + \Delta W$.

(2) In an isolated system the entropy does not decrease. Thus if during some process a system absorbs heat ΔQ, then its entropy changes by $\Delta S = \Delta Q/T + \Delta S_{\text{internal}} \geq \Delta Q/T$.

(3) The entropy at absolute zero temperature is zero, or it is independent of other parameters, so can be set to zero. Another form is that the ground state of a quantum system has finite multiplicity (or at least not exponential with N).

A simple consequence is that the heat capacity vanishes at absolute zero temperature:

$$C_X = \left.\frac{\partial Q}{\partial T}\right|_X = T \left.\frac{\partial S}{\partial T}\right|_X \to 0 \quad \text{if } T \to 0,$$

where the subscript X corresponds to the quantity held fixed, it can be anything suitable, e.g. volume or pressure.

Let us suppose now that we bring a thermodynamic system from one state to another. Certain thermodynamic quantities, like work or heat do depend on which path is taken, while others like total energy, entropy, free energy, temperature, pressure, etc., are path-independent. We call the latter ones *state variables*.

Now we return to free energies. Consider a system kept at fixed temperature T, which undergoes some change. If it is otherwise isolated from its environment (canonical ensemble), then its energy changes only by the absorbed heat: $\Delta E = \Delta Q$. The change in entropy is

$$\Delta S = \frac{\Delta Q}{T} + \Delta S_{\text{int}} \geq \frac{\Delta Q}{T},$$

since the second law of thermodynamics states that $\Delta S_{\text{int}} \geq 0$. Introducing the Helmholtz free energy as $A(T) = E - TS$, its change is

$$\Delta A = \Delta E - T\Delta S \leq 0\,,$$

so the free energy never increases during any change or transition. Then it follows that at stable equilibrium it must be minimal. This is a very important point, which makes free energies central in thermodynamics. The corresponding observation in statistical mechanics, where there are always fluctuations, is that the probability of a macroscopic state is proportional to $\exp\left(-\frac{A(T)}{k_B T}\right)$. For a large system the difference between free energies of different states are large: $\Delta A \gg k_B T$, so only the state with the lowest free energy is observed.

The above observations hold for any ensemble, when the appropriate free energy is used. To illustrate this we will have another example. Consider now a system kept at constant temperature T and pressure p, so exchange of heat and volume with its environment is allowed. When undergoing a change, its energy changes by $\Delta E = \Delta Q - p\Delta V$. The change in entropy is

$$\Delta S = \frac{\Delta Q}{T} + \Delta S_{\text{int}} \geq \frac{\Delta Q}{T}.$$

If we now introduce the Gibbs free energy: $G(T,p) = E - TS + pV$, its change is

$$\Delta G = \Delta E - T\Delta S + p\Delta V \leq 0,$$

which is non-positive again. The Gibbs free energy is the relevant quantity for constant T, p environments, like in biochemical reactions.

4.5 Phase transitions

As we mentioned in the beginning of this chapter, complex systems display a nontrivial collective behaviour of the constituents. Phase transitions are the most spectacular of these emergent phenomena. They correspond to sudden change in behaviour by a small change in the controlling parameters.

Probably the first examples to come to mind are the melting of ice and the boiling of water – those in the domain of physics are the best understood, since the microscopic interactions are completely known. There are many other, potentially more contested examples as well, like the fall of communism in East Europe around 1989.

Coming back to physics, the transition occurs between different phases. The possible phases include the familiar examples of solids, liquids and gases, but there are many more, like magnets, a whole zoo of liquid crystals, superconductors, superfluids, different crystal structures of the same material, etc.

When considering phases, *symmetry* is an important concept. It can be thought of as a collection of operations that map a system back to itself. If two states have different symmetry, these are necessarily different phases. However, it is possible to have two different phases with the same symmetry, like both liquids and gases are homogeneous (translation invariant in a statistical sense) and isotropic (all directions are equivalent). When state A and state B corresponds to phases of different symmetry, any path connecting A and B has at some point a jump in symmetry, thus it necessarily involves a phase transition. However, when the phases are of the same symmetry, it is possible, as in Fig. 4.1, to move from A to B without a phase transition.

Order parameter field

The first step to describe a phase transition is to define an *order parameter field* [2]. This is an extraction of the relevant quantities from a large number of degrees of freedom. This is not straightforward to do, and represents a very important step in understanding the phase transition. A few examples are given below:

- *Liquids and gases*: the order parameter field is the density ρ: it is a *scalar* assigned to each point in space.
- *Magnets*: the order parameter field is the locally averaged magnetisation: it is a *vector* $\mathbf{v}$ at each space point.

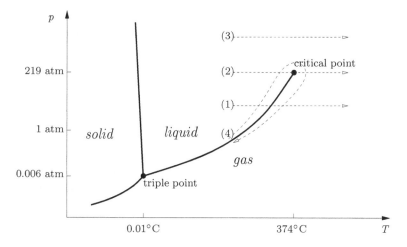

Figure 4.1 Simplified phase diagram of water. Path (1) corresponds to an abrupt phase transition between the liquid and gas phases, path (2) is a continuous phase transition, while path (3) involves no phase transition as it crosses neither a coexistence line nor a critical point. Since liquid and gas has the same symmetry, it is possible, as in path (4), to move from one side of the coexistence line to the other without a phase transition.

With the extra restriction of fixed temperature, $|\mathbf{v}|$ becomes fixed: corresponding to a point on the surface of a *sphere*.

- *Nematic liquid crystals*: these consist of thin rod molecules, which have an orientational order. The order parameter field is the locally averaged orientation $\mathbf{d}$. Since the two ends of the molecules are equivalent, a 180° rotation does not change the orientation. Mathematically this property can be expressed as $\mathbf{d} = -\mathbf{d}$: these objects are called *directors*.

  With the extra restriction of fixed temperature, $|\mathbf{d}|$ becomes fixed, corresponding to a point on a *hemisphere*.

- *Crystals*: the order parameter field is the local translation needed to return an atom to a perfect lattice site. Since all perfect lattice sites are equivalent, this means that our object does not change if we add integer times a lattice vector. For example, in two dimensions this means $\mathbf{u} = \mathbf{u} + ma_x\hat{x} + na_y\hat{y}$, where a_x and a_y are the lattice constants and $\hat{x}$ and $\hat{y}$ are the lattice directions. These objects are often called *wrapped vectors*. They can be represented by a rectangle, where the pairs of opposite sides are identified, which is topologically equivalent to the surface of a torus.

- *Superconductors*: the order parameter field is a complex number Ψ, corresponding to the quantum mechanical phase of the condensed state. At fixed temperature $|\Psi|$ is fixed, ie. a point on a *circle*.

Topological defects

A phase is often not perfect: the order parameter field is not uniform everywhere. Often these are small fluctuations that can return easily to the perfect state, but sometimes this is not the case. We call *defect* a "tear" (some sort of singularity or discontinuity) in the order parameter field, and *topological defect* is a "tear which cannot be patched". This means that no continuous local deformation of the order parameter field can remove the defect.

The simplest example is a vortex line superconductor at fixed T. If we consider a closed path encircling the line, the order parameter goes around the circle in the order parameter space. The lack of defect would correspond to the order parameter being constant as we follow the closed path. Since it is not possible to deform continuously the order parameter field from the initial state where following the closed path we go around the circle in the order parameter space to a final state where we don't move in the order parameter space, this is a topological defect. A rigorous treatment of this phenomena belongs to the *topology* branch of mathematics. It follows that the vortex line cannot end (in reality it can only end at the boundary of the sample).

Another example is a crystal, where in half of the domain there is an extra layer of atoms. The edge of this layer is a line defect: encircling it corresponds to moving on a closed loop on the torus of the order parameter space, which cannot be contracted continuously to a single point.

Our last example is nematic liquid crystals at fixed T. The order parameter space is a hemisphere, where one can select a closed loop which cannot be contracted continuously to a single point. This means that nematic liquid crystals can have line defects. This in in contrast with three-dimensional magnets, which do not have line defects: there the order parameter space is a full sphere, where any circle can be contracted to a point, for example by moving them to the north pole.

Abrupt phase transitions

Phase transitions can be grouped into two classes: *abrupt phase transitions*, where the order parameter has a discontinuity; and *continuous phase transitions*, where the order parameter has a singularity but still continuous.

Abrupt phase transitions are sometimes called first order, and the continuous one as second order. This terminology originates from Ehrenfest's classification, which considered the lowest derivative of the free energy that is discontinuous at the transition. Ehrenfest's approach is no longer used, as it turned out that different higher order phase transitions are not fundamentally different, some even do not fit in (e.g. divergent derivatives).

First let us consider the abrupt phase transition, and our example will be water held at fixed T and p. As illustrated on Fig. 4.1, if the parameters (T and p) are varied on a path which crosses a phase coexistence curve, a sudden jump in properties occur. This can be captured by the order parameter, which is in this case the density ρ.

What underlies an abrupt phase transition is that when the external parameters are varied, the current phase, which used to be stable (lowest free energy), becomes metastable, as another phase, which might be very far in configuration space, becomes lower in free energy. In the phase diagram (representing the lowest free energy phase as a function of parameters like T and p) one crosses from a domain A where the free energy of phase A is lower to domain B where that of phase B is lower.

When the coexistence curve (of equal free energies) is crossed, the system does not necessarily recognise immediately that some other state would be lower in free energy: this happens via fluctuations. Suppose that by fluctuations a small sphere of radius R of the lower free energy phase is formed. To fix notation let us consider cooling down a gas to form liquid at fixed pressure. Then the Gibbs free energy of the small liquid droplet is

$$G_{\text{droplet}}(R) = c_1 R^2 - c_2 R^3 \Delta T ,$$

see Fig. 4.2. The first term corresponds to surface tension: it costs free energy to create an interface between the two phases. This is positive and quadratic in R, dominating the expression for small R. The origin of the second term is the difference between the bulk free energies: this is proportional to the volume of the droplet ($\sim R^3$). The prefactor is proportional to ΔT (the difference between the coexistence temperature and the actual temperature), as the bulk free energies are smooth functions of the temperature, therefore their difference can be approximated to be linear in ΔT. This term is negative, and dominates at large R. The function $G_{\text{droplet}}(R)$ has a maximum, which is easy to obtain

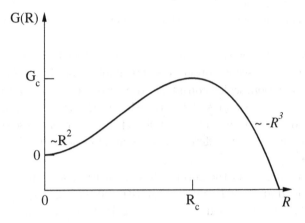

Figure 4.2 The Gibbs free energy of a droplet as a function of its radius.

by differentiation: the maximum can be denoted by G_c, which is taken at finite R_c. Due to the temperature dependence of the volume term, $G_c \sim \Delta T^{-2}$.

The free energy of a large droplet is negative, therefore this is the thermodynamically stable state. However, the system needs to discover that there is indeed a phase with lower free energy than the current one, which it does via fluctuations. In the simplest picture the system attempts to overcome the free energy barrier many times, and each attempt succeeds with probability $\exp(-\beta G_c)$. The rate by which the attempts are made (denoted by "prefactor" below) is determined by fundamental frequencies of the system, and is often very hard to predict, but the strong temperature dependence is contained mostly in the Boltzmann factor. So the rate by which the droplets form spontaneously, often called nucleation rate, is given by

$$\Gamma = (\text{prefactor}) \; e^{-\frac{G_c}{k_B T}} = (\text{prefactor}) \; e^{-\frac{c_3}{k_B T} \frac{1}{\Delta T^2}} .$$

This means *very* small rate for small ΔT, that is why a metastable phase has a macroscopic lifetime for small enough undercooling.

All the above relates to spontaneously forming droplets in free space, or "bulk nucleation". In practice there might be other channels to form droplets, like on the surface of the container, around an impurity, etc., which all have their individual barrier height; the smallest of these provides the dominant contribution to the overall nucleation rate.

Continuous phase transitions

When a system is brought through a path in parameter space which goes through a critical point (see path (2) in Fig. 4.1), the order parameter stays continuous, but at this point has a singularity: the derivative is infinite similarly to the square root function. In fact in this type of phase transition typically a large number of physical quantities have singularities, or in other words have non-trivial power law dependencies.

One such quantity is the parameter dependence (e.g. temperature dependence) of the jump of the order parameter between the two phases near the critical point. For the liquid–gas transition of water, the order parameter can be approximated by

$$\frac{\rho}{\rho_c} = 1 + s \left(1 - \frac{T}{T_c}\right) \pm \rho_0 \left(1 - \frac{T}{T_c}\right)^\beta. \tag{4.20}$$

Another example is the uniaxial magnet, where the order parameter, the magnetisation, has a nonzero value below the Curie temperature, which can be either positive or negative, and its absolute value scales as

$$M \propto \left(1 - \frac{T}{T_c}\right)^\beta.$$

Surprisingly the value of the exponent β for both the liquid–gas transition of water and the uniaxial magnet is the same, an often quoted value is $\beta = 0.325 \pm 0.005$. This is not coincidence: the scaling exponents (the exponents of the power laws for the singular quantities) are the same for a number of other physical quantities as well.[4]

The above examples illustrate the concept of *universality*: the behaviour near a continuous phase transition point is independent of the microscopic details. The various systems can be classified into a small number of groups. These groups are called *universality classes*: collections of systems with the same singular properties at the critical point. For example in thermal systems the scaling behaviour (the scaling exponent and the shape of the scaling function) of the susceptibility, specific heat, correlation length, etc., are identical within a universality class, while other quantities like the parameters (e.g. temperature) corresponding to the critical point, and also the prefactors are different. The

---

[4] In some sense the uniaxial magnet is simpler, as the phase coexistence curve falls on the $T = 0$ line. For the liquid–gas transition of water the coexistence curve is parallel to neither the T nor the p axis in the phase diagram, giving rise to the linear term in (4.20). The singular term, which is the important one, however, is the same.

universality class typically depends only on fundamental properties, like the symmetry of the order parameter, and the dimensionality.

As we have seen, the liquid–gas transition of water (and other materials!) and uniaxial magnets are in the same universality class. This is quite surprising at first sight, but on closer inspection one can see that the order parameter is of the same type (scalar), the dimensionality (three dimensions) is the same, and in fact it is possible to find a mapping between the two systems.

Universality is a very powerful concept, since one member of a universality class fully represents all other members. Thus to learn everything about a particular universality class it is enough to find a simple enough member which is tractable theoretically or numerically. This is the reason why physicists often study very simple "toy" models, which genuinely represent the more complex members of their universality class in all important aspects.

Fluctuations, which were exponentially suppressed in abrupt phase transitions, are very prevalent in continuous phase transitions. The characteristic spatial size of the fluctuations (the correlation length) diverges at the critical point.

4.6 Surface growth[5]

A large class of processes in nature can be described as growth processes, where one phase grows and invades a region previously occupied by another phase. These processes are often not in equilibrium (in fact far from equilibrium), where a persistent driving force brings the process forward, resulting in a propagating rough front. The growth front is not progressing uniformly, it is not smooth, because of some noise or inherent disorder in the system.

To give some concrete examples, we can think of the progress of the wetting of a paper or table cloth, or the burning front of a slowly burning paper. An example from the nanosciences is molecular beam epitaxy, where atoms are deposited in vacuum onto the surface of a crystal, which due to the shot noise of the deposited atoms creates a rough surface after a number of deposited layers. The growth of bacterial colonies can also be in this class of processes, especially when the growth substrate is relatively dry so the bacteria don't swim, and the nutrients are abundant.

---

[5] A detailed treatment of the statistical mechanics aspects of surface growth can be found in [3].

(It must be mentioned that the bacterial colony growth is much more complex phenomena, including sometimes even genetic shifts. And even in more simple cases a different physical phenomenon can dominate, like in the scarce nutrients regime, where the constant driving is replaced by growth dominated by the diffusion of nutrients. In the latter case instead of the rough growth front a fine branching fractal pattern emerges.)

As usual, we only would like to know aggregate information about the process, like how "non-smooth" the front is. We consider the surface growth problem on an initially flat substrate; denote the coordinates along the substrate as x, and the distance of the interface from the substrate is h (height). We can define the average height as

$$\bar{h}(t) = \langle h(x,t) \rangle_x, \qquad \bar{h}(t) \sim t,$$

which typically grows linearly in time. The *roughness* or *width* of the surface is defined as the root-mean-square deviation of h from $\bar{h}$:

$$w(L,t) = \sqrt{\left\langle \left(h(x,t) - \bar{h}(t) \right)^2 \right\rangle_x},$$

where L is the linear size (length) of the substrate.

In typical surface growth processes the width initially grows as a power of time, then at some crossover time (which depends on L) it saturates:

early times, $t \ll t_\times$: $\quad w(L,t) \sim t^\beta$ $\qquad \beta$: growth exponent

late times, $t \gg t_\times$: $\quad w(L,t) \sim w_{\mathrm{sat}}(L) \sim L^\alpha$ $\quad \alpha$: roughness exponent

crossover time: $\qquad t_\times \sim L^z$ $\qquad\qquad\qquad z$: dynamic exponent

When plotting width against time, we can achieve data collapse if the width is rescaled by the saturation width, and time by the crossover time. This way the surface width can be expressed by a single-argument scaling function:

$$w(L,t) \sim L^\alpha f\left(\frac{t}{L^z} \right) \qquad\qquad f(u) \sim \begin{cases} u^\beta, & \text{if } u \ll 1, \\ \text{const}, & \text{if } u \gg 1, \end{cases}$$

which is called the Family–Vicsek scaling relation.

Evaluating w at the crossover time where approximately both the early and the late time behaviour holds, we can obtain a relation between the exponents:

$$\left. \begin{array}{l} w(L,t_\times) \quad \sim L^\alpha \\ \qquad\qquad \sim t_\times^\beta \sim L^{z\beta} \end{array} \right\} \qquad z = \frac{\alpha}{\beta}.$$

The surface generated by the growth process has no intermediate characteristic length scales, thus displays some kind of self-similarity. In this case the x and the h directions are not equivalent, so different magnification in the x and h direction yields statistically similar objects:

$$h(x) \quad \text{similar to} \quad b^\alpha h\left(\frac{x}{b}\right).$$

Such functions are called self-similar with self-similarity exponent α. One such familiar function is the graph of random walk (note here t plays the role of x):

$$\langle (y(t+\Delta t) - y(t))^2 \rangle_t \sim \Delta t \quad \Rightarrow \quad \alpha = \frac{1}{2}.$$

These surface growth processes can be illustrated by simple models, for example:

- Random deposition: unit square blocks are released above integer positions of a substrate, and they just land on top of previously dropped blocks. For this model α is undefined, and $\beta = 1/2$.
- Random deposition with surface relaxation: as for random deposition, but the blocks are allowed to jump to the nearest neighbour substrate position to achieve lowest position.
- Restricted solid-on-solid (RSOS) model: only those growth events are allowed which keep local slope bounded: maintain $|h(x)-h(x+1)| \leq 1$

Another approach to understand the surface growth processes is to consider continuum equations:

$$\frac{\partial h}{\partial t} = G[h(\cdot), x, t] + \eta,$$

where the noise term is often unbiased and delta-correlated, for example:

$$\eta(x,t): \quad \langle \eta(x,t) \rangle = 0, \quad \langle \eta(x,t)\eta(x',t') \rangle = 2D\delta(x-x')\delta(t-t').$$

These equations have the following desired symmetries:

$t \to t + \Delta t$
$h \to h + \Delta h$
$x \to x + \Delta x$
$x \to -x$, or rotation
$h \to -h$ (in certain cases, e.g. if in equilibrium)

The simplest such equation is the Edwards–Wilkinson equation:

$$\frac{\partial h}{\partial t} = \nabla^2 h + \eta(x,t) \qquad \alpha = 1 - \frac{d}{2}, \quad \beta = \frac{1}{2} - \frac{d}{4}, \quad z = 2.$$

The simplest one breaking the $h \to -h$ symmetry is the Kardar–Parisi–Zhang (KPZ) equation:

$$\frac{\partial h}{\partial t} = \nabla^2 h + \frac{\lambda}{2}(\nabla h)^2 + \eta(x, t) \qquad \text{for } d = 1: \quad \alpha = \frac{1}{2}, \ \beta = \frac{1}{3}, \ z = \frac{3}{2},$$

which describes e.g. the RSOS model. The continuum equations and discrete models can be classified into universality classes (as we have seen in continuous phase transitions), where the scaling exponents are the same across a class.

4.7 Collective biological motion: flocking

In this section we will consider a model, which is a first step towards understanding the motion of groups of animals: for example schools of fish, herds of quadruples, flocks of birds (of which flocks of starlings is particularly impressive), as well as cooperative motion in bacterial colonies.

The model consists of a collection of self-propelled particles (representing e.g. the individual birds), which move in continuous space with constant speed but varying direction [4]. This arrangement is quite unlike typical physical systems made of passive particles, as momentum and energy is not conserved; but much more typical in biological systems where the individuals are more advanced, and active, having their own energy supply.

The particles move for time Δt in a ballistic motion, then select a new direction for their velocity. For simplicity we restrict the particles in two dimensions, where the direction is given by an angle θ. The update rules of the ith particle's position, velocity and direction is

$$\mathbf{x}_i(t + \Delta t) = \mathbf{x}_i(t) + \mathbf{v}_i(t)\,\Delta t,$$
$$\mathbf{v}_i(t) = v(\cos\theta_i(t)\,\hat{x} + \sin\theta_i(t)\,\hat{y}),$$
$$\theta_i(t + \Delta t) = \langle \theta_j(t) \rangle_{j: |\mathbf{x}_j - \mathbf{x}_i| < R} + \underbrace{\Delta\theta}_{\text{uniform in } [\eta/2, \eta/2]}.$$

So at the end of each Δt timestep each particle takes the average direction of all particles within radius R, and adds a noise of controlled amplitude.

What are the relevant parameters of this model? For any model we should take care to reduce the number of parameters as much as possible (but no further!) in order not to be lost in a high dimensional parameter space.

Without loss of generality we can set $\Delta t = 1$ and $R = 1$, i.e. these will be our units of time and space. For the speed v of the particles we already fixed the units, so need to consider its actual value. The $v \to 0$ limit would mean stationary objects (which would correspond to the XY model of magnets), while $v \to \infty$ limit would mean that each timestep a given particle meets completely new ones, corresponding to complete mixing or mean-field models. For our flocking model in the relevant, intermediate range, say $0.003 < v < 0.3$, the actual value of v turns out not to affect the behaviour. So as long as the value of v is moderate, its value does not matter: it is not a relevant parameter either. There is one more symbol above, the amplitude of the noise, $0 < \eta < 2\pi$, which is indeed one of the relevant parameters. The second relevant parameter is a bit hidden: it is the average density of particles, ρ. Finally we have the system size: either the number of particles N or the side L of the two-dimensional box (the two are related by $\rho = N/L^2$), which does matter but we will strive to see everything in the thermodynamic limit $N \to \infty$.

We see that for some values of the parameters the particles align the direction of their velocities globally, creating an ordered phase, while for other values there is no global order either because each particle moves randomly or there are small coherent groups but these are uncorrelated with each other. To quantify this order, we define the order parameter as the average velocity over the whole system normalised by v, or to follow [4] we take the absolute value:

$$\Phi = \frac{1}{Nv} \left| \sum_i \mathbf{v}_i \right| .$$

As the parameters are varied, in the $N \to \infty$ limit Φ changes from nonzero to zero value. On the ordered (nonzero) side we see power law scaling:

fixed ρ: $\Phi \sim (\eta_c(\rho) - \eta)^\beta$,
fixed η: $\Phi \sim (\rho - \rho_c(\eta))^{\beta'}$,

where numerically it was found that $\beta \approx 0.45$ and $\beta' \approx 0.35$ [4].

This system is not an equilibrium system, the phase transition is purely due to kinetic effects; this is an example of kinetic phase transitions.

It turns out that as for many other continuous phase transitions, the order parameter, which is now function of two parameters $\Phi(\eta, \rho)$, can in fact be written as a scaling function of as single variable:

$$\Phi(\eta, \rho) = \tilde{\Phi}\left(\frac{\eta}{\eta_c(\rho)}\right), \qquad \tilde{\Phi}(u) \sim \begin{cases} (1-u)^\beta, & \text{if } u < 1, \\ 0, & \text{if } u > 1. \end{cases}$$

With the help of the scaling function we can now look at the scaling exponents β and β'. For fixed η and varying ρ, we defined β' as $\Phi \sim (\rho - \rho_c(\eta))^{\beta'}$. Denoting $\epsilon = \rho - \rho_c(\eta)$:

$$\Phi(\eta, \rho_c(\eta) + \epsilon) = \tilde{\Phi}\left(\frac{\eta}{\eta_c(\rho_c(\eta) + \epsilon)}\right) \approx \tilde{\Phi}\left(\frac{\eta}{\eta + \epsilon \left.\frac{d\eta_c}{d\rho}\right|_{\rho=\rho_c(\eta)}}\right)$$

$$\approx \tilde{\Phi}\left(1 - \frac{\epsilon}{\eta}\frac{d\eta_c}{d\rho}\right) \sim \epsilon^\beta. \tag{4.21}$$

This shows [5] that $\beta = \beta'$, despite the original different numerical estimates.

There is recent ongoing research (as of 2010–2011) on the dynamics of flocks of starlings in Italy. The current understanding is that while the above model is a good starting point, real birds don't really consider a fixed radius neighbourhood, but instead watch a fixed number of neighbours and keep them oriented by trying to maintain fixed relative angles between them.

References

[1] E. T. Jaynes, *Probability Theory: The logic of science.* Cambridge University Press (2003).

[2] J. P. Sethna, *Statistical Mechanics: Entropy, order parameters, and complexity.* Oxford University Press (2006).

[3] A.-L. Barabási, H. E. Stanley, *Fractal Concepts in Surface Growth.* Cambridge University Press (1995).

[4] T. Vicsek, A. Czirók, E. Ben-Jacob, I. Cohen, O. Shocket, *Phys. Rev. Lett.* **75**, 1226 (1995)

[5] A. Czirók, H. E. Stanley, T. Vicsek; *J. Phys. A: Math. Gen.* **30**, 1375 (1997)

5

Numerical simulation of continuous systems

Colm Connaughton

5.1 Introduction

5.1.1 Partial differential equations in complexity science

Partial differential equations (PDEs), that is to say equations relating partial derivatives of functions of more than one variable, are part of the bedrock of most quantitative disciplines and complexity science is no exception. They invariably arise when continuous fields are introduced into models. The classic example is the distribution of heat in a thermal conductor which typically varies continuously with time and with position. The continuous field in this case is $T(\mathbf{x}, t)$ the temperature at position $\mathbf{x}$ and time t which, in this case, satisfies a linear PDE called the diffusion equation.

In complexity science, PDEs often result from the process of coarse-graining whereby microscopically discrete processes are averaged over small scales to produce an effective continuous description of larger scales. Since much of complexity science focuses on the emergent properties of such coarse-grained descriptions, the analysis of PDEs is a key part of our complexity science toolkit. Some examples of coarse-graining as applied to interacting particle systems were discussed in Chapter 3. Another well-known example is the effective description of traffic flow using a coarse-grained fluid description described by a PDE known as Burgers' equation and variants of it, cf. Chapter 4. We will revisit this application in more detail later. Real fluids also provide a wealth of examples of complex behaviour, all described by the well-known Navier-Stokes equations or variants of them which are famous for being among the most mathematically intractable PDEs of classical physics. Phenomena such as pattern formation, turbulence, phase separation and chaotic mixing are all encoded within these equations and play a key role in understanding the dynamics of a diverse range of natural complex systems from the human heart to the global climate to clusters of galaxies. Not all applications come from the physical sciences of course. Another famous PDE which has played a key role in theoretical economics is the Black–Scholes equation, which provided one of the first systematic models for option pricing. A final important application of PDEs in complexity science relates to their connection with stochastic processes such as the simple random walk models studied in Chapter 1. For a given stochastic model of a complex system it is often advantageous to derive the so-called Fokker–Planck equation, which is a deterministic PDE describing the time evolution of the probability density of the stochastic process. For example, the Fokker–Planck equation for a random walker is the diffusion equation.

5.1.2 What is this chapter about?

Despite the enormous amount of attention they have received over the centuries in pure and applied mathematics, most PDEs arising in modelling applications in complexity science can easily become impractically difficult or even impossible to solve by analytical methods. Therefore numerical methods of solution are invariably required in practice. This set of notes does not attempt to seriously address the multitude of specialist techniques which exist for the analytic solution of PDEs and jumps straight to the numerical techniques. Even within the numerical domain, we barely scratch the surface of what is possible, choosing to focus primarily on linear equations and finite difference methods. This chapter is intended only as a departure point to hopefully equip the reader with the vocabulary and knowledge of the underlying issues to allow him/her to have an idea of how to begin when his/her research requires resort to numerical techniques to solve some particular PDE.

Although ordinary differential equations and dynamical systems have been discussed extensively in Chapter 2 we shall revisit them here from a numerical perspective. In order to make the transition into the world of numerical algorithms for PDEs one must first have a good understanding of the concept of timestepping. Thus we shall warm up by giving detailed consideration in Section 5.2 to the numerical algorithms which produced the pictures of Chapter 2 learning along the way about the essential concepts of discretisation, finite difference approximations and timestepping algorithms. Next in Section 5.3 we shall look at a number of special scenarios in which partial differential equations can be reduced to ordinary differential equations which are amenable to the methods introduced in Section 5.2. We will learn about the method of characteristics and how it can be converted into a rudimentary numerical solver. We will also meet some particular cases where symmetries of the problem permit a reduction of a PDE to an ordinary differential equation, such as travelling waves and similarity solutions. These solution methods are not general and break immediately when the underlying symmetry is broken by, say, initial or boundary conditions. Nevertheless they are very useful benchmarks for checking the correctness of a numerical code. Next in Section 5.4 we will look at several different finite-difference approaches to the diffusion equation. Along the way we will confront issues of consistency and stability, the trade-offs between implicit and explicit methods and address the crucial issue of how to implement boundary conditions numerically. Finally in Section 5.5 we will adopt a rather similar program for hyperbolic PDEs, principally the wave equation, where questions of stability are complicated by the presence of conservation laws and finite difference methods must contend with the difficulties and opportunities presented by the phenomenon of numerical dissipation.

5.2 Ordinary differential equations (ODEs)

5.2.1 Ordinary differential equations

Differential equations are the principal mathematical tool for mathematical modelling of complex systems. An *ordinary* differential equation (ODE) is an equation involving one or more derivatives of a function of a single variable. The word "ordinary" serves to distinguish it from a *partial* differential equation (PDE) which involves one or more partial derivatives of a function of several variables. Although this course is ostensibly about solving PDEs, it makes a lot of sense to start by looking at ODEs.

Definition 5.2.1 Let $u : \mathbb{R} \to \mathbb{R}$ be a real-valued function of a single real variable, t, and $F : \mathbb{R}^{m+1} \to \mathbb{R}$ be a real-valued function of a vector formed from t, $u(t)$ and its first $m-1$ derivatives. The equation

$$\frac{d^m u}{dt^m}(t) = F\left(t, u(t), \frac{du}{dt}(t), \dots, \frac{d^{m-1}u}{dt^{m-1}}(t)\right) \tag{5.1}$$

is an ordinary differential equation of order m.

The general solution of an mth order ODE involves m constants of integration. Here are some simple examples which you already know:

Example 5.2.2 (Exponential growth equation)

$$\frac{du}{dt}(t) = \lambda u(t) \qquad\qquad \lambda > 0 \qquad\qquad (5.2)$$

has general solution

$$u(t) = c_1 \exp(\lambda t).$$

Example 5.2.3 (Exponential decay equation)

$$\frac{du}{dt}(t) = -\lambda u(t) \qquad\qquad \lambda > 0 \qquad\qquad (5.3)$$

has general solution

$$u(t) = c_1 \exp(-\lambda t).$$

Example 5.2.4 (Simple harmonic motion equation)

$$\frac{d^2 u}{dt^2}(t) = -\omega^2 u(t) \qquad\qquad \omega > 0 \qquad\qquad (5.4)$$

has general solution

$$u(t) = c_1 \sin(\omega t) + c_2 \cos(\omega t).$$

Often, physical problems correspond to a particular solution with specific values of the constants. These are determined by requiring that the solution satisfy particular initial or boundary conditions. Determining appropriate initial or boundary conditions is part of the modelling. There are three archetypal problems which often come up in modelling:

- **Initial value problems**: The value of $u(t)$ and a sufficient number of derivatives are specified at an initial time, $t = t_0$ and we require the values of $u(t)$ for $t > t_0$. For example, taking $t_0 = 0$, find the particular solution of Example 5.2.4 which satisfies:

$$u(0) = 0 \qquad\qquad \frac{du}{dt}(0) = 1.$$

- **Boundary value problems**: Boundary value problems arise when information on the behaviour of the solution is specified at two different points, x_1 and x_2, and we require the solution, $u(x)$, on the interval $[x_1, x_2]$. Choosing to call the dependent variable x instead of t is a nod to the fact that boundary value problems often arise from problems having spatial dependence rather than temporal dependence. For example, find the solution of

$$\frac{d^2 u}{dx^2}(x) = 0$$

on the interval $[x_1, x_2]$ given that

$$u(x_1) = U_1 \qquad u(x_2) = U_2.$$

- **Eigenvalue problems**: Eigenvalue problems are boundary value problems involving a free parameter, λ. The objective is to find the value(s) of λ, the eigenvalue(s), for which the problem has a solution and to determine the corresponding eigenfunction(s), $u_\lambda(x)$. For example, find the values of λ for which

$$\frac{d}{dx}\left[(1 - x^2)\frac{du}{dx}(x)\right] + \lambda u(x) = 0$$

has solutions satisfying the boundary conditions

$$u(0) = 1 \qquad \frac{du}{dt}(1) = 1,$$

and find the associated eigenfunctions, $u_\lambda(x)$.

In this course we will principally be interested in initial value problems and boundary value problems.

5.2.2 Dynamical systems

Systems of ODEs arise very often in applications. They can arise as a direct output of the modelling. Also, more importantly for this text, when we discretise the spatial part of a PDE, we are really replacing the PDE with a (usually large) system of ODEs.

Definition 5.2.5 Let $\mathbf{u} : \mathbb{R} \to \mathbb{R}^n$ be a vector-valued function of a single real variable, t. Let $\mathbf{F} : \mathbb{R}^{nm+1} \to \mathbb{R}^n$ be a vector-valued function of the vector $\left(t, \mathbf{u}(t), \frac{d\mathbf{u}}{dt}(t), \ldots, \frac{d^{m-1}\mathbf{u}}{dt^{m-1}}(t)\right)$ taking values in $\mathbb{R}^n$. The equation

$$\frac{d^m \mathbf{u}}{dt^m}(t) = \mathbf{F}\left(t, \mathbf{u}(t), \frac{d\mathbf{u}}{dt}(t), \ldots, \frac{d^{m-1}\mathbf{u}}{dt^{m-1}}(t)\right) \qquad (5.5)$$

is a system of mth-order ordinary differential equations of dimension n.

This is rather complicated and we rarely need to consider such general systems of equations. In practice, we often deal only with systems of ODEs of first order. This is a result of the fact that any ODE of order m (and by extension any system of equations of order m) can be rewritten as a system of first-order ODEs of dimension m.

Example 5.2.6 (Reduction of order)   Consider the general mth-order ODE given by Eq. (5.1). Define a vector $\mathbf{u}(t)$ in $\mathbb{R}^m$ as follows:

$$\mathbf{u}(t) = \left(u(t), \frac{dt}{du}(t), \ldots, \frac{d^{m-1}t}{du^{m-1}}(t) \right).$$

Define the function $\mathbf{F} : \mathbb{R}^{m+1} \to \mathbb{R}^m$ as

$$\mathbf{F}(t, \mathbf{u}(t)) = (u_2(t), \ldots, u_m(t), F(t, \mathbf{u}(t))).$$

By construction, Eq. (5.1) is equivalent to the first-order system of ODEs given by

$$\frac{d\mathbf{u}}{dt}(t) = \mathbf{F}\left(t, \mathbf{u}(t)\right). \tag{5.6}$$

Another distinction which we can do away with using the language of first-order systems is that between autonomous and non-autonomous systems. A system of ODEs is called *autonomous* if the independent variable (t in most of our examples) does not appear explicitly on the right-hand side and is called *non-autonomous* otherwise. A non-autonomous system of dimension n can be re-written as an autonomous system of dimension $m + 1$ by augmenting it with an additional equation: $\dot{t} = 1$.

Example 5.2.7 (Reduction to an autonomous system)   Consider the non-autonomous m-dimensional first-order system of ODEs given by Eq. (5.6). Define a vector $\mathbf{v}(t)$ in $\mathbb{R}^{m+1}$ as follows: $\mathbf{v}(t) = (\mathbf{u}(t), v_{m+1}(t))$. Define the function $\mathbf{G} : \mathbb{R}^{m+1} \to \mathbb{R}^{m+1}$ as

$$\mathbf{G}(\mathbf{v}(t)) = (\mathbf{F}(v_{m+1}(t), (v_1(t), \ldots, v_m(t)), 1))).$$

By construction, Eq. (5.6) is equivalent to the first-order autonomous system of ODEs given by

$$\frac{d\mathbf{v}}{dt}(t) = \mathbf{G}\left(\mathbf{v}(t)\right). \tag{5.7}$$

While the notation might be clumsy, the principles are illustrated clearly by the following example:

Example 5.2.8 (The driven damped pendulum)   The angle, $\theta(t)$, made with the vertical by a damped pendulum subject to an external harmonic forcing satisfies the following ODE:

$$\frac{d^2\theta}{dt^2} + \alpha\frac{d\theta}{dt} + \beta\sin(\omega_0^2\,\theta) = \delta\cos(\omega_f\,t + \phi). \qquad (5.8)$$

where α, β, γ, ω_0, ω_f and ϕ are parameters. This equation is equivalent to the following first-order system of dimension 3:

$$\frac{dv_1}{dt} = v_2, \qquad (5.9)$$

$$\frac{dv_2}{dt} = -\alpha v_2 - \beta\sin(\omega_0^2\,v_1) + \delta\cos(\omega_f\,v_3 + \phi), \qquad (5.10)$$

$$\frac{dv_3}{dt} = 1. \qquad (5.11)$$

By deriving methods of solving Eq. (5.7) we can actually solve a lot of different ODE IVPs and, as we shall see, lots of PDE IVPs too.

5.2.3 Approximation of derivatives by finite differences

Computers are finite machines and cannot represent the real numbers. When faced with a function of a continuous variable, the best we can hope to do with a computer is to approximate it by its values at a finite number of points. The process of replacing a function by an array of values at particular points is called *discretisation*. We replace $v(t)$ with $t \in [t_1, t_2]$ by a list of values, $\{v_i \ : \ i = 1 \ldots N\}$ where $v_i = v(t_1 + ih)$ where $h = (t_2 - t_1)/N$. There is no reason why the discretisation should be on a set of uniformly spaced points but we will mostly deal with such uniform discretisations. Intuitively, derivatives of v at a point can be approximated by differences of the v_i at adjacent points:

$$\frac{dv}{dt}(t_i) \approx \frac{v_{i+1} - v_i}{h}.$$

This intuition is made precise using Taylor's theorem:

Theorem 5.2.9 (Reminder of Taylor's theorem)   *If $v(t)$ is a real-valued function which is differentiable $n + 1$ times on the interval $[t, t + h]$ then there exists a point, ξ, in $[t, t + h]$ such that*

$$v(t + h) = v(t) + \frac{1}{1!}\,h\,\frac{dv}{dt}(t) + \frac{1}{2!}\,h^2\,\frac{d^2v}{dt^2}(t) + \cdots$$
$$+ \frac{1}{n!}\,h^n\,\frac{d^nv}{dt^n}(t) + h^{n+1}R_{n+1}(\xi), \qquad (5.12)$$

where

$$R_{n+1}(\xi) = \frac{1}{(n+1)!} \frac{d^{n+1}v}{dt^{n+1}}(\xi).$$

Note that the theorem does not tell us the value of ξ.

We will use Eq. (5.12) a lot. Taylor's theorem to first order ($n = 1$) tells us that

$$v(t+h) = v(t) + h \frac{dv}{dt}(t) + h^2 R_2(\xi)$$

for some $\xi \in [t, t+h]$. In the light of the above discussion of approximating derivatives using differences, we can take $t = t_i$ and re-arrange this to give

$$\frac{dv}{dt}(t_i) = \frac{v_{i+1} - v_i}{h} - R_2(\xi)\,h. \tag{5.13}$$

This is called the first-order forward difference approximation to the derivative of v. Taylor's theorem makes the sense of this approximation precise: we make an error proportional to h (hence the terminology "first order") when we make this approximation. Thus it is ok when h is "small". (Compared to what?)

We could equally well use the Taylor expansion of $v(t-h)$:

$$v(t-h) = v(t) - h \frac{dv}{dt}(t) + h^2 R_2(\xi)$$

for some $\xi \in [t-h, t]$. Rearranging this gives what is called the first-order backwards difference approximation:

$$\frac{dv}{dt}(t_i) = \frac{v_i - v_{i-1}}{h} + R_2(\xi)\,h. \tag{5.14}$$

It also obviously makes an error of order h. To get a first-order approximation of the derivative requires that we know the values of v at two points. The key idea of more general finite difference approximations is that by using more points we can get better approximations. In what sense do we mean better? The error made in the approximation is proportional to a higher power of h. An important example is the (second-order) centred difference formula which we can construct using three values, v_{i-1}, v_i and v_{i+1}. We have two Taylor expansions:

$$v_{i+1} = v_i + h \frac{dv}{dt}(t_i) + \frac{1}{2!} h^2 \frac{d^2v}{dt^2}(t_i) + h^3 R_3^+, \tag{5.15}$$

$$v_{i-1} = v_i - h \frac{dv}{dt}(t_i) + \frac{1}{2!} h^2 \frac{d^2v}{dt^2}(t_i) - h^3 R_3^-. \tag{5.16}$$

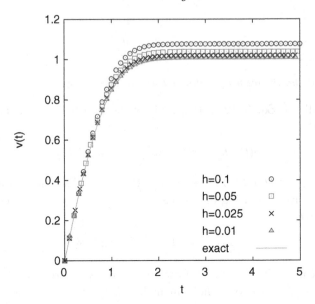

Figure 5.1 Comparison between the numerical solution of Eq. (5.20) obtained using the forward Euler method and the exact solution, Eq. (5.21) for a range of values of the stepsize, h.

Subtracting Eq. (5.16) from Eq. (5.15) and rearranging gives:

$$\frac{dv}{dt}(t_i) = \frac{v_{i+1} - v_{i-1}}{2h} + \frac{1}{2}(R_3^+ + R_3^-)h^2. \qquad (5.17)$$

By using three points, we can get an approximation that is second-order accurate. What about even higher order approximations? What would happen if the discretisation is not uniform?

5.2.4 Timestepping algorithms 1: Euler methods

Forward Euler method

We have seen how, if we know v_i and v_{i+1}, Eq. (5.13) allows us to approximate $\frac{dv}{dt}(t_i)$ in a controlled way. We can turn this on its head: if we know v_i and $\frac{dv}{dt}(t_i)$ then we can approximate v_{i+1}:

$$v_{i+1} = v_i + h\frac{dv}{dt}(t_i) + R_2 h^2. \qquad (5.18)$$

For the system of ODEs given by Eq. (5.7) this gives:

$$\mathbf{v}_{i+1} = \mathbf{v}_i + h\,\mathbf{G}_i + O(h^2). \qquad (5.19)$$

Here we have adopted the obvious notation $\mathbf{G}_i = \mathbf{G}(\mathbf{v}(t_i))$. Furthermore, from now on, unless we explicitly need to, we shall stop carrying around the constants preceeding the error term and simply denote an nth order error term by $O(h^n)$. Starting from the initial condition, for Eq. (5.7) and knowledge of $\mathbf{G}(\mathbf{v})$) we obtain an approximate value for $\mathbf{v}$ a time h later. We can then iterate the process to obtain an approximation to the values of $\mathbf{v}$ at all subsequent times, t_i. Such an iteration might be possible to carry out analytically in some simple cases (for which we are likely to be able to solve the original equation anyway) but, in general, is ideal for computer implementation. This iteration process is called *timestepping*. Eq. (5.19) provides the simplest possible timestepping method. It is called the forward Euler method.

In Fig. 5.1, this method is applied to the following problem:

$$\frac{d^2v}{dt^2} + 2t\frac{dv}{dt} - av = 0, \tag{5.20}$$

for the particular case of $a = 1$, $v(0) = 0$, $\frac{dv}{dt}(0) = \frac{2}{\sqrt{\pi}}$. For this value of the parameter, a, and these initial conditions, Eq. (5.20) has a simple exact solution:

$$v(t) = \mathrm{Erf}(t), \tag{5.21}$$

where $\mathrm{Erf}(x)$ is defined as

$$\mathrm{Erf}(x) = \frac{2}{\sqrt{\pi}} \int_0^x e^{-y^2}\, dy. \tag{5.22}$$

Figure 5.2 compares this method with the other timestepping algorithms covered in this chapter.

Backward Euler method

We could equally well have used the backward difference formula, Eq. (5.14) to derive a timestepping algorithm in which case we would have found

$$\mathbf{v}_{i+1} = \mathbf{v}_i + h\,\mathbf{G}_{i+1} + O(h^2). \tag{5.23}$$

$\mathbf{v}_{i+1}$ (the quantity we are trying to find) enters on both sides of the equation. This means we are not, in general in a position to write down an explicit formula for $\mathbf{v}_{i+1}$ in terms of $\mathbf{v}_i$ as we did for the forward Euler method, Eq. (5.19). Rather we are required to solve a (generally nonlinear) equation to find the value of $\mathbf{v}$ at the next timestep. For non-trivial $\mathbf{G}$ this can be quite hard.

Step-wise vs global error

In Eq. (5.19), we refer to R_2h^2 as the *step-wise error*. This is distinct from the *global error*, which is the total error that occurs in the

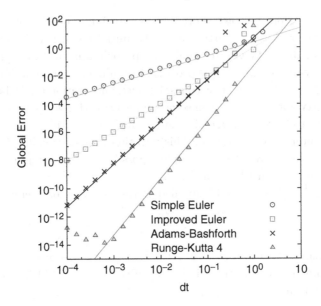

Figure 5.2 Log-log plot showing the behaviour of the global error as a function of step size for a selection of the timestepping algorithms which we shall study in this chapter applied to Eq. (5.20). The straight lines show the theoretically expected error, h^n, where $n = 1$ for the simple Euler method, $n = 2$ for the improved Euler method, $n = 3$ for the Adams–Bashforth method and $n = 4$ for the fourth-order Runge–Kutta method.

approximation of $\mathbf{v}(t_2)$ if we integrate the equation from t_1 to t_2 using a particular timestepping algorithm. If we divide the time domain into N intervals of length $h = (t_2 - t_1)/N$ then $N = (t_2 - t_1)/h$. If we make a step-wise error of $O(h^n)$ in our integration routine, then the global error is therefore $O((t_2 - t_1)h^{n-1})$. Hence the global error for the forward Euler method is $O(h)$. This is very poor accuracy. This is one reason why the forward Euler method is almost never used in practice.

Explicit vs implicit methods

The examples of the forward and backward Euler methods illustrates a very important distinction between different timestepping algorithms: *explicit* vs *implicit*. Forward Euler is an explicit method. Backward Euler is implicit. Although they both have the same degree of accuracy – a step-wise error of $O(h^2)$ – the implicit version of the algorithm typically requires a lot more work per timestep. However, it has superior stability properties.

Improving on the Euler method

We would like to do better than $O(h)$ for the global error. One way is to keep more terms from the Taylor expansion in the derivation of the algorithm. Using Eq. (5.12) and Eq. (5.7):

$$\mathbf{v}_{i+1} = \mathbf{v}_i + h\frac{d\mathbf{v}}{dt}(t_i) + \frac{1}{2}h^2\frac{d^2\mathbf{v}}{dt^2}(t_i) + O(h^3)$$

$$= v_i + hG_i + \frac{1}{2}h^2\frac{d\mathbf{G}}{dt}(\mathbf{v}(t_i)) + O(h^3) \qquad (5.24)$$

Using the chain rule together with Eq. (5.7), the kth component of $\frac{d^2\mathbf{v}}{dt^2}(t_i)$ evaluated at time t_i is

$$\frac{dG^{(k)}}{dt}(\mathbf{v}_i) = \sum_{j=1}^{m}\frac{dG^{(k)}}{dv^{(j)}}\bigg|_{t=t_i}\; G_i^{(j)}.$$

This is starting to look messy and indeed, although the algorithm Eq. (5.24) now has a step-wise error of $O(h^3)$, for nontrivial $\mathbf{G}$ the pain of calculating higher order derivatives of the components of $\mathbf{G}$ with respect to the components of $\mathbf{v}$ means that this approach is not often used. Better alternatives are available to which we now turn.

5.2.5 Timestepping algorithms 2: predictor–corrector methods

Implicit trapezoidal method

So far we have thought of timestepping algorithms based on approxima-tion of the solution by its Taylor series. A complementary way to think about it is to begin from the formal solution of Eq. (5.7):

$$\mathbf{v}_{i+1} = \mathbf{v}_i + \int_{t_i}^{t_i+h} \mathbf{G}(\mathbf{v}(\tau))\, d\tau, \qquad (5.25)$$

and think of how to approximate the integral,

$$I = \int_{t_i}^{t_i+h} \mathbf{G}(\mathbf{v}(\tau))\, d\tau. \qquad (5.26)$$

The simplest possibility is to use the rectangular approximations familiar from the definition of the Riemann integral. We can use either a left or right Riemann approximation:

$$I \approx h\mathbf{G}(\mathbf{v}(t_i)), \qquad (5.27)$$

$$I \approx h\mathbf{G}(\mathbf{v}(t_{i+1})). \qquad (5.28)$$

These approximations obviously give us back the forward and backward Euler methods which we have already seen. A better approximation would be to use the trapezoidal rule:

$$I \approx \frac{1}{2}h\left[\mathbf{G}(\mathbf{v}(t_i)) + \mathbf{G}(\mathbf{v}(t_{i+1}))\right],$$

which would give us the timestepping algorithm

$$\mathbf{v}_{i+1} = \mathbf{v}_i + \frac{h}{2}(\mathbf{G}_i + \mathbf{G}_{i+1}). \tag{5.29}$$

This is known as the implicit trapezoidal method. The step-wise error in this approximation is $O(h^3)$. To see this, we Taylor expand the terms sitting at later terms in Eq. (5.29) and compare the resulting expansion to the true Taylor expansion.

Example 5.2.10 (Accuracy of the implicit trapezoidal method) Let us assume that we have a scalar equation,

$$\frac{dv}{dt} = G(v).$$

(The extension of the analysis to the vector case is straightforward but indicially messy.) The true solution at time t_{i+1} up to order h^3 is, from Taylor's theorem and the chain rule:

$$v_{i+1} = v_i + h\frac{dv}{dt}(t_i) + \frac{h^2}{2}\frac{d^2v}{dt^2}(t_i) + R_3\,h^3$$

$$= v_i + hG_i + \frac{1}{2}h^2 G_i\, G_i' + R_3\,h^3.$$

Let us denote our approximate solution by $\tilde{v}(t)$. From Eq. (5.29) we get

$$\tilde{v}_{i+1} = v_i + \frac{h}{2}\left[G(v_i) + G(v(t_{i+1}))\right].$$

We can write

$$G(v(t_{i+1})) = G(v(t_i)) + h\frac{dG}{dt}(v(t_i)) + R_2 h^2 = G_i + hG_i'G_i + R_2\,h^2.$$

Substituting this back gives

$$\tilde{v}_{i+1} = v_i + hG + \frac{1}{2}h^2 G_i\, G_i' + \frac{1}{2}h^3 R_2.$$

We now see that $\tilde{v}_{i+1} - v_{i+1} = O(h^3)$. Hence the implicit trapezoidal method has an $O(h^3)$ step-wise error.

I have written this example out in some detail since it is a standard way of deriving the step-wise error for any timestepping algorithm.

The principal drawback of the implicit trapezoidal method is that it is implicit and hence computationally expensive and tricky to code. Can we get higher order accuracy with an explicit method?

Improved Euler method

A way to get around the necessity of solving implicit equations is try to "guess" the value of v_{i+1} and hence estimate the value of G_{i+1} to go into the RHS of Eq. (5.29). How do we "guess"? One way is to use a less accurate explicit method to predict v_{i+1} and then use a higher order method such as Eq. (5.29) to correct this prediction. Such an approach is called a predictor–corrector method. Here is the simplest example:

- Step 1: Make a prediction, v_{i+1}^* for the value of v_{i+1} using the forward Euler method:

$$v_{i+1}^* = v_i + h G_i,$$

  and calculate

$$G_{i+1}^* = G(v_{i+1}^*).$$

- Step 2: Use the trapezoidal method to correct this guess:

$$v_{i+1} = v_i + \frac{h}{2}\left[G_i + G_{i+1}^*\right]. \tag{5.30}$$

Equation (5.30) is called the improved Euler method. It is explicit and has a step-wise error of $O(h^3)$. The price to be paid is that we now have to evaluate G twice per timestep. If G is a very complicated function, this might be worth considering but typically this is a considerable improvement over the forward Euler method.

There are two common approaches to making further improvements to the error:

1. Use more points to better approximate the integral, Eq. (5.26). This leads to class of algorithms known as multistep methods.
2. Use predictors of the solution at several points in the interval $[t_i, t_{i+1}]$ and combine them in clever ways to cancel errors. This approach leads to a class of algorithms known as Runge–Kutta methods.

Multistep methods

Theorem 5.2.11 (Reminder of Lagrange interpolation)   *Consider the n + 1 points, (t_0, G_0), $(t_1, G_1) \ldots (t_n, G_n)$ constructed from evaluating a function, $G(t)$ at $n + 1$ points, t_0, $t_1 \ldots t_n$. The unique polynomial of degree n passing through these points is the Lagrange interpolant:*

$$L_n(t) = \sum_{i=0}^{n} G_i \, p_i(t), \tag{5.31}$$

where

$$p_i(t) = \Pi_{j=0 i \neq j}^{n} \frac{t - t_j}{t_i - t_j}. \tag{5.32}$$

Using these interpolation formulae it is possible to use previously computed values of the solution to more accurately approximate the integral in Eq. (5.26). Suppose that we have equally spaced grid-points, 0, $-h$, $-2h \ldots$. Then the first few Lagrange interpolants are

$$L_1(t) = \left(-\frac{t}{h}\right) G_{-1} + \frac{t+h}{h} G_0, \tag{5.33}$$

$$L_2(t) = \frac{t(t+h)}{2h^2} G_{-2} - \frac{t(t+2h)}{h^2} G_{-1} + \frac{(t+h)(t+2h)}{2h^2} G_0, \tag{5.34}$$

$$L_3(t) = -\frac{t(t+h)(t+2h)}{6h^3} G_{-3} + \frac{t(t+h)(t+3h)}{2h^3} G_{-2} \tag{5.35}$$
$$- \frac{t(t+2h)(t-3h)}{2h^3} G_{-1} + \frac{(t+h)(t+2h)(t+3h)}{6h^3} G_0.$$

While straightforward to calculate, the formulae become increasingly horrible as n increases. We will derive a three-step algorithm using $n = 2$ (quadratic interpolation). We shall assume a scalar equation – extension to the vector case is done component-wise. Suppose we know (from previous timesteps) the values of v_{i-2}, v_{i-1} and v_i and hence the values of G_{i-2}, G_{i-1} and G_i. The requirement to know the solution at several previous steps explains the term "multistep" used to describe these algorithms.

Adams–Bashforth methods

Use the known data points $(-2h, G_{-2})$, $(-h, G_{-1})$ and $(0, G_i)$ and Eq. (5.34) to calculate an approximation, $\widetilde{G}(t)$, to $G(t)$ over the interval $[-2h, 0]$. Use $\widetilde{G}(t)$ and Eq. (5.26) to calculate an extrapolated value for v_1:

$$v_1 = \int_0^h G(\tau)\,d\tau$$

$$v_1 \approx v_0 + \int_0^h \widetilde{G}(\tau)\,d\tau$$

$$v_1 = v_0 + h\left(\frac{5}{12}G_{-2} - \frac{4}{3}G_{-1} + \frac{23}{12}G_0\right). \tag{5.36}$$

Equation (5.36) is the three-step Adams–Bashforth Method.

Adams–Moulton methods

A more sophisticated approach is to use the Adams–Bashforth Method, Eq. (5.36), as a predictor for a more refined estimate of the integral Eq. (5.26):

- **Step1 – Predictor:**

$$v_1^* = v_0 + h\left(\frac{5}{12}G_{-2} - \frac{4}{3}G_{-1} + \frac{23}{12}G_0\right). \tag{5.37}$$

From v_1^* we estimate $G_1^* = G(v_1^*)$.

- **Step 2 – Corrector:** Now use the known data points $(-h, G_{-1})$ and $(0, G_0)$ with the predicted data point, (h, G_1^*) to calculate an approximation, $\widetilde{G}^*(t)$, to $G(t)$, over the interval $[-h, h]$ which we then insert into Eq. (5.26). The Lagrange polynomial passing through the points $(-h, G_{-1}), (0, G_0), (h, G_1)$ is

$$L_2'(t) = \frac{t(t-h)}{2h^2}G_{-1} - \frac{(t-h)(t+h)}{h^2}G_0 + \frac{t(t+h)}{2h^2}G_1. \tag{5.38}$$

We use our predicted value, G_1^*, for G_1, in Eq. (5.38) and substitute the new interpolation into Eq. (5.26) and integrate to obtain the corrected value:

$$v_1 = v_0 + \int_0^h \widetilde{G}^*(\tau)\,d\tau$$

$$= v_0 + h\left(-\frac{1}{12}G_{-1} + \frac{2}{3}G_0 + \frac{5}{12}G_1^*\right). \tag{5.39}$$

Be careful: the interpolating polynomial in the predictor and corrector stages are *different* so the resulting coefficients are not the same. Eqs. (5.37) and (5.39) together constitute the three-step Adams–Moulton method. It is explicit, multistep and has a step-wise error of $O(h^3)$. More accurate Adams–Bashforth and Adams–Moulton methods can be derived using higher order Lagrange interpolants. For example, an Adams–Bashforth method with $O(h^4)$ step-wise error can be derived using knowledge of three previous steps together with Eq. (5.35). The formulae for the first few are listed at [1]. The big disadvantage of multistep methods is that they are obviously not self-starting and must rely on a comparably accurate single-step method to calculate enough initial points from the initial data to get going.

5.2.6 Timestepping algorithms 3: Runge–Kutta methods

Runge–Kutta methods aim to retain the accuracy of the multistep predictor–corrector methods described in Section 5.2.5 but without having to use more than the current value to predict the next one – i.e. they are *self-starting*. The idea is roughly to make several predictions of the value of the solution at several points in the interval $[t_1, t_{i+1}]$ and then weight them cleverly so as to cancel errors.

An nth order Runge–Kutta method for Eq. (5.7) looks as follows. We first calculate n estimated values of $\mathbf{G}$ which are somewhat like the predictors used in Section 5.2.5:

$$\mathbf{g}_1 = \mathbf{G}(\mathbf{v}_i),$$
$$\mathbf{g}_2 = \mathbf{G}(\mathbf{v}_i + a_{21}\, h\, \mathbf{g}_1),$$
$$\mathbf{g}_3 = \mathbf{G}(\mathbf{v}_i + a_{31}\, h\, \mathbf{g}_1 + a_{32}\, h, \mathbf{g}_2),$$
$$\vdots$$
$$\mathbf{g}_n = \mathbf{G}(\mathbf{v}_i + a_{n1}\, h\, \mathbf{g}_1 + \ldots + a_{n\,n-1}\, h, \mathbf{g}_{n-1}).$$

We then calcalate v_{i+1} as

$$\mathbf{v}_{i+1} = \mathbf{v}_i + h\,(b_1\mathbf{g}_1 + b_2\mathbf{g}_2 + \ldots + b_n\mathbf{g}_n). \tag{5.40}$$

The art lies in choosing the a_{ij} and b_i such that Eq. (5.40) has a step-wise error of $O(h^{n+1})$. The way to do this is by comparing Taylor expansions as we did to determine the accuracy of the improved Euler method in Example 5.2.10 and choose the values of the constants such that the requisite error terms vanish. It turns out that this choice is not unique so that there are actually parametric families of Runge–Kutta methods of a given order.

We shall derive a second-order method explicitly for a scalar equation. Derivations of higher order methods provide nothing new conceptually but require a lot more algebra. Extension to the vector case, as usual, is straightforward but requires care with indices. Let us denote our general two-stage Runge–Kutta algorithm by

$$g_1 = G(v_i),$$
$$g_2 = G(v_i + a\,h\,g_1),$$
$$v_{i+1} = v_i + h\,(b_1 g_1 + b_2 g_2).$$

Taylor expand g_2 up to second order in h:

$$g_2 = G(v_i) + a\,h\,G_i\,\frac{dG}{dv}(v_i) + R_2\,h^2$$
$$= G_i + a\,h\,G_i\,G_i' + R_2\,h^2.$$

Our approximate value of v_{i+1} is then

$$\tilde{v}_{i+1} = v_i + (b_1 + b_2)hG_i + ab_2 h^2 G_i\,G_i' + R_2\,h^3.$$

If we compare this to the usual Taylor expansion of v_{i+1} we see that we can make the two expansions identical up to $O(h^3)$ if we choose

$$b_1 + b_1 = 1,$$
$$ab_2 = \frac{1}{2}.$$

The method then has a step-wise error of $O(h^3)$. Note that we have two equations for three unknowns so there is a one-parameter family of Runge–Kutta algorithms of this order. A popular choice is $b_1 = b_2 = \frac{1}{2}$ and $a = 1$. This gives, what is often considered the "standard" second-order Runge–Kutta method:

$$g_1 = G(v_i)$$
$$g_2 = G(v_i + h\,g_1) \tag{5.41}$$
$$v_{i+1} = v_i + \frac{h}{2}\,(g_1 + g_2).$$

Perhaps it looks familiar. The standard fourth-order Runge–Kutta method, with a step-wise error of $O(h^5)$, is really the workhorse of numerical integration since it has a very favourable balance of accuracy, stability and efficiency properties. It is often the standard choice. We shall not derive it but you would be well advised to use it in your day-to-day life. It takes the following form:

$$\mathbf{g}_1 = \mathbf{G}(\mathbf{v}_i),$$
$$\mathbf{g}_2 = \mathbf{G}(\mathbf{v}_i + \frac{h}{2}\,\mathbf{g}_1),$$
$$\mathbf{g}_3 = \mathbf{G}(\mathbf{v}_i + \frac{h}{2}\,\mathbf{g}_2), \qquad (5.42)$$
$$\mathbf{g}_4 = \mathbf{G}(\mathbf{v}_i + h\,\mathbf{g}_3),$$
$$\mathbf{v}_{i+1} = \mathbf{v}_i + \frac{h}{6}(\mathbf{g}_1 + 2\mathbf{g}_2 + 2\mathbf{g}_3 + \mathbf{g}_4).$$

5.2.7 Adaptive timestepping

Up until now we have talked a lot about the behaviour of numerical algorithms as $h \to 0$. In practice we need to operate at a finite value of h. How do we choose it? Ideally we would like to choose the timestep such that the error per timestep is less than some threshold, ϵ. We measure the error by comparing the numerical solution at a grid point, $\tilde{\mathbf{v}}_i$, to the exact solution, $\mathbf{v}(t_i)$, assuming it is known. Two criteria are commonly used:

$$E_a(h) = |\tilde{\mathbf{v}}_i - \mathbf{v}_i| \le \epsilon \quad \text{absolute error threshold},$$
$$E_r(h) = \frac{|\tilde{\mathbf{v}}_i - \mathbf{v}_i|}{|\mathbf{v}_i|} \le \epsilon \quad \text{relative error threshold}.$$

Intuitively, and mathematically from Taylor's theorem, the error is largest when the solution is rapidly varying. Often the solution does not vary rapidly everywhere. By setting the timestep in this situation so that the error threshold is satisfied during the intervals of rapid variation, we end up working very inefficiently during the intervals of slower varia-tion where a much larger timestep would have easily satisfied the error threshold. Here is a two-dimensional example which you already know: a relaxation oscillator when $\mu \gg 1$:

$$\frac{dx}{dt} = \mu(y - (\frac{1}{3}x^3 - x)),$$
$$\frac{dy}{dt} = -\frac{1}{\mu}x. \qquad (5.43)$$

The solution of this system has two widely separated timescales – jumps which proceed on a time of $O(\frac{1}{\mu})$ (very fast when $\mu \gg 1$) and crawls which proceed on a timescale of μ (very slow when $\mu \gg 1$) (see Fig. 5.3). To integrate Eqs. (5.43) efficiently for a given error threshold, we need

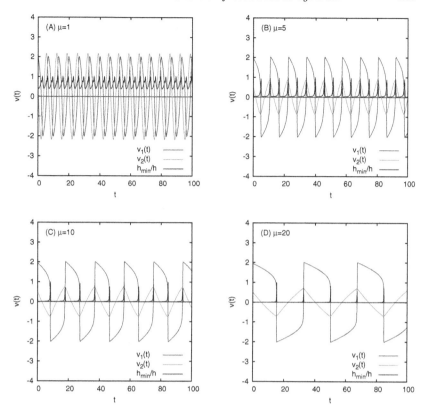

Figure 5.3 Solutions of Eq. (5.43) for several different values of μ obtained using the forward Euler method with adaptive timestepping. Also shown is how the timestep varies relative to its global minimum value as the solution evolves.

to take small steps during the jumps and large ones during the crawls. This process is called *adaptive timestepping* (see Fig. 5.4).

The problem, of course, is that we don't know the local error for a given timestep since we do not know the exact solution. So how do we adjust the timestep if we do not know the error? A common approach is to use trial steps: at time i we calculate one trial step starting from the current solution, $\mathbf{v}_i$ using the current value of h and obtain an estimate of the solution at the next time which we shall call $\mathbf{v}_{i+1}^{B}$. We then calculate a second trial step starting from $\mathbf{v}_i$ and taking *two* steps, each of length $h/2$, to obtain a second estimate of the solution at the next time which

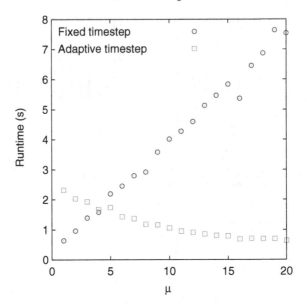

Figure 5.4 Comparison of the performance of the simple Euler method with adaptive and fixed timestepping strategies applied to Eq. (5.43) for several different values of μ.

we shall call $\mathbf{v}_{i+1}^{\mathrm{S}}$. We can then estimate the local error as

$$\Delta = \left| \mathbf{v}_{i+1}^{\mathrm{B}} - \mathbf{v}_{i+1}^{\mathrm{S}} \right|.$$

We can monitor the value of Δ to ensure that it stays below the error threshold.

If we are using a method with a step-wise error of order h^n, we know that $\Delta = ch^n$ for some c. The most efficient choice of step is that for which $\Delta = \epsilon$ (remember ϵ is our error threshold. Thus we would like to choose the new timestep, $\tilde{h}$ such that $c\tilde{h}^n = \epsilon$. But we know from the trial step that $c = \Delta/h^n$. From this we can obtain the following rule to get the new timestep from the current step, the required error threshold and the local error estimated from the trial steps:

$$\tilde{h} = \left(\frac{\epsilon}{\Delta} \right)^{\frac{1}{n}} h. \tag{5.44}$$

It is common to include a "safety factor", $\sigma_1 < 1$, to ensure that we stay a little below the error threshold:

$$\tilde{h}_1 = \sigma_1 \left(\frac{\epsilon}{\Delta} \right)^{\frac{1}{n}} h. \tag{5.45}$$

Equally, since the solutions of ODE's are usually smooth, it is often sensible to ensure that the timestep does not increase or decrease by more than a factor of σ_2 (2 say) in a single step. To do this we impose the second constraint:

$$\tilde{h} = \max\left\{\tilde{h}_1, \frac{h}{\sigma_2}\right\},$$

$$\tilde{h} = \min\left\{\tilde{h}_1, \sigma_2 h\right\}.$$

5.2.8 Snakes in the grass

Equipped with an accurate, stable integration algorithm and an adaptive timestepping strategy you might now feel confident in your ability to obtain the solution to any problem which applications might throw at you. Of course problems arise. We will mention two common ones that you should be aware of – singular problems and stiff problems.

Singularities

A singularity occurs at some time, $t = t^*$ if the solution of an ODE tends to infinity as $t \to t^*$. Naturally this will cause problems for any numerical integration routine. A properly functioning adaptive stepping strategy should reduce the timestep to zero as a singularity is approached. The following is a simple example on which you might like to test some of your integration routines.

Example 5.2.12 (A simple singularity)   Consider the equation,

$$\frac{dv}{dt} = \lambda v^2.$$

with initial data $v(0) = v_0$. This equation is separable and has solution

$$v(t) = \frac{v_0}{1 - \lambda v_0 t}.$$

This clearly has a singularity at $t^* = (\lambda v_0)^{-1}$.

Stiffness

Stiffness is an unpleasant property of some problems which causes adaptive algorithms to require very small timesteps even though the solution is not changing very quickly. There is no uniformly accepted definition of stiffness. Whether a problem is stiff or not depends on the equation, the initial condition, the numerical method being used and the interval of integration. The common feature of stiff problems elucidated in a nice article by Moler [2], is that the required solution is slowly varying but "nearby" solutions vary rapidly. The archetypal stiff problem is the decay equation, Eq. (5.3), with $\lambda \gg 1$. Efficient solution of stiff problems typically require implicit algorithms. Explicit algorithms with proper adaptive stepping will work but usually take an unfeasibly long time.

Here is a more interesting nonlinear example taken from [2]:

Example 5.2.13 (A stiff problem)   Consider the equation,

$$\frac{dv}{dt} = v^2 - v^3.$$

with initial data $v(0) = \epsilon$. Find the solution on the time interval $[0, 2/\epsilon]$.

Its solution is shown in Fig. 5.5(a). The problem becomes stiff as ϵ is decreased. We see, for a given error tolerance (in this case, a relative error threshold of 10^{-5}), that if $\epsilon \ll 1$ the implicit backward Euler method can compute the latter half of the solution (the stiff part) with enormously larger timesteps than the corresponding explicit method. An illuminating animation and further discussion of stiffness with references is available on *Scholarpedia* [3].

5.3 Partial differential equations (PDEs)

5.3.1 Introduction to PDEs and their mathematical classification

The function to be determined, $v(x, t)$, is now a function of several variables (two for us). The choice of notation suggests that one of these variables should be interpreted as a spatial variable and the other as a temporal variable. This is often, but certainly not always, the case in applications. A partial differential equation (PDE) relates partial derivatives of v. Many modelling problems lead to first- or second-order PDEs. We shall deal only with these two cases.

- **First-order PDEs:** We shall consider first-order PDEs of the form

$$a(v, x, t)\frac{\partial v}{\partial t} + b(v, x, t)\frac{\partial x}{\partial t} = c(v, x, t). \tag{5.46}$$

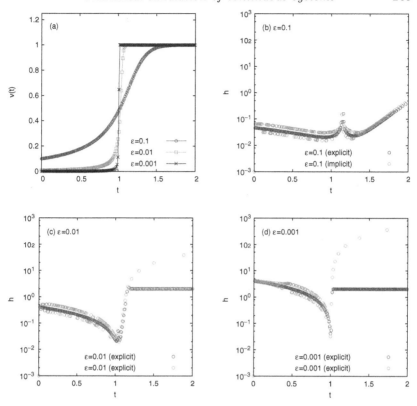

Figure 5.5 Comparison between behaviour of the forward and backward Euler methods with adaptive timestepping applied to the problem of Example 5.2.13 for a range of values of ϵ.

This is called a *quasi-linear* equation because, although the functions a, b and c can be nonlinear, there are no powers of partial derivatives of v higher than 1.

• **General second-order linear PDE:** A general second-order linear PDE takes the form

$$A\frac{\partial^2 v}{\partial t^2} + 2B\frac{\partial^2 v}{\partial x \partial t} + C\frac{\partial^2 v}{\partial x^2} + D\frac{\partial v}{\partial t} + E\frac{\partial v}{\partial x} + Fv + G = 0, \quad (5.47)$$

where the coefficients A to G are generally functions of x and t.

Classification of second-order PDEs

Linear second-order PDEs are grouped into three classes – elliptic, parabolic and hyperbolic – according to the following:

- $B^2 - 4AC < 0$: elliptic ("equilibrium");
- $B^2 - 4AC = 0$: parabolic ("diffusive propagation");
- $B^2 - 4AC > 0$: hyperbolic ("wave propagation").

Unlike for ODEs there are no general methods for solving PDEs. Identifying the class of problem will guide you in your choice of appropriate tools for solving it. This classification has limited usefulness in practice. General PDEs, particularly nonlinear ones, may not fall into any of these classes. Among linear second-order PDEs, there are archetypal examples of each category of equation:

- **Poisson equation (elliptic)**: Given $\rho(x, y)$, find $v(x, y)$ satisfying

$$\frac{\partial^2 v}{\partial x^2} + \frac{\partial^2 v}{\partial y^2} = \rho(x, y) \tag{5.48}$$

  in some region, $\Gamma \in \mathbb{R}^2$. Usually posed as a boundary value problem with behaviour specified on the boundary of Γ.

- **Diffusion equation (parabolic)**: Find $v(x, t)$ satisfying

$$\frac{\partial v}{\partial t} = D \frac{\partial^2 v}{\partial x^2} \tag{5.49}$$

  in the domain $(x, t) \in [x_L, x_R] \times [0, \infty)$. This is solved as an initial value problem with initial data $v(x, 0) = V(x)$ and boundary conditions specified at $x = x_L$ and $x = x_R$.

- **wave equation (hyperbolic)**: Find $v(x, t)$ satisfying

$$\frac{\partial^2 v}{\partial t^2} = c^2 \frac{\partial^2 v}{\partial x^2} \tag{5.50}$$

  in the domain $(x, t) \in [x_L, x_R] \times [0, \infty)$. This is also solved as an initial value problem with initial data $v(x, 0) = V(x)$ and boundary conditions specified at $x = x_L$ and $x = x_R$.

Here are some other examples which arise in applications. Some fit into the above classification and some don't:

- **The advection equation:**

$$\frac{\partial v}{\partial t} + c\frac{\partial v}{\partial x} = 0. \tag{5.51}$$

- **The inviscid Burgers' equation:**

$$\frac{\partial v}{\partial t} + v\frac{\partial v}{\partial x} = 0. \tag{5.52}$$

- **Burgers' equation:**

$$\frac{\partial v}{\partial t} + v\frac{\partial v}{\partial x} = \nu\frac{\partial^2 v}{\partial x^2}. \tag{5.53}$$

- **The telegraph equation:**

$$\frac{\partial^2 v}{\partial t^2} = c^2\frac{\partial^2 v}{\partial x^2} - k\frac{\partial v}{\partial t} - bv. \tag{5.54}$$

- **The Black–Scholes equation:**

$$\frac{\partial v}{\partial t} + rx\frac{\partial v}{\partial x} + \frac{1}{2}\sigma^2 x^2\frac{\partial^2 v}{\partial x^2} - rv = 0. \tag{5.55}$$

Types of boundary conditions

Determining the correct boundary conditions to impose on the solution of a PDE are as important a part of the modelling process as the derivation of the PDE itself. Some of the most common boundary conditions which arise in practice have names. Here we have in mind parabolic and hyperbolic problems in one spatial dimension having spatial domain $[x_L, x_R]$ (x_L and/or x_R could be ∞) where the boundary conditions should be imposed at $x = x_L$ and $x = x_R$. For elliptic problems, the boundary conditions should be specified along a line in the $x - y$ plane.

1. **Dirichlet boundary conditions:** The behaviour of $v(x, t)$ itself is specified on the boundaries:

$$v(x_L, t) = V_L(t) \qquad v(x_R, t) = V_R(t). \tag{5.56}$$

2. **Neumann boundary conditions:** The behaviour of the partial derivative of $v(x, t)$ is specified on the boundaries:

$$\frac{\partial v}{\partial x}(x_L, t) = V_L(t) \qquad \frac{\partial v}{\partial x}(x_R, t) = V_R(t). \tag{5.57}$$

3. **Periodic boundary conditions:** The boundaries "wrap around" so that the solution satisfies:

$$v(x + L, t) = v(x, t), \tag{5.58}$$

where $L = x_R - x_L$ is the size of the domain.

4. **Boundary conditions at infinity**

In some problems involving infinite domains, it is sufficient to provide the asymptotic behaviour of the solution as $x \to \infty$. For example, $v(x, t) \to 0$ as $x \to \pm\infty$.

5.3.2 Non-dimensionalisation

Modelling applications almost always produce PDEs where the quantities involved have actual physical units (dimensions). Mathematically we *never* want to work with dimensional quantities since functions of dimensional quantities make no sense: if x is a length, measured in metres, then e^x is meaningless. When presented with a PDE coming from a modelling problem, it is important to be able to regroup all dimensional quantities into dimensionless combinations thus identifying the essential combinations of parameters which control the solution. This process, known as *non-dimensionalisation*, is almost entirely trivial from a theoretical point of view and and hence is usually neglected in textbooks.

There is a tremendous temptation, theoretically, to "choose units" in which various parameters in a problem are equal to one so as to simplify formulae and get going with the "real" work of solving the problem mathematically. Often this replaces a proper consideration of how to non-dimensionalise the problem. The result is then that you may find it difficult to restore the original units. Nothing aggravates experimentalists more than to be presented with a theoretical curve without units!

Here is a simple example from physics which, from a mathematician's point of view, is littered with annoying physical units.

Example 5.3.1 (One-dimensional diffusion of heat)   A rod of length 2.0 m has the following temperature profile initially:

$$v(x, 0) = \begin{cases} 10\ \text{K} & 0 < x < 1.0\ \text{m}, \\ 0\ \text{K} & 1.0\ \text{m} < x < 2.0\ \text{m}. \end{cases}$$

The left end is maintained at a constant temperature of 10 K and the right end is maintained at 0 K. If thermal diffusivity of the material is $\pi \times 10^{-2}\ \text{m}^2\text{s}^{-1}$, what is the temperature profile in the rod after 25 s?

This problem is described by the diffusion equation, Eq. (5.49), on the interval $[0, 2.0]$ with $D = \pi \times 10^{-2}$ and an initial condition as described above. We want to know what is the solution at $t = 25$ s. However, we never want to work with such dimensional quantities. Let us denote dimensional variables with overlines: $\bar{x}$, $\bar{v}$, $\bar{t}$, etc. The dimensional equation is

$$\frac{\partial \bar{v}}{\partial \bar{t}} = \bar{D}\frac{\partial^2 \bar{v}}{\partial \bar{x}^2}.$$ (5.59)

Let us introduce non-dimensional variables, x, t, v defined by:

$$\bar{x} = x\,L, L = \text{a characteristic length},$$
$$\bar{t} = t\,T, T = \text{a characteristic time},$$ (5.60)
$$\bar{v} = v\,V, V = \text{a characteristic temperature}.$$

Equation (5.59) then trivially leads to

$$\frac{\partial v}{\partial t} = \left(\frac{\bar{D}T}{L^2}\right)\frac{\partial^2 v}{\partial x^2}.$$

If we define the non-dimensional diffusivity,

$$D = \frac{\bar{D}T}{L^2},$$

then we are arrive at a version of the diffusion equation,

$$\frac{\partial v}{\partial t} = D\frac{\partial^2 v}{\partial x^2},$$ (5.61)

where everything is dimensionless. This is the equation we would like to handle mathematically or numerically. We are free to choose L, T and V in Eqs. (5.60). Appropriate choices are suggested by the problem (not the only possibility):

$$V = 10.0 \text{ K},$$
$$L = 2.0 \text{ m},$$
$$T = 25.0 \text{ s}.$$

With these choices we then require to solve the non-dimensional equation, Eq. (5.59), on the interval $x \in [0, 1]$, with $D = \frac{\pi \times 10^{-2} \times 25}{2^2} = 0.196$ and initial condition

$$v(x, 0) = \begin{cases} 1 & 0 < x < \frac{1}{2}, \\ 0 & \frac{1}{2} \leq x < 1, \end{cases}$$

and find the solution at $t = 1$. Once we have done this, we use Eqs. (5.60) to restore the original units.

We now get to the business of solving some PDEs. As is the case for ODEs, some are exactly solvable by various analytic techniques but the vast majority are not. One could devote an entire course to the analytic tools which have been developed for certain PDEs. We shall not touch on this huge subject here and focus on numerical techniques which are appropriate for "typical" PDEs – those which are not analytically solvable. We shall start by looking at situations where PDEs can be reduced to ODEs which may then be tackled using the methods that we have already discussed in Section 5.2. The first such situation occurs for first order PDEs which may be reduced to an ODE using the method of characteristics. The second such situation occurs when some symmetry of a problem means that the independent variables enter the solution only in a certain combination. This gives rise to a class of solutions known as similarity solutions.

5.3.3 First-order PDEs: method of characteristics

Definition 5.3.2 (Reminder: Geometry of curves and surfaces in $\mathbb{R}^3$:) Let us label the coordinates in $\mathbb{R}^3$ by (x, t, z) instead of the usual (x, y, z) to adapt our notation to the PDE application below.

- A function,

$$\mathcal{C} \; : \; \mathbb{R} \to \mathbb{R}^3 \; : \; s \to (x(s), t(s), z(s)),$$

  defines a curve in $\mathbb{R}^3$.
- Given a curve, $\mathcal{C}$, the 3-vector, $\mathbf{T}(s)$ defined by

$$\mathbf{T}(s) = \left(\frac{dx}{ds}, \frac{dt}{ds}, \frac{dz}{ds} \right) \tag{5.62}$$

  is tangent to $\mathcal{C}$ at the point $\mathcal{C}(s)$.
- A function of two variables, $v(x, t)$, can be used to define a surface, $\mathcal{S}$, in $\mathbb{R}^3$ as follows:

$$\mathcal{S} \; : \; \mathbb{R}^2 \to \mathbb{R}^3 \; : \; (x, t) \to (x, t, v(x, t)). \tag{5.63}$$

- The 3-vector, $\mathbf{N}(x, t)$, defined by

$$\mathbf{N}(x, t) = \left(\frac{\partial v}{\partial x}, \frac{\partial v}{\partial t}, -1 \right), \tag{5.64}$$

  is normal to the surface $\mathcal{S}$ at the point $\mathcal{S}(x, t)$.

- Any 3-vector, **T** satisfying

$$\mathbf{T} \cdot \mathbf{N}(x, t) = 0$$

is in the tangent plane to the surface, $\mathcal{S}$, at the point $\mathcal{S}(x, t)$.

Consider the first-order quasi-linear PDE

$$a(x, t, v)\frac{\partial v}{\partial x} + b(x, t, v)\frac{\partial v}{\partial t} = c(x, t, v), \tag{5.65}$$

given "initial data"

$$v(x(\sigma), t(\sigma)) = \mathcal{V}(\sigma), \tag{5.66}$$

along a curve, $\Gamma(\sigma) = (x(\sigma), t(\sigma))$ in the (x, t) in plane. If we could find the solution of Eq. (5.65), $v(x, t)$, then it would define a surface $\mathcal{S}(x, t)$ as in Eq. (5.63). Since the solution satisfies the boundary condition, the surface, $\mathcal{S}$ includes the curve $\mathcal{C}(\sigma) : \sigma \to (x(\sigma), t(\sigma), \mathcal{V}(\sigma))$. Therefore, if we can construct the surface, $\mathcal{S}$, starting from the curve, $\mathcal{C}(\sigma)$, then we solve Eq. (5.65).

The normal to $\mathcal{S}$ at the point (x, t) is

$$\mathbf{N}(x, t) = \left(\frac{\partial v}{\partial x}, \frac{\partial v}{\partial t}, -1\right). \tag{5.67}$$

Now consider the vector,

$$\mathbf{T}(x, t) = (a(x, t, v(x, t)), \ b(x, t, v(x, t)), \ c(x, t, v(x, t))). \tag{5.68}$$

Then using Eq. (5.65).

$$\mathbf{N} \cdot \mathbf{T} = a\frac{\partial v}{\partial x} + b\frac{\partial v}{\partial t} - c$$
$$= 0$$

Thus the vector **T** defined as in Eq. (5.68) is always tangent to the solution surface, $\mathcal{S}$. Here's the key to the method of characteristics: if we define a curve, $\mathcal{C}(s)$ in $\mathbb{R}^3$ by the equation

$$\frac{d\mathbf{x}}{ds} = \mathbf{T}(\mathbf{x}(s)), \tag{5.69}$$

then the tangent to this curve at a given point is **T** (by Eq. (5.62)). Thus if the **x** starts on the solution surface, $\mathcal{S}$ then it stays there. If we start such a curve from each point on the boundary curve, $\mathcal{C}(\sigma)$ then the union of all of these curves is the solution surface. There is a mathematical caveat here: the curve, $\mathcal{C}(\sigma)$, defined by the boundary data must be nowhere parallel to $\mathbf{T}(x(\sigma), t(\sigma))$. Boundary data satisfying this condition is called characteristic data.

This somewhat abstract discussion shows that in principle one can construct the solution surface from the boundary data provided it is characteristic. In fact, we have a concrete method of computing the solution. Let us suppose, for concreteness, that the boundary data is space-like: $C(X) = (X, 0, V(X))$ so that it corresponds to an honest initial condition. The method of characteristics requires that we solve the equations

$$\frac{dx}{ds} = a(x(s), t(s), z(s))x(0) = X,$$

$$\frac{dt}{ds} = b(x(s), t(s), z(s))t(0) = 0, \tag{5.70}$$

$$\frac{dz}{ds} = c(x(s), t(s), z(s))z(0) = V(X).$$

The curve in $\mathbb{R}^2$ given by $(x(s), t(s)$ emanating from the boundary point $(X, 0)$ is called a *characteristic curve*. Solving the characteristic equations, Eqs. (5.70), propagates the initial data along the characteristic curves. To find the solution at time T, we take a discrete set of initial values of x and for each such point, we integrate Eqs. (5.70) until $t(s) = T$. Examples of the results obtained by doing this are shown in Figs. 5.6, 5.7 and 5.8.

Example 5.3.3 (The advection equation)   Solve the equation,

$$\frac{\partial v}{\partial t} + c\frac{\partial v}{\partial x} = 0. \tag{5.71}$$

with initial data $v(x, 0) = V(x)$. Comparing with Eq. (5.65),

$$a(x, t, v) = c,$$

$$b(x, t, v) = 1,$$

$$c(x, t.v) = 0.$$

From Eq. (5.70), the characteristic curve emanating from $(X, 0)$ satisfies

$$\frac{dx}{ds} = c \qquad x(0) = X,$$

$$\frac{dt}{ds} = 1 \qquad t(0) = 0, \tag{5.72}$$

$$\frac{dz}{ds} = 0 \qquad z(0) = V(X).$$

These are easily solved:

$$x(s) = cs + X,$$

$$t(s) = s,$$

$$z(s) = V(X).$$

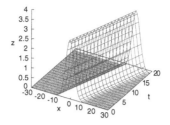

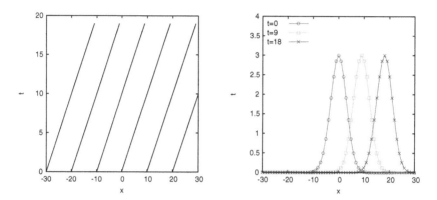

Figure 5.6 Solution of Eq. (5.71) over the time interval $[0:20]$ with $V(x) = 3.0\exp(-0.1*x^2)$. Shown are the global solution, the characteristic curves in the $x-t$ plane and some snapshots of $v(x,t)$ at different times.

Eliminating s in favour of t, the characteristic curve emanating from $(X,0)$ is $x = ct + X$. On this curve, $v(x,t) = V(X)$. Putting these together, the solution is given by

$$v(x,t) = V(x - ct).\tag{5.73}$$

Thus the initial data simply propagates along the characteristic curves. The numerical solution is shown in Fig. 5.6.

Example 5.3.4 (A nonlinear advection–reaction equation)   Solve the equation,

$$\frac{\partial v}{\partial t} + c\frac{\partial v}{\partial x} = -\lambda v^2.\tag{5.74}$$

with initial data $v(x,0) = \cos^2(x)$. The characteristic curve emanating from $(X, 0)$ satisfies

$$\frac{dx}{ds} = c \qquad x(0) = X,$$

$$\frac{dt}{ds} = 1 \qquad t(0) = 0, \qquad (5.75)$$

$$\frac{dz}{ds} = -\lambda z^2 \qquad z(0) = \cos^2(X).$$

Solving these equations gives:

$$x(s) = cs + X$$

$$t(s) = s$$

$$z(s) = \frac{\cos^2(X)}{1 + \lambda t \cos^2(X)}.$$

Eliminating s in favour of t, the characteristic curve emanating from $(X, 0)$ is again $x = ct + X$. The solution is given by $\frac{\cos^2(X)}{1 + \lambda t \cos^2(X)}$.

$$v(x, t) = \frac{\cos^2(x - ct)}{1 + \lambda t \cos^2(x - ct)} \qquad (5.76)$$

Provided $\lambda > 0$, the initial data propagates along the characteristic curves as shown in Fig. 5.7. If $\lambda < 0$, the initial data grows and forms a singularity at time $t^* = 1/\lambda$.

Example 5.3.5 (An advection equation with density dependent speed) Solve the equation,

$$\frac{\partial v}{\partial t} + c(v)\frac{\partial v}{\partial x} = 0. \qquad (5.77)$$

with initial data $v(x, 0) = V(x)$. The characteristic curve emanating from $(X, 0)$ satisfies

$$\frac{dx}{ds} = c(z) \qquad x(0) = X$$

$$\frac{dt}{ds} = 1 \qquad t(0) = 0 \qquad (5.78)$$

$$\frac{dz}{ds} = 0 \qquad z(0) = V(X).$$

We can easily integrate the last two equations:

$$t(s) = s,$$

$$z(s) = V(X).$$

We find that z is constant along the characteristic curve. Thus we can integrate the first equation to obtain

$$x(s) = c(V(X))s + X.$$

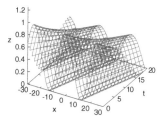

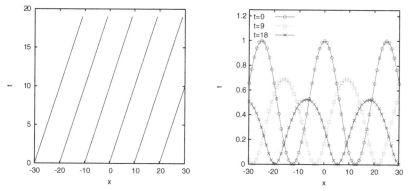

Figure 5.7 Solution of Eq. (5.74) over the time interval $[0:20]$ with $V(x) = \cos^2(\frac{4*\pi x}{x_R - x_L})$. Shown are the global solution, the characteristic curves in the $x - t$ plane and some snapshots of $v(x, t)$ at different times.

Eliminating s in favour of t, the characteristic curve emanating from $(X, 0)$ is therefore $x = c(V(X))t + X$. On this curve, $v(x, t) = $. The solution is given by

$$v(x, t) = V(x - c(V(X))t. \tag{5.79}$$

The characteristic curves are again straight lines but the slope, $\frac{dt}{dx} = 1/c(V(X))$ depends on the initial point. We cannot say more without specifying $c(v)$. Numerical results are shown in Fig. 5.8 for

$$c(v) = 2.0(1 - \frac{v}{3}). \tag{5.80}$$

5.3.4 Similarity solutions and travelling waves

Another way to reduce PDEs to ODEs is to look for solutions where the independent variables, x and t only enter the solution in some combination, ξ. If this occurs, then the equation written in terms of ξ is ordinary. We have already seen this happen in a simple way for the advection equation.

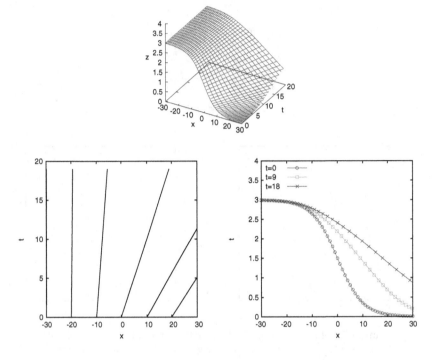

Figure 5.8 Solution of Eq. (5.77) over the time interval $[0 : 20]$ with $V(x) = 1.5 * (1 - \tanh(0.1 * x))$. Shown are the global solution, the characteristic curves in the $x - t$ plane and some snapshots of $v(x, t)$ at different times.

5.3.5 Travelling waves

A travelling wave solution is of the form

$$v(x,t) = F(\xi) \qquad\qquad \text{where } \xi = x - ut, \qquad (5.81)$$

for some u. If $u > 0$, this corresponds to a solution with a spatial profile described by $F(\xi)$ which translates to the right in time at a speed u. If $u < 0$, the profile translates to the left. We usually require $F(\xi)$ to be bounded. The issue is for what values of u, if any, is such a solution possible.

Travelling wave solution of the advection equation

Consider again the advection equation, Eq. (5.71), which we solved in Example 5.3.3 using the method of characteristics. Let us seek a solution in the form of Eq. (5.81). By the chain rule:

$$\frac{\partial v}{\partial t} = \frac{dF}{d\xi}\frac{\partial \xi}{\partial t} = -u\frac{dF}{d\xi},$$

$$\frac{\partial v}{\partial x} = \frac{dF}{d\xi}\frac{\partial \xi}{\partial x} = \frac{dF}{d\xi}.$$

Equation (5.71) then reduces to

$$(-u + c)F'(\xi) = 0.$$

Thus Eq. (5.81) solves Eq. (5.71) for *any* $F(\xi)$ provided we choose $u = c$. By choosing $F = V$, we match the initial data and obtain the solution

$$v(x,t) = V(x - ct),$$

as in Example 5.3.3.

Travelling wave solution of the linear wave equation

Above is the simplest example of a travelling wave solution of a PDE. A slightly more complicated example is the linear wave equation, Eq. (5.50). Let us seek a solution in the form of Eq. (5.81). Two applications of the chain rule gives

$$\frac{\partial^2 v}{\partial t^2} = u^2 \frac{d^2 F}{d\xi^2},$$

$$\frac{\partial^2 v}{\partial x^2} = \frac{d^2 F}{d\xi^2},$$

so that Eq. (5.50) reduces to

$$(u^2 - c^2)F''(\xi) = 0.$$

Thus Eq. (5.81) solves Eq. (5.50) for *any* $F(\xi)$ provided we choose $u = \pm c$. We have two possible wave solutions, one travelling to the left and the other to the right. Equation (5.50) is linear so the general solution involves both. One way to match the initial data is to choose

$$v(x,t) = \frac{1}{2}[V(x - ct) + V(x + ct)].$$

Travelling wave solutions of linear hyperbolic equations have a propagation speed determined from the equation and have a lot of freedom in choosing their shape. This allows us to match the initial data. In general, however, there is not really an "initial" condition for a travelling wave – it propagates from $t = -\infty$ to $t = +\infty$ over the whole real line. Travelling wave solutions of nonlinear equations, when they exist, are much more special as the following example shows.

A nonlinear travelling wave: Burgers' equation

Consider Burgers' equation, Eq. (5.53), with the following boundary conditions at infinity:

$$v(x,t) \to v_L \qquad x \to -\infty,$$
$$v(x,t) \to v_R \qquad x \to \infty.$$

Assuming a solution in the form of Eq. (5.81) and using the chain rule reduces the equation to the following:

$$-uF' + FF' - \nu F'' = 0, \qquad (5.82)$$
$$F(\xi) \to v_L \qquad \xi \to -\infty,$$
$$F(\xi) \to v_R \qquad \xi \to \infty.$$

Unlike in the previous linear problems, we now have an honest ODE to solve and the wave speed is to be determined as part of the solution. Equation (5.82) is not an IVP like those we studied in Section 5.2. Rather it is a nonlinear eigenvalue problem – it is not obvious that there are values of u for which a trajectory leaving v_L at $\xi = -\infty$ will tend to v_R as $\xi \to \infty$. Determining whether such trajectories are possible is best done using the dynamical systems techniques that you learned in Chapter 2 (this is a much easier example than the ones you looked at there!).

We can proceed a little further analytically before resorting to qualitative or numerical techniques. Observing that $\frac{d}{d\xi}\left(\frac{1}{2}F^2\right) = F\frac{dF}{d\xi}$ allows us to write Eq. (5.82) in a form which can be integrated once:

$$\frac{d}{d\xi}\left[-uF + \frac{1}{2}F^2 - \nu F'\right] = 0$$

$$\Rightarrow -uF + \frac{1}{2}F^2 - \nu F' = c_1$$

$$\Rightarrow 2\nu\frac{dF}{d\xi} = F^2 - 2uF - 2c_1$$

$$\Rightarrow 2\nu\frac{dF}{d\xi} = (F - F_+)(F - F_-), \qquad (5.83)$$

where

$$F_\pm = u \pm \sqrt{u^2 + 2c_1}.$$

Let us rescale, $\xi = 2\nu\xi'$, to remove ν:

$$\frac{dF}{d\xi} = (F - F_+)(F - F_-). \qquad (5.84)$$

The fixed points are at $F = F_\pm$. Looking at the phase portrait, we can see the form of the solution:

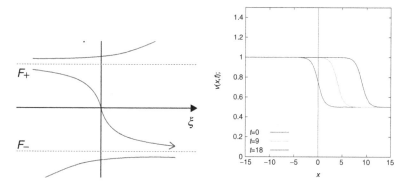

Figure 5.9 Phase portrait for Eq. (5.83) and plots of the resulting travelling wave for $v_R = 1.0$ and $v_L = 0.5$.

The travelling wave solution is a heteroclinic orbit connecting the unstable fixed point at F_+ to the stable fixed point at F_-. It is now clear that we need to choose u and c_1 so that $F_+ = v_L$ and $F_- = v_R$. These conditions lead us to choose

$$u = \frac{v_L + v_R}{2},$$
$$c = -\frac{v_L v_R}{2}. \tag{5.85}$$

Note that, unlike in the linear case, the propagation speed is determined by the boundary conditions, not by the equation itself. In the case of Burgers' equation, it is possible to solve Eq. (5.84) exactly (the equation is separable):

$$\frac{dF}{(F - v_L)(F - v_R)} = d\xi'$$

$$\Rightarrow \left(-\frac{1}{(v_L - v_R)(F - v_R)} + \frac{1}{(v_L - v_R)(F - v_L)} \right) dF = d\xi'$$

$$\Rightarrow \left(\frac{1}{(F - v_R)} - \frac{1}{(F - v_L)} \right) dF = -(v_L - v_R) \, d\xi'$$

$$\Rightarrow \log\left(\frac{F - v_R}{F - v_L} \right) = -(v_L - v_R)\xi' + \log(c_2)$$

$$\Rightarrow \frac{F - v_R}{F - v_L} = c_2 \, e^{-(v_L - v_R)\xi'}$$

$$\Rightarrow F(\xi') = \frac{v_R - v_L\, c_2\, e^{-(v_L - v_R)\,\xi'}}{1 - c_2\, e^{-(v_L - v_R)\,\xi'}}.$$

If we now impose the condition that the trajectory should pass through the centre point, $F(0) = \frac{v_L + v_R}{2}$, then we are led to choose $c_2 = -1$ and restoring $\xi = \frac{\xi'}{2\nu}$ we obtain the travelling wave profile

$$F(\xi) = \frac{v_R + v_L\, e^{-\alpha\xi}}{1 + e^{-\alpha\xi}} \qquad \text{with } \alpha = \frac{v_L - v_R}{2\nu}. \qquad (5.86)$$

In general, we are not so lucky and the ODE describing a travelling wave must be solved numerically. Note that for the nonlinear travelling wave, there is no freedom to adjust the solution to fit particular initial data.

5.3.6 Similarity solutions

Travelling waves are solutions of PDEs which are invariant under translation in space. A related concept is a similarity solution which is invariant under a rescaling of space and time. As was the case for travelling waves, only certain equations admit such solutions, and even for those that do, they are only realisable when the initial and boundary conditions are compatible with the rescaling symmetry. The standard strategy is to seek a solution of the form

$$v(x, t) = t^a\, F(\xi) \qquad \text{where } \xi = xt^b, \qquad (5.87)$$

for some exponents a and b which should be determined from the equation.

Let us consider the diffusion equation as an example:

$$\frac{\partial v}{\partial t} = D\frac{\partial^2 v}{\partial x^2}. \qquad (5.88)$$

Look for a solution of the form Eq. (5.87). Then

$$\frac{\partial v}{\partial t} = at^{a-1}F(\xi) + t^a\frac{dF}{d\xi}\frac{d\xi}{dt}$$
$$= t^{a-1}\left(aF + b\xi F'\right),$$
$$\frac{\partial v}{\partial x} = t^a\frac{dF}{d\xi}\frac{d\xi}{dx}$$
$$= t^{a+b}F',$$
$$\frac{\partial^2 v}{\partial x^2} = t^{a+2b}F''.$$

Comparing the left and right sides of the equation, for consistency we must choose

$$a - 1 = a + 2b.$$

Thus we must choose $b = -\frac{1}{2}$. a remains arbitrary and the profile of the similarity solution is determined from by solving

$$aF - \frac{1}{2}\xi F' - DF'' = 0.$$

We can hide the dependence on D by rescaling $\xi \to \sqrt{D}\xi'$ and dropping the primes immediately:

$$aF - \frac{1}{2}\xi F' - F'' = 0. \tag{5.89}$$

This is tricky to solve for general a. We shall look at a couple of special cases:

1. **A diffusing interface:** $a = 0$

   We can use an integrating factor, $e^{\frac{\xi^2}{4}}$:

$$\frac{1}{2}\xi F' + F'' = 0$$

$$\Rightarrow e^{\frac{\xi^2}{4}}\xi F' + e^{\frac{\xi^2}{4}}F'' = 0$$

$$\Rightarrow \frac{d}{d\xi}\left[e^{\frac{\xi^2}{4}}F'\right] = 0$$

$$\Rightarrow F'(\xi) = c_1 e^{-\frac{\xi^2}{4}}$$

$$\Rightarrow F(\xi) = c_1 \int_{-\infty}^{\xi} e^{-\frac{\eta^2}{4}} d\eta + c_2$$

If we recall the definition of $\mathrm{Erf}(x)$:

$$\mathrm{Erf}(x) = \frac{2}{\sqrt{\pi}} \int_0^x e^{-y^2} dy,$$

we obtain

$$F(\xi) = c_1 \sqrt{\pi}\, \mathrm{Erf}\left(\frac{\xi}{2}\right) + c_2. \tag{5.90}$$

Restoring $\xi = \sqrt{D}\xi'$ and recalling that $\xi = \frac{x}{\sqrt{t}}$ we finally get

$$v(x,t) = \sqrt{\pi}c_1 \mathrm{Erf}\left(\frac{x}{\sqrt{4Dt}}\right) + c_2. \tag{5.91}$$

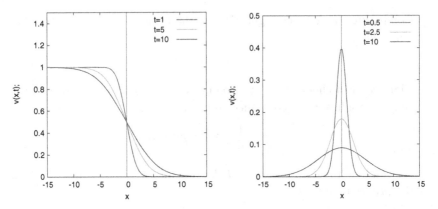

Figure 5.10 Two similarity solutions for the diffusion equation.

We can now choose c_1 and c_2 such that

$$v(x,t) \to v_L \qquad x \to -\infty,$$
$$v(x,t) \to v_R \qquad x \to \infty,$$

then

$$v(x,t) = \frac{v_L + v_R}{2} - \left(\frac{v_L - v_R}{2}\right) \mathrm{Erf}\left(\frac{x}{\sqrt{4Dt}}\right). \qquad (5.92)$$

This solution describes a diffusing interface as can be seen from Fig. 5.10.

2. **A diffusing pulse:** $a = -\frac{1}{2}$

$$F'' + \frac{1}{2}\xi F' + \frac{1}{2}F = 0$$

has a Gaussian solution

$$Ae^{-\frac{\xi^2}{4}}$$

by inspection. Restoring the dependence on D we can write

$$F(\xi) = Ae^{-\frac{\xi^2}{4D}}.$$

We can fix A by setting the total mass equal to 1: $\int_{-\infty}^{\infty} F(\xi)d\xi = 1$. Noting that

$$\int_{-\infty}^{\infty} e^{-\frac{\xi^2}{4D}}\, d\xi = \sqrt{4\pi D},$$

we are led to choose $A = \frac{1}{2\sqrt{D\pi}}$. In terms of x and t we obtain the familiar solution describing a diffusing pulse:

$$v(x,t) = \frac{1}{\sqrt{4\pi Dt}}e^{-\frac{x^2}{4Dt}}. \qquad (5.93)$$

In general, similarity solutions of nonlinear PDEs lead to ODEs which cannot be solved explicitly as in these examples and must be solved numerically.

5.4 The diffusion equation

5.4.1 Forward time centred space (FTCS) method

In this section we shall focus on methods of solving the diffusion equation with source term:

$$\frac{\partial v}{\partial t} = D\frac{\partial^2 v}{\partial x^2} + f(x,t), \tag{5.94}$$

in the domain $(x,t) \in [x_L, x_R] \times [0,T]$. This is to be solved with initial data $v(x,0) = V(x)$ and boundary conditions specified at $x = x_L$ and $x = x_R$. We shall denote the size of the spatial domain by $L = v_R - v_L$.

The similarity solutions which we constructed in Section 5.3 do not generalise to cases with a finite domain, general initial condition or source term (even though they may exhibit behaviour which is typical in an asymptotic sense). We need a more robust method. This leads us to the topic of numerical PDEs proper. We used finite difference approximations of derivatives to build numerical algorithms capable of solving ODEs. We shall do the same here although life is complicated a bit by the fact that we now have derivatives with respect to space and time.

Let us discretise space first, ignoring the source term in Equation (5.94) for the time being. We divide the spatial domain into $N - 1$ intervals of length $\Delta x = \frac{L}{N-1}$ using N equally spaced points, $x_i = x_L + i\Delta x$. With this definition, $x_0 = x_L$ and $x_{N-1} = x_R$. We create a vector, $\mathbf{v}(t) \in \mathbb{R}^N$ from the values, $v_i(t) = v(x_i, t)$, of $v(x,t)$ at the N grid points, x_i:

$$\mathbf{v}(t) = (v_0(t), v_1(t), \dots, v_{N-1}(t)).$$

Equation (5.94) tells us how each component of $\mathbf{v}$ evolves in time:

$$\frac{\partial v_i}{\partial t}(t) = D\frac{\partial^2 v}{\partial x^2}(x_i, t).$$

We can approximate the second derivative on the RHS at a given time with a finite difference:

$$\frac{\partial v_i}{\partial t}(t) = \frac{D}{(\Delta x)^2}[v_{i+1}(t) - 2v_i(t) + v_{i-1}(t)] + O(\Delta x).$$

We are immediately confronted with the question of what to do at the boundaries. For now lets assume periodic boundary conditions (the solution wraps around at the end of the spatial domain):

$$v_N(t) = v_0(t),$$
$$v_{-1}(t) = v_{N-1}(t).$$

We will return to the question of how to handle other boundary conditions in Section 5.4.2. Note that we now have an approximation of the form

$$\frac{d\mathbf{v}}{dt} = \mathbf{G}(\mathbf{v}). \qquad (5.95)$$

Although the dimension, N, of this first-order system is considerably larger in practice than those which we solved in Section 5.2, in principle everything we learned there is applicable now. It is clear that $\mathbf{G}(\mathbf{v})$ is rather simple here:

$$\mathbf{G}(\mathbf{v}) = A\mathbf{v}, \qquad (5.96)$$

where is an $N \times N$ matrix of the form (for $N = 5$):

$$A = \frac{D}{(\Delta x)^2} \begin{pmatrix} -2 & 1 & 0 & 0 & 1 \\ 1 & -2 & 1 & 0 & 0 \\ 0 & 1 & -2 & 1 & 0 \\ 0 & 0 & 1 & -2 & 1 \\ 1 & 0 & 0 & 1 & -2 \end{pmatrix}. \qquad (5.97)$$

With the exception of the effects of the periodic boundary conditions, A, is a tridiagonal matrix. Tridiagonal matrices and banded matrices in general are attractive in designing numerical algorithms since there exist very efficient algorithms for performing matrix operations with such matrices.

To solve our original problem we must now select a timestepping algorithm to advance Eq. (5.95). The simplest choice would be to use the forward Euler method. We choose a time increment, h, and create temporal gridpoints, $t_j = j\,h$. The forward Euler method is

$$\mathbf{v}(t_{j+1}) = \mathbf{v}(t_j) + hA\mathbf{v}(t_j). \qquad (5.98)$$

If we now adopt the notation

$$v_{i,j} = v(x_i, t_j),$$

we can see that the forward Euler method gives us an explicit timestepping algorithm for Eq. (5.94) without the source term:

$$v_{i,j+1} = v_{i,j} + \delta\left[v_{i+1,j} - 2v_{i,j} + v_{i-1,j}\right], \qquad (5.99)$$

where

$$\delta = \frac{Dh}{(\Delta x)^2}. \tag{5.100}$$

Equation (5.99) is called the forward time centred space (FTCS) algorithm. It can be written as a simple matrix multiplication:

$$\mathbf{v}_{j+1} = B\mathbf{v}_j, \tag{5.101}$$

where

$$B = \begin{pmatrix} 1 - 2\delta & \delta & 0 & 0 & \delta \\ \delta & 1 - 2\delta & \delta & 0 & 0 \\ 0 & \delta & 1 - 2\delta & \delta & 0 \\ 0 & 0 & \delta & 1 - 2\delta & \delta \\ \delta & 0 & 0 & \delta & 1 - 2\delta \end{pmatrix}. \tag{5.102}$$

For this reason, there are strong links between linear algebra and numerical analysis of PDEs. One of the main differences between the algorithms which we studied in Section 5.2 and Eq. (5.99) is that there are now two sources of approximation – a spatial and temporal discretisation. These can interact to make questions of stability and convergence more delicate.

5.4.2 Source terms and boundary conditions in the FTCS method

Including a source term in the FTCS method

It is not hard to incorporate the source term in Eq. (5.94) into the FTCS method. We create vector, $\mathbf{f}(t) \in \mathbb{R}^N$, by evaluating the source term at each spatial gridpoint:

$$\mathbf{f}(t) = (f_0(t), f_1(t), \dots, f_{N-1}(t)),$$

where $f_i(t) = f(x_i, t)$. The analogue of Eq. (5.95) is then

$$\frac{d\mathbf{v}}{dt} = A(\mathbf{v}) + \mathbf{f}(t), \tag{5.103}$$

with the matrix A still given by Eq. (5.97). Note that the problem is now potentially non-autonomous. We could increase the dimension by one to get an autonomous system as we learned to do in Section 5.2. If we did that, however, the $\mathbf{G}$ operator for the augmented system would no longer be banded or (unless the source was very special) linear. This would be bad news for large N since we would no longer be able to use all those fast algorithms for performing linear operations on banded matrices. Instead, for PDE applications, we generally prefer to work directly with the non-autonomous system and write our forward Euler method with the source included explicitly:

$$\mathbf{v}(t_{j+1}) = \mathbf{v}(t_j) + hA\mathbf{v}(t_j) + h\mathbf{f}(t_j). \qquad (5.104)$$

A single timestep of the FTCS algorithm, Eq. (5.101), now requires a vector addition in addition to a matrix multiplication:

$$\mathbf{v}_{j+1} = B\mathbf{v}_j + \mathbf{b}_j, \qquad (5.105)$$

where $\mathbf{b}_j = h\mathbf{f}_j$. One thing to watch out for here is not to use more sophisticated integrators which have been designed for non-autonomous systems.

Imposing dirichlet boundary conditions in the FTCS method
Dirichlet conditions:

$$v(x_L, t) = D_L(t), \qquad (5.106)$$
$$v(x_R, t) = D_R(t).$$

(D_L and D_R could be constant or zero!) With these boundary conditions we know the time evolution of v_0 and v_{N-1}. Thus we only need to compute the solution for the points on the interior of the domain, $[x_L + \Delta x, x_R - \Delta x]$. From a conceptual point of view, it is easier to think of $[x_L + \Delta x, x_R - \Delta x] = [\tilde{x}_L . \tilde{x}_R]$ as a new domain to be discretised. The boundary points are considered to be external to the domain (sometimes called "false points"). Thus we divide the new domain $[\tilde{x}_L . \tilde{x}_R]$ into $\tilde{N} - 1 = N - 3$ intervals of length Δx using $\tilde{N} = N - 2$ equally spaced points $x_i = \tilde{x}_L + i\Delta x$. In the new domain, the boundary conditions are applied at nodes x_{-1} and $x_{\tilde{N}}$.

The FTCS method is the same as before for the interior points $i = 1, \ldots \tilde{N} - 2$:

$$v_{i,j+1} = v_{i,j} + \delta \left[v_{i+1,j} - 2v_{i,j} + v_{i-1,j} \right] \qquad (5.107)$$

but at the boundary points, $i = 0$ and $i = \tilde{N} - 1$ we have

$$v_{0,j+1} = v_{0,j} + \delta \left[v_{1,j} - 2v_{0,j} + D_L(t_j) \right],$$
$$v_{\tilde{N}-1,j+1} = v_{\tilde{N}-1,j} + \delta \left[D_R(t_j) - 2v_{\tilde{N}-1,j} + v_{\tilde{N}-2,j} \right]. \qquad (5.108)$$

Equations (5.107) and (5.108) are equivalent to the $\tilde{N} \times \tilde{N}$ linear system

$$\mathbf{v}_{j+1} = B\mathbf{v}_j + \mathbf{b}_j, \qquad (5.109)$$

where (for $\tilde{N} = 5$)

$$
B = \begin{pmatrix}
1 - 2\delta & \delta & 0 & 0 & 0 \\
\delta & 1 - 2\delta & \delta & 0 & 0 \\
0 & \delta & 1 - 2\delta & \delta & 0 \\
0 & 0 & \delta & 1 - 2\delta & \delta \\
0 & 0 & 0 & \delta & 1 - 2\delta
\end{pmatrix}, \qquad (5.110)
$$

$$
\mathbf{b} = \begin{pmatrix}
\delta D_L(t_j) \\
0 \\
0 \\
0 \\
\delta D_R(t_j)
\end{pmatrix}. \qquad (5.111)
$$

Imposing Neumann boundary conditions in the FTCS method

Neumann conditions:

$$
\frac{\partial v}{\partial x}(x_L, t) = N_L(t), \qquad (5.112)
$$

$$
\frac{\partial v}{\partial x}(x_R, t) = N_R(t).
$$

To implement these boundary conditions, we again use "false points", x_{-1} and x_N, which are external points. We use a centred difference to approximate $\frac{\partial v}{\partial x}(x_L, t)$ and set it equal to the desired boundary condition:

$$
\frac{\partial v}{\partial x}(x_L, t) = \frac{v_1 - v_{-1}}{2\Delta x} + O(\Delta x^2) = N_L(t).
$$

From this we can determine v_{-1}:

$$
v_{-1}(t) = v_1(t) - 2\,\Delta x\, N_L(t). \qquad (5.113)
$$

Similarly at the right boundary we determine v_N:

$$
v_N(t) = v_{N-1}(t) + 2\,\Delta x\, N_R(t). \qquad (5.114)
$$

The FTCS method is the same as before for the interior points $i = 1, \ldots \tilde{N} - 2$:

$$
v_{i,j+1} = v_{i,j} + \delta \left[v_{i+1,j} - 2v_{i,j} + v_{i-1,j}\right], \qquad (5.115)
$$

but at the boundary points, $i = 0$ and $i = N - 1$ we have

$$
v_{0,j+1} = v_{0,j} + 2\delta \left[v_{1,j} - v_{0,j}\right] - \frac{2Dh}{\Delta x} N_L(t_j),
$$

$$
v_{N-1,j+1} = v_{N-1,j} + 2\delta \left[-v_{N-1,j} + v_{N-2,j}\right] + \frac{2Dh}{\Delta x} N_R(t_j). \qquad (5.116)
$$

Equations (5.115) and (5.116) are equivalent to the $N \times N$ linear system

$$\mathbf{v}_{j+1} = B\mathbf{v}_j + \mathbf{b}_j, \qquad (5.117)$$

where (for $N = 5$)

$$B = \begin{pmatrix} 1 - 2\delta & 2\delta & 0 & 0 & 0 \\ \delta & 1 - 2\delta & \delta & 0 & 0 \\ 0 & \delta & 1 - 2\delta & \delta & 0 \\ 0 & 0 & \delta & 1 - 2\delta & \delta \\ 0 & 0 & 0 & 2\delta & 1 - 2\delta \end{pmatrix}, \qquad (5.118)$$

$$\mathbf{b} = \begin{pmatrix} -\frac{2Dh}{\Delta x} N_L(t_j) \\ 0 \\ 0 \\ 0 \\ \frac{2Dh}{\Delta x} N_R(t_j) \end{pmatrix}. \qquad (5.119)$$

Figure 5.11 shows some numerical solutions to the diffusion equation with gaussian initial conditions obtained using the FTCS method. Although Dirichlet boundary conditions have been imposed, Fig. 5.11 shows the evolution at early times before the solution starts to feel the boundaries. The solution is therefore very well approximated by the self-similar solution, Eq. (5.93), obtained in Chapter 2. The solution on the

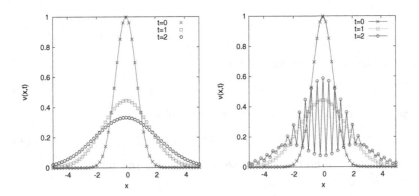

Figure 5.11 Snapshots of the numerical solution of Eq. (5.94) with $f = 0$ and $D = 1$ computed using the FTCS method, Eq. (5.99). The spatial interval was $[-20 : 20]$ (only the range $[-5 : 5]$ is shown) with Dirichlet boundary conditions $v(-20, t) = v(20, t) = 0$ imposed at the edges. The initial condition was $v(x, 0) = e^{-x^2}$. For the computation on the left, $h = 0.02$. For the computation on the right, $h = 0.0266$. Solid lines show the approximate analytic solution.

left has a timestep of $h = 2.00 \times 10^{-2}$. The solution on the right has a timestep of $h = 2.66 \times 10^{-2}$. It is clear that something goes catastrophically wrong with the FTCS method under certain circumstances. The oscillatory behaviour captured in the right panel of Fig. 5.11 is the leading edge of an exponentially growing instability which, within a few more timesteps completely engulfs the entire solution rendering the numerical solution useless.

Is this instability absent in the left panel or would the left-hand computation fall victim to the same instability if we waited for slightly longer time? The question of when the numerical solution obtained from a given numerical algorithm converges to the exact solution is one of the central questions of numerical analysis. We now turn to this issue.

5.4.3 Consistency and stability of the FTCS method

Formal definitions of consistency and stability

We now look at the concepts of consistency and stability which allow us to understand when a numerical solution to a PDE converges to the exact solution. We shall then apply them to the FTCS method.

Definition 5.4.1 (Consistency) A finite difference method is *consistent* if the numerical solution computed after a fixed number of steps converges to the exact solution as h and Δx tend to zero.

Consistency ensures that the finite difference equation converges to the original PDE.

Definition 5.4.2 (Stability) A finite difference method is *stable* if the numerical solution computed after a fixed time remains bounded as $h \to 0$.

Stability ensures that the numerical solution at a finite time does not blow up as the timestep is reduced to zero. Consistency and stability together ensure convergence of the numerical solution according to the following theorem:

Theorem 5.4.3 (Lax–Richtmeyer theorem) *A finite difference approximation to a well-posed linear IVP converges to the exact solution as h and Δx tend to zero if and only if it is consistent and stable.*

Consistency of the FTCS method

We want to compare the numerical and exact solution after a fixed number of timesteps, say $j + 1$. Let us denote the numerically computed solution at space-time point x_i, t_j by $\widetilde{v}_{i,j}$ to distinguish it from corresponding value of the exact solution of Eq. (5.94), which we denote by $v_{i,j}$. For simplicity, we shall assume that $f = 0$. Recall how $\widetilde{v}_{i,j+1}$ is computed using the FTCS method:

$$\widetilde{v}_{i,j+1} = \delta\widetilde{v}_{i-1,j} + (1 - 2\delta)\widetilde{v}_{i,j} + \delta\widetilde{v}_{i+1,j}, \tag{5.120}$$

where $\delta = \frac{Dh}{(\Delta x)^2}$. The error at each point is

$$\varepsilon_{i,j} = v_{i,j} - \widetilde{v}_{i,j}.$$

From Eq. (5.120), we easily obtain

$$\varepsilon_{i,j+1} = \delta(\varepsilon_{i-1,j} + \varepsilon_{i+1,j}) + (1 - 2\delta)\varepsilon_{i,j} + \delta(v_{i-1,j} + v_{i+1,j})$$
$$+ (1 - 2\delta)v_{i,j} - v_{i,j+1}. \tag{5.121}$$

By Taylor's theorem

$$v_{i+1,j} = v_{i,j} + (\Delta x)\frac{\partial v}{\partial x}(x_i, t_j) + \frac{1}{2}(\Delta x)^2\frac{\partial^2 v}{\partial x^2}(\eta_i, t_j),$$

$$v_{i-1,j} = v_{i,j} - (\Delta x)\frac{\partial v}{\partial x}(x_i, t_j) + \frac{1}{2}(\Delta x)^2\frac{\partial^2 v}{\partial x^2}(\mu_i, t_j),$$

$$v_{i,j+1} = v_{i,j} + h\frac{\partial v}{\partial t}(x_i, \tau_j),$$

for some $\eta_i \in [x_i, x_{i+1}]$, $\mu_i \in [x_{i-1}, x_i]$ and $\tau_j \in [t_j, t_{j+1}]$. Putting these into Eq. (5.121) and doing some algebra we obtain

$$\varepsilon_{i,j+1} = \delta(\varepsilon_{i-1,j} + \varepsilon_{i+1,j}) + (1 - 2\delta)\varepsilon_{i,j}$$
$$+ h\left[\frac{D}{2}\left(\frac{\partial^2 v}{\partial x^2}(\eta_i, t_j) + \frac{\partial^2 v}{\partial x^2}(\mu_i, t_j)\right) - \frac{\partial v}{\partial t}(x_i, \tau_j)\right]. \tag{5.122}$$

Now, the solution of the PDE exists so the absolute value of term in the square brackets has a maximum value over all the (i, j) in the computational grid which we denote by $\mathcal{M}$:

$$\mathcal{M} = \max_{i,j}\left|\frac{D}{2}\left(\frac{\partial^2 v}{\partial x^2}(\eta_i, t_j) + \frac{\partial^2 v}{\partial x^2}(\mu_i, t_j)\right) - \frac{\partial v}{\partial t}(x_i, \tau_j)\right|. \tag{5.123}$$

Let us denote by E_j the maximum over the spatial points of absolute value of the error at a fixed time-slice, j:

$$E_j = \max_i |\varepsilon_{i,j}|. \tag{5.124}$$

From Eq. (5.122), provided that $1 - 2\delta > 0$,

$$|\varepsilon_{i,j+1}| \leq \delta[E_j + E_j] + (1 - 2\delta)E_j + \mathcal{M}h$$
$$= E_j + h\mathcal{M}.$$

Let us now assume that $1 - 2\delta > 0$ and take the maximum over i:

$$E_{j+1} \le E_j + h\mathcal{M}. \tag{5.125}$$

We can iterate this argument:

$$E_{j+1} \le E_j + h\mathcal{M} \le E_{j-1} + 2h\mathcal{M} \cdots \le E0 + jh\mathcal{M} = jh\mathcal{M}, \tag{5.126}$$

since $E_0 = 0$. Now let $h \to 0$ and $\Delta x \to 0$. E_{j+1} will tend to zero provided that $\mathcal{M}$ remains finite. Clearly as $h \to 0$ and $\Delta x \to 0$, $\eta_i \to x_i$, $\mu_i \to x_i$ and $\tau_j \to t_j$. Then from Eq. (5.123) and Eq. (5.94) (recall we are taking $f = 0$) we see that $\mathcal{M} \to 0$. Hence the error after $j + 1$ steps, E_{j+1} tends to zero. We conclude that the FTCS method, Eq. (5.120), is consistent if

$$\delta = \frac{Dh}{(\Delta x)^2} < \frac{1}{2}. \tag{5.127}$$

Notice that in this argument, we cannot take h and Δx to zero independently.

Stability of the FTCS method

Theorem 5.4.4 (Finding the largest eigenvalue of a matrix) *If A is an $N \times N$ diagonalisable matrix with eiganvalues $\lambda_1 > \lambda_2 \ge \lambda_3 \ge \cdots \ge \lambda_N$. The for any vector $\mathbf{b}_0$ having nonzero component in the direction of the eigenvector associated with λ_1, (iteratively) define*

$$\mathbf{b}_{k+1} = \frac{A\,\mathbf{b}_k}{|A\,\mathbf{b}_k|}.$$

Then $\mathbf{b}_{k+1} \to \lambda_1 \mathbf{b}_k$ as $k \to \infty$. This is called the power method.

Let us consider the case of Dirichlet conditions for concreteness. Equation (5.109) shows that the FTCS method, Eq. (5.120), is naturally expressed as a linear system,

$$\tilde{\mathbf{v}}_{j+1} = B\tilde{\mathbf{v}}_j + \mathbf{b}_j.$$

In reality, we have some error relative to the exact solution, which means

$$\mathbf{v}_{j+1} + \varepsilon_{j+1} = B(\mathbf{v}_j + \varepsilon_j) + \mathbf{b}_j$$
$$\Rightarrow \varepsilon_{j+1} = B\varepsilon_j$$
$$\Rightarrow \varepsilon_{j+n} = B^n \varepsilon_j.$$

The propagation of errors in the numerical method is controlled by the matrix B. Let us fix a time, T, and compute the solution using a timestep of $h = T/m$. As $h \to 0$,

$$|\varepsilon_m| \sim |\lambda_1|^m |\varepsilon_1|,$$

where λ_1 is the largest eigenvalue of the matrix B given by Eq. (5.110). Clearly the FTCS method will be stable if $|\lambda_1| \leq 1$. Computing the eigenvalues of Eq. (5.110) is somewhat technical so I will just quote the answer:

$$\lambda_n = 1 - 4\delta \sin^2 \left(\frac{n\pi}{2(N+1)} \right) \qquad n = 1, \ldots, N. \qquad (5.128)$$

For stability we need $|\lambda_n| \leq 1$. $\delta > 0$ so we automatically have $\lambda_n \leq 1$. We need to ensure that $\lambda_n \geq -1$. Since $\sin^2(x) \leq 1$ it is sufficient to require $1 - 4\delta < -1$, which translates into

$$\delta = \frac{Dh}{(\Delta x)^2} < \frac{1}{2}. \qquad (5.129)$$

We conclude that the FTCS method is conditionally stable. From this analysis and the Lax–Richtmyer Theorem quoted above, our numerical solution will converge to the analytical solution as $h \to 0$ and $\Delta x \to 0$ provided that we keep $h < \frac{(\Delta x)^2}{2D}$. From a theoretical perspective, this is exactly what we want. From a practical perspective, the fact that we have to decrease our timestep as the square of the spatial grid spacing in order to maintain stability as we increase the spatial resolution is a severe constraint on the efficiency of the method. We shall fix this problem in the next section.

Von Neumann stability analysis

This method of determining the stability of the FTCS method requires that we find the largest eigenvalue of the evolution matrix, a task which is mathematically difficult in general. We now introduce another approach to determinining numerical stability which, although less general, is easier in practice. The idea is to study the growth of a trial solution taking the form of a periodic wave.

Consider the FTCS method:

$$v_{n\,m+1} = v_{n\,m} + \delta \left(v_{n+1\,m} - 2v_{n\,m} + v_{n-1\,m} \right) \qquad (5.130)$$

with a trial solution $v(x,t) = a(t)\,e^{i\,k\,x}$. This trial solution leads to

$$a(t_{m+1})e^{i\,k\,x_n} = a(t_m)\,e^{i\,k\,x} \left[1 + \delta \left(e^{i\,k\,\Delta x} - 2 + e^{-i\,k\,\Delta x} \right) \right]. \qquad (5.131)$$

It is then easy to show that

$$\frac{a(t_{m+1})}{a(t_m)} = 1 + \delta \left(2 \cos(k \, \Delta x) - 2 \right).$$

For stability we require

$$\left| \frac{a(t_{m+1})}{a(t_m)} \right| \leq 1,$$

or alternatively

$$-1 \leq 1 - 2\delta \left(1 - \cos(k \, \Delta x) \right) \leq 1.$$

The latter inequality is clearly satisfied since $0 \leq 1 - \cos(x) \leq 2$. The first inequality requires that $\delta < \frac{1}{2}$. This is the same stability criterion which we obtained via the matrix method previously.

5.4.4 The Crank–Nicholson method

The Crank–Nicholson method is an improvement on the FTCS method which is *unconditionally* stable. Of course there is a price to pay: the method is implicit. The idea is to base the finite difference scheme on the point $(x_i, t_j + \frac{h}{2})$. That is, we approximate the equation

$$\frac{\partial v}{\partial t} \left(x_i, t_j + \frac{h}{2} \right) = D \frac{\partial^2 v}{\partial x^2} \left(x_i, t_j + \frac{h}{2} \right). \tag{5.132}$$

We use a centred difference formula for the time derivative and approximate the spatial derivative by the average of the second-order difference approximation at t_j and t_{j+1}:

$$\frac{\partial v}{\partial t}(x_i, t_j + \frac{h}{2}) = \frac{v_{i,j+1} - v_{i,j}}{h} + O(h^2)$$

$$\frac{\partial^2 v}{\partial x^2}(x_i, t_j + \frac{h}{2}) = \frac{D}{2} \left[\frac{v_{i+1,j} - 2v_{i,j} + v_{i-1,j}}{(\Delta x)^2} \right.$$
$$\left. + \frac{v_{i+1,j+1} - 2v_{i,j+1} + v_{i-1,j+1}}{(\Delta x)^2} \right] + O((\Delta x)^2)$$

Note that the Crank–Nicholson method is also more accurate than the FTCS method. We thus arrive at the finite difference equation

$$\frac{v_{i,j+1} - v_{i,j}}{h} = \frac{D}{2} \left[\frac{v_{i+1,j} - 2v_{i,j} + v_{i-1,j}}{(\Delta x)^2} \right.$$
$$\left. + \frac{v_{i+1,j+1} - 2v_{i,j+1} + v_{i-1,j+1}}{(\Delta x)^2} \right].$$

We can rearrange this to give a set of equations relating the $v_{i,j+1}$ to the $v_{i,j}$:

$$-\frac{\delta}{2}v_{i+1,j+1} + (1+\delta)v_{i,j+1} - \frac{\delta}{2}v_{i-1,j+1} = \frac{\delta}{2}v_{i+1,j} + (1-\delta)v_{i,j} + \frac{\delta}{2}v_{i-1,j},$$
$$(5.133)$$

where $\delta = \frac{Dh}{(\Delta x)^2}$. The method is implicit since we must solve a set of N simultaneous equations in order to obtain the $v_{i,j+1}$ from the $v_{i,j}$. What about the boundaries? Suppose we impose Dirichlet conditions on the spatial boundaries: $x = x_L$ and $x = x_R$:

$$v(x_L, t) = D_L(t),$$
$$v(x_R, t) = D_R(t).$$

We introduce fictitious points, x_{-1} and x_N as we did in Section 5.4.2 and impose the boundary conditions on these points. At $i = 0$ we have the approximation

$$\frac{v_{0,j+1} - v_{0,j}}{h} = \frac{D}{2}\left[\frac{v_{1,j} - 2v_{0,j} + D_L(t_j)}{(\Delta x)^2}\right.$$
$$\left. + \frac{v_{1,j+1} - 2v_{0,j+1} + D_L(t_{j+1})}{(\Delta x)^2}\right].$$

At $i = N - 1$ we have the approximation

$$\frac{v_{N-1,j+1} - v_{N-1,j}}{h} = \frac{D}{2}\left[\frac{D_R(t_j) - 2v_{N-1,j} + v_{N-2,j}}{(\Delta x)^2}\right. \qquad (5.134)$$
$$\left. + \frac{D_R(t_{j+1}) - 2v_{N-1,j+1} + v_{N-2,j+1}}{(\Delta x)^2}\right].$$

Equations (5.134) and (5.134) can be arranged to give the appropriate boundary equations to supplement Eq. (5.133):

$$-\frac{\delta}{2}v_{1,j+1} + (1+\delta)v_{0,j+1} = \frac{\delta}{2}v_{1,j} + (1-\delta)v_{0,j} + \frac{\delta}{2}D_L(t_{j+1})$$
$$+ \frac{\delta}{2}D_L(t_j)$$

$$(1+\delta)v_{N-1,j+1} - \frac{\delta}{2}v_{N-2,j+1} = (1-\delta)v_{N-1,j} + \frac{\delta}{2}v_{N-2,j}$$
$$+ \frac{\delta}{2}D_R(t_{j+1}) + \frac{\delta}{2}D_R(t_j). \quad (5.135)$$

After multiplying across by 2 for convenience, Eq. (5.133) together with Eqs. (5.135) can be concisely expressed as a linear system:

$$A\mathbf{v}_{j+1} = B\mathbf{v}_j + \mathbf{b}_{j+1} + \mathbf{b}_j, \qquad (5.136)$$

where

$$
A = \begin{pmatrix}
2(1+\delta) & -\delta & 0 & 0 & 0 \\
-\delta & 2(1+\delta) & -\delta & 0 & 0 \\
0 & -\delta & 2(1+\delta) & -\delta & 0 \\
0 & 0 & -\delta & 2(1+\delta) & -\delta \\
0 & 0 & 0 & -\delta & 2(1+\delta)
\end{pmatrix}, \quad (5.137)
$$

$$
\mathbf{b}_{j+1} = \begin{pmatrix}
\delta D_L(t_{j+1}) \\
0 \\
0 \\
0 \\
\delta D_R(t_{j+1})
\end{pmatrix}, \quad (5.138)
$$

and

$$
B = \begin{pmatrix}
2(1-\delta) & \delta & 0 & 0 & 0 \\
\delta & 2(1-\delta) & \delta & 0 & 0 \\
0 & \delta & 2(1-\delta) & \delta & 0 \\
0 & 0 & \delta & 2(1-\delta) & \delta \\
0 & 0 & 0 & \delta & 2(1-\delta)
\end{pmatrix}, \quad (5.139)
$$

$$
\mathbf{b}_j = \begin{pmatrix}
\delta D_L(t_j) \\
0 \\
0 \\
0 \\
\delta D_R(t_j)
\end{pmatrix}. \quad (5.140)
$$

We can solve the required set of equations at each step as follows:

$$
\mathbf{v}_{j+1} = (A^{-1}B)\mathbf{v}_j + A^{-1}\mathbf{b}_{j+1} + A^{-1}\mathbf{b}_j. \quad (5.141)
$$

In this example, we only need to invert the matrix A once at the beginning of the calculation since it does not change from one timestep to the next. In more complicated problems, for example if D were time-dependent or if adaptive stepping were used (in which case, δ would vary in time), then A would be different at each step. In these cases, a full matrix inversion is required at each step. This is potentially expensive but, as mentioned already, fast algorithms exist for performing inversions of the kind of banded matrices which result from discretisation of differential operators like $\frac{\partial^2}{\partial x^2}$.

A similar set of steps can be followed to implement Neumann boundary conditions within the Crank–Nicholson method.

Figure 5.12 shows some snapshots of the numerical solutions obtained with the Crank–Nicholson method with the same set of parameters for

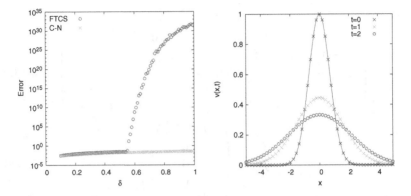

Figure 5.12 Left panel: comparision of the numerical error made by the FTCS and Crank–Nicholson methods as a function of δ for the problem described in Fig. 5.11. The conditional stability of the FTCS method is clearly evident whereas the Crank–Nicholson method is suggested to be unconditionally stable. Right Panel: explicit snapshots of the numerical results obtained using the Crank–Nicholson method for $\delta = 0.665$. Compare the corresponding results for the FTCS method in the right panel of Fig. 5.11.

which the FTCS method was unstable (see Fig. 5.11). The left panel shows how the error varies as a function of δ for the two methods. The plot confirms our analysis of the previous section that the FTCS method is conditionally stable (stable if $\delta < \frac{1}{2}$) and suggests that the Crank–Nicholson method is unconditionally stable. We now perform a mathematical analysis which confirms that this is the case.

5.4.5 Stability of the Crank–Nicholson method

The consistency of the Crank–Nicholson method is a rather lengthy piece of analysis which does not differ significantly from that performed in Section 5.4.3 for the FTCS method. We omit it here.

The stability of the Crank–Nicholson method requires that we understand the structure of the matrix $A^{-1} B$ in Eq. (5.141) with A and B given by Eqs. (5.137) and (5.139). We just quote the result for the eigenvalues of $A^{-1} B$:

$$\lambda_n = \frac{2 - 4\delta \sin^2\left(\frac{n\pi}{2N}\right)}{2 + 4\delta \sin^2\left(\frac{n\pi}{2N}\right)}. \qquad (5.142)$$

Note that for $\delta = 0$, $\lambda_n = 1$ for all values of n. As $\delta \to \infty$, $\lambda_n \to -1$ for

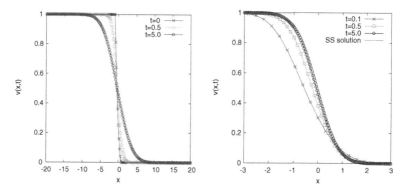

Figure 5.13 Snapshots of the numerical solution to Eq. (5.94) with the initial conditions given by Eq. (5.143) and the Dirichlet boundary conditions. Left panel shows the raw numerical solution and the right panel shows the numerical solution rescaled according to Eq. (5.144).

all values of n. In general,

$$
\begin{aligned}
|\lambda_n| &= \frac{\left|2 - 4\delta \sin^2\left(\frac{n\pi}{2N}\right)\right|}{2 + 4\delta \sin^2\left(\frac{n\pi}{2N}\right)} \\
&\leq \frac{2 + 4\delta \sin^2\left(\frac{n\pi}{2N}\right)}{2 + 4\delta \sin^2\left(\frac{n\pi}{2N}\right)} \\
&= 1.
\end{aligned}
$$

Hence the Crank–Nicholson method is *unconditionally* stable.

5.4.6 Similarity solutions as attractors: rescaling

In our discussion of similarity solutions we mentioned briefly that, although such solutions are special, they are often attracting in an asymptotic sense. That is, lots of initial conditions converge to the similarity solution at large times provided that there are no boundary conditions or terms in the equation which break the scaling symmetry. In Chapter 2 we lacked the technology to be able to verify this claim since we could only find the self-similar solution in isolation and could not match it to arbitrary initial conditions.

Now, however, equipped with the FTCS or Crank–Nicholson methods, we can solve the diffusion equation for arbitrary initial conditions and check if this is indeed the case. Let us reconsider the diffusing interface problem which we earlier described using the self-similar solution

Eq. (5.92). Now suppose that our initial interface does not have a step function profile but something asymmetric:

$$v(x,0) = \begin{cases} 1 & x < -1, \\ \frac{1}{4}(1-x)^2 & -1 \le x \le 1, \\ 0 & x > 1. \end{cases} \qquad (5.143)$$

This clearly does not "fit" with the self-similar $\mathrm{Erf}(x)$ profile. However, that does not pose any problems to the methods that we have developed in the last few sections. If the self-similar solution is attracting, however, we should find that at large times our numerical solution, $\widetilde{v}(x,t)$, should behave as

$$\widetilde{v}(x,t) \approx t^a F(xt^b), \qquad (5.144)$$

with a and b determined as before. Suppose that we plot $\widetilde{v}(x,t)/t^a$ as a function of xt^b then the data should collapse onto the single curve $F(x)$ as t gets large. This procedure is illustrated in Fig. 5.13. The numerical solution has been computed using the Crank–Nicholson method starting with the above initial data on the spatial intervale $[-20:20]$ with Dirichlet boundary conditions $v(-20,t) = 1$ and $v(20,t) = 0$. Note that we have set things up so that the solution can evolve for a relatively long time without feeling the boundaries. This gives the solution time to converge to the similarity solution before the boundaries start to break the scaling symmetry. The conclusion to be drawn from Fig. 5.13 is that the solutions quickly adopt the $\mathrm{Erf}(x)$ profile predicted by the similarity analysis as we claimed would be the case.

5.5 Hyperbolic PDEs

5.5.1 The advection equation revisited

The linear wave equation is the archetypal hyperbolic equation. In this chapter we want to devise methods to solve

$$\frac{\partial^2 v}{\partial t^2} = c^2 \frac{\partial^2 v}{\partial x^2} \qquad (5.145)$$

on the interval $[x_L, x_R]$ subject to periodic, Dirichlet or Neumann boundary conditions as we did for parabolic equations in Section 5.4. Since Eq. (5.145) is second order in time, we require two pieces of initial data to specify the problem:

$$v(x,0) = V(x),\tag{5.146}$$

$$\frac{\partial v}{\partial x}(x,0) = V'(x).$$

Hyperbolic equations are generally much tougher to solve numerically than their parabolic counterparts for reasons which we will now explore. Let us begin by writing Eq. (5.145) as a system of first-order PDEs. Define

$$v_1(x,t) = \frac{\partial v}{\partial t},$$

$$v_2(x,t) = c\frac{\partial v}{\partial x}.$$

It then follows that

$$\frac{\partial v_1}{\partial t} = \frac{\partial^2 v}{\partial t^2} = c^2\frac{\partial^2 v}{\partial x^2} = c\frac{\partial}{\partial x}\left(c\frac{\partial v}{\partial x}\right) = c\frac{\partial v_2}{\partial x},$$

$$\frac{\partial v_2}{\partial t} = c\frac{\partial^2 v}{\partial t\partial x} = c\frac{\partial}{\partial x}\left(c\frac{\partial v}{\partial t}\right) = c\frac{\partial v_1}{\partial x},$$

so that the two can be compactly written as

$$\frac{\partial}{\partial t}\begin{pmatrix} v_1 \\ v_2 \end{pmatrix} = -\frac{\partial}{\partial x}\left[\begin{pmatrix} 0 & -c \\ -c & 0 \end{pmatrix}\begin{pmatrix} v_1 \\ v_2 \end{pmatrix}\right].\tag{5.147}$$

In many applications, Eq. (5.147) are actually the primary equations from which Eq. (5.145) is derived. The initial conditions are then specified for v_1 and v_2. We shall adopt this perspective since it will side-step unnecessary complications relating v_1 and v_2 to v. From this point of view, the wave equation provides a simple example of a (vector-valued) conservation law:

$$\frac{\partial \mathbf{v}}{\partial t} = -\frac{\partial}{\partial x}\left[\mathbf{F(v)}\right],\tag{5.148}$$

where $\mathbf{v} = (v_1, v_2)$ and the flux, $\mathbf{F(v)}$, is given by

$$\mathbf{F(v)} = C\,\mathbf{v},$$

where C is a matrix

$$C = \begin{pmatrix} 0 & -c \\ -c & 0 \end{pmatrix}.$$

Many hyperbolic equations can be thought of as conservation laws, so we will study them from this perspective. Although not strictly a hyperbolic equation, the scalar advection equation, Eq. (5.51), which we studied in

Section 5.3, is an even simpler example of a conservation law. It can be written as

$$\frac{\partial v}{\partial t} = -\frac{\partial F(v)}{\partial x} \qquad F(v) = c\, v. \qquad (5.149)$$

Many of the issues which arise in the numerical solution of general hyperbolic equations also manifest themselves at the level of Eq. (5.149) so we shall, for clarity and simplicity, derive algorithms which solve this equation and then extend them to include vector-valued cases like Eq. (5.147).

An obvious starting point (forgetting, for now, that we know the method of characteristics) is to derive the analogue of the FTCS method for Eq. (5.149). As in Section 5.4.1 we use a centred difference approximation for the spatial derivative and a forward difference approximation for the time derivative:

$$-\frac{\partial F}{\partial x}(x_i, t_j) \approx -\frac{F_{i+1,j} - F_{i-1,j}}{2\,\Delta x} = -\frac{c}{2\,\Delta x}\left(v_{i+1,j} - v_{i-1,j}\right),$$

$$\frac{\partial v}{\partial t}(x_i, t_j) \approx \frac{v_{i,j+1} - v_{i,j}}{h}.$$

Putting these together in Eq. (5.149) we arrive at the FTCS method for the advection equation:

$$v_{i,j+1} = v_{i,j} - \frac{1}{2}\gamma\left(v_{i+1,j} - v_{i-1,j}\right), \qquad (5.150)$$

where

$$\gamma = \frac{c\,h}{\Delta x}. \qquad (5.151)$$

What is the pointwise accuracy of this method? As usual, we can determine this by comparing the Taylor expansions of the left-hand and right-hand sides of Eq. (5.150) and seeing at what order they differ. Beginning with the left-hand side:

$$v_{i,j+1} = v_{i,j} + h\left.\frac{\partial v}{\partial t}\right|_{i,j} + \frac{1}{2}h^2\left.\frac{\partial^2 v}{\partial t^2}\right|_{i,j} + O(h^3)$$

$$= v_{i,j} - c\,h\left.\frac{\partial v}{\partial x}\right|_{i,j} + \frac{1}{2}c^2h^2\left.\frac{\partial^2 v}{\partial x^2}\right|_{i,j} + O(h^3). \qquad (5.152)$$

Now the right-hand side:

$$v_{i,j} - \frac{ch}{2\Delta x}\left(v_{i+1,j} - v_{i-1,j}\right)$$

$$= v_{i,j} - \frac{ch}{2\Delta x}\left(v_{i,j} + \Delta x\left.\frac{\partial v}{\partial x}\right|_{i,j} + \frac{(\Delta x)^2}{2}\left.\frac{\partial^2 v}{\partial x^2}\right|_{i,j} + \frac{(\Delta x)^3}{6}\left.\frac{\partial^3 v}{\partial x^3}\right|_{i,j}\right.$$

$$+ ((\Delta x)^4) - \left[v_{i,j} - \Delta x\left.\frac{\partial v}{\partial x}\right|_{i,j} + \frac{(\Delta x)^2}{2}\left.\frac{\partial^2 v}{\partial x^2}\right|_{i,j}\right.$$

$$\left.\left. - \frac{(\Delta x)^3}{6}\left.\frac{\partial^3 v}{\partial x^3}\right|_{i,j} + ((\Delta x)^4)\right]\right)$$

$$= v_{i,j} - ch\left.\frac{\partial v}{\partial x}\right|_{i,j} + \frac{1}{6}c(\Delta x)^2 h\left.\frac{\partial^3 v}{\partial x^3}\right|_{i,j} + O(h(\Delta x)^3). \qquad (5.153)$$

Comparing Eq. (5.152) and Eq. (5.153), we conclude that the pointwise errors for the FTCS method, Eq. (5.150), are $O(h^2)$ and $O((\Delta x)^2 h)$.

Recall that the FTCS method derived for the diffusion equation was conditionally stable. We were required to choose $\delta < \frac{1}{2}$ in order to maintain stability. What is the corresponding stability condition for Eq. (5.150)? Let us perform a Neumann stability analysis. Consider a trial solution

$$v(x,t) = a(t)\,e^{i\,k\,x},$$

and substitute it into the finite difference method. We obtain

$$a(t_{j+1})\,e^{i\,k\,x} = a(t_j)\,e^{i\,k\,x} - \frac{1}{2}\gamma\,a(t_j)\left[e^{i\,k\,(x_i+\Delta x)} - e^{i\,k\,(x_i-\Delta x)}\right]$$

$$\Rightarrow \frac{a(t_{j+1})}{a(t_j)} = 1 - \frac{1}{2}\gamma\left[e^{i\,k\,\Delta x} - e^{-i\,k\,\Delta x}\right]$$

$$= 1 - \gamma\,i\,\sin(k\,\Delta x).$$

For stability we need

$$\left|\frac{a(t_{j+1})}{a(t_j)}\right| < 1$$

$$\Rightarrow |1 - \gamma\,i\,\sin(k\,\Delta x)| < 1$$

$$\Rightarrow 1 + \gamma^2\sin^2(k\,\Delta x) < 1.$$

This is clearly impossible for any real γ. We conclude therefore that the FTCS method applied to the advection equation is unconditionally unstable. The results of attempting to apply the method is shown in Fig. 5.14. As was the case for the diffusion equation, the numerical

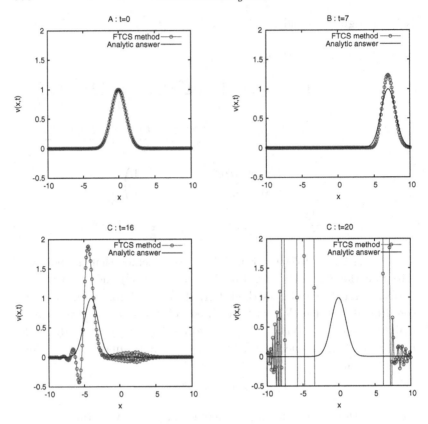

Figure 5.14 Instability of the FTCS method applied to the advection equation.

method produces unphysical oscillations which grow exponentially in time and swamp the solution. The problem can be postponed by taking smaller timesteps but ultimately renders the FTCS method unsuitable for solving hyperbolic problems.

5.5.2 Lax method and CFL criterion

The problem of unconditional instability for the FTCS method is easily fixed although fixing it does, as we shall see, introduce new difficulties. The simplest fix is to use the Lax method. The Lax method modifies the FTCS method, Eq. (5.150), in a seemingly trivial way: the $v_{i,j}$ term on the RHS is replaced by the average, $(v_{i+1,j} + v_{i-1,j})/2$, of the solution at the two adjacent grid points. The resulting method is

$$v_{i,j+1} = \frac{1}{2}(v_{i+1,j} + v_{i-1,j}) - \frac{1}{2}\gamma(v_{i+1,j} - v_{i-1,j}). \qquad (5.154)$$

A similar analysis to that which we applied to the FTCS method in the previous section shows that the leading pointwise errors for this method are $O(h^2)$ and $O((\Delta x)^2)$. Thus, on the face of it, the Lax method, Eq. (5.154), is *less* accurate than Eq. (5.150). Its stability properties are much improved, however. To perform a Neumann stability analysis, we consider a trial solution of the usual form, $v(x,t) = a(t)\,e^{i\,k\,x}$, and substitute it into Eq. (5.154). After some manipulations we find

$$\frac{a(t_{j+1})}{a(t_j)} = \cos(k\,\Delta x) - \gamma\,i\,\sin(k\,\Delta x).$$

For stability we require

$$\left|\frac{a(t_{j+1})}{a(t_j)}\right| < 1$$

$$\Rightarrow \cos^2(k\,\Delta x) + \gamma^2\,\sin^2(k\,\Delta x) < 1$$

$$\Rightarrow (\gamma^2 - 1)\,\sin^2(k\,\Delta x) < 0$$

$$\Rightarrow \gamma = \frac{c\,h}{\Delta x} < 1. \qquad (5.155)$$

The condition, $\frac{c\,h}{\Delta x} < 1$, is known as the CFL condition (Courant–Friedrichs–Lewy). It is a very common constraint arising in explicit numerical method for hyperbolic equations: a finer spatial grid requires smaller timesteps in order to maintain stability. Note, however, that the restriction is not as bad as for the FTCS method for the diffusion equation where we required $h \sim (\Delta x)^2$ to maintain stability. Remember that hyperbolic equations generally describe the propagation of waves (in the advection equation, the propagation speed is c). The CFL condition has an intuitive interpretation: the timestep must be sufficiently small that the distance, $c\,h$, travelled by the wave in a single timestep does not exceed the grid spacing, Δx.

Figure 5.15 shows snapshots of the results obtained by applying Eq. (5.154) to Eq. (5.149) with a gaussian initial condition for three different values of γ set by fixing $c = 1$, $\Delta x = 0.1$ and varying h. For $\gamma = 1.25$, the numerical method is clearly unstable, producing exponentially growing oscillations similar to those in Fig. 5.14. For $\gamma = 1$, the method seems to work well as we would expect from our stability analysis. For $\gamma = 0.75$ we observe, perhaps surprisingly, that the method again

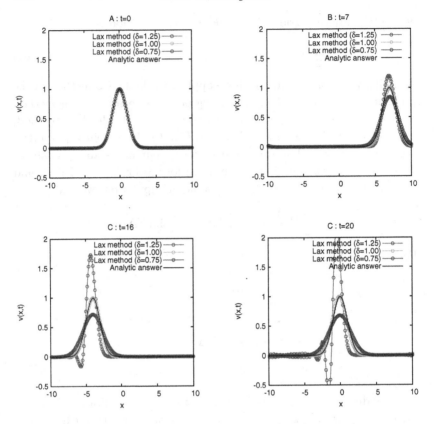

Figure 5.15 Lax method applied to the advection equation.

fails. The numerical solution decays in time whereas the analytic solution does not. This runs counter to our expectation that a smaller timestep should always produce a more accurate solution. What is going on?

With a little thought, it is clear that the Lax method, Eq. (5.154), is equivalent to the FTCS method applied to a modified equation:

$$\frac{\partial v}{\partial t} = -c\frac{\partial v}{\partial x} + D\frac{\partial^2 v}{\partial x^2} \qquad D = \frac{(\Delta x)^2}{2\,h}. \qquad (5.156)$$

Thus the Lax method works by introducing an effective diffusion term into the advection method. This diffusion damps out the oscillations which would otherwise grow and destroy the solution. The artificial diffusion which occurs in the Lax method is called *numerical diffusion*. The amount of numerical diffusion is inversely proportional to h. This explains why the Lax method fails when $\gamma \ll 1$: the numerical diffusion is so strong that it damps out the solution. The price to be paid for stabilising the FTCS method is that we must accept some artificial spreading and damping of the numerically computed wave – an effect known as *numerical dispersion*. The trade-off between numerical stability and dispersion is a common issue in the numerical study of hyperbolic problems. The Lax method works best when $\gamma = 1$, which is the least amount of diffusion sufficient to stabilise the solution. Note that the diffusion is not zero in this case – the numerical solution will decay eventually even if it is not very evident from Fig. 5.15.

5.5.3 Conservation laws and Lax–Wendroff method

The Lax–Wendroff method will be our third and final numerical algorithm to derive for the advection equation. It builds on the success of the Lax method in suppressing the instability of the FTCS method via numerical diffusion but, by keeping more terms in the relevant Taylor expansion, it reduces the rate at which the numerical solution is damped out by the artificial diffusion.

We will derive Lax–Wendroff for a general conservation law,

$$\frac{\partial v}{\partial t} = -\frac{\partial F(v)}{\partial x}, \tag{5.157}$$

since the thinking which underpins the method is clearer there. We shall then specialise our formulae to the particular case of Eq. (5.149). The idea behind the method is very similar to the Taylor series methods for ODEs which we briefly discussed for ODEs in section 5.2.4. We start from the Taylor expansion for $v_{i,j+1}$:

$$v_{i,j+1} = v_{i,j} + h\left.\frac{\partial v}{\partial t}\right|_{i,j} + \frac{1}{2}h^2\left.\frac{\partial^2 v}{\partial t^2}\right|_{i,j} + O(h^3). \tag{5.158}$$

We now use Eq. (5.157) to express the time derivatives in terms of the flux, F:

$$\left.\frac{\partial v}{\partial t}\right|_{i,j} = -\left.\frac{\partial F}{\partial x}\right|_{i,j} \tag{5.159}$$

$$\left.\frac{\partial^2 v}{\partial t^2}\right|_{i,j} = -\frac{\partial}{\partial t}\left[\frac{\partial F}{\partial x}\right]\Big|_{i,j}$$

$$= -\frac{\partial}{\partial x}\left[\frac{dF}{dv}\frac{\partial v}{\partial t}\right]\Big|_{i,j}$$

$$= \frac{\partial}{\partial x}\left[F'(v)\frac{\partial F}{\partial x}\right]\Big|_{i,j} \tag{5.160}$$

Putting Eqs. (5.159) and (5.160) together with Eq. (5.158) we have the essence of the Lax–Wendroff method:

$$v_{i,j+1} = v_{i,j} - h \left.\frac{\partial F}{\partial x}\right|_{i,j} + \frac{1}{2} h^2 \frac{\partial}{\partial x}\left[F'(v)\frac{\partial F}{\partial x}\right]\Bigg|_{i,j}. \tag{5.161}$$

The tricky bit is discretising the spatial derivatives on the RHS of this formula. The first term is easy – we just use a centred difference approximation as we have done for the FTCS and Lax methods previously:

$$\left.\frac{\partial F}{\partial x}\right|_{i,j} \approx \frac{v_{i+1,j} - v_{i-1,j}}{2\Delta x} \tag{5.162}$$

For the second term, we discretise the outer derivative using a centred difference formula involving the points $x_i - \frac{\Delta x}{2}$, x_i and $x_i + \frac{\Delta x}{2}$:

$$\frac{\partial}{\partial x}\left[F'(v)\frac{\partial F}{\partial x}\right]\Bigg|_{i,j} = \frac{1}{\Delta x}\left[F'(v)|_{i+\frac{1}{2},j}\left.\frac{\partial F}{\partial x}\right|_{i+\frac{1}{2},j}\right.$$

$$\left. - F'(v)|_{i-\frac{1}{2},j}\left.\frac{\partial F}{\partial x}\right|_{i-\frac{1}{2},j}\right]. \tag{5.163}$$

We now estimate $F'(v)$ at the intermediate gridpoints by linear interpolation:

$$F'(v)|_{i+\frac{1}{2},j} = F'\left(\frac{1}{2}[v_{i+1,j} + v_{i,j}]\right) \equiv F'_{i+\frac{1}{2},j},$$

$$F'(v)|_{i-\frac{1}{2},j} = F'\left(\frac{1}{2}[v_{i,j} + v_{i-1,j}]\right) \equiv F'_{i-\frac{1}{2},j},$$

where we have introduced the notation $F'_{i+\frac{1}{2},j}$ and $F'_{i-\frac{1}{2},j}$ to simplify the subsequent formulae. We approximate the inner derivatives again with centred difference formulae with spacing $\Delta x/2$:

$$\left.\frac{\partial F}{\partial x}\right|_{i+\frac{1}{2},j} \approx \frac{F_{i+1,j} - F_{i,j}}{\Delta x},$$

$$\left.\frac{\partial F}{\partial x}\right|_{i-\frac{1}{2},j} \approx \frac{F_{i,j} - F_{i-1,j}}{\Delta x}.$$

Putting all this together we arrive at the general Lax–Wendroff method:

$$v_{i,j+1} = v_{i,j} - \frac{h}{2\Delta x}[F_{i+1,j} - F_{i-1,j}]$$

$$+ \frac{1}{2}\frac{h^2}{\Delta x}\left[F'_{i+\frac{1}{2},j}\left(\frac{F_{i+1,j} - F_{i,j}}{\Delta x}\right) - F'_{i-\frac{1}{2},j}\left(\frac{F_{i,j} - F_{i-1,j}}{\Delta x}\right)\right]. \tag{5.164}$$

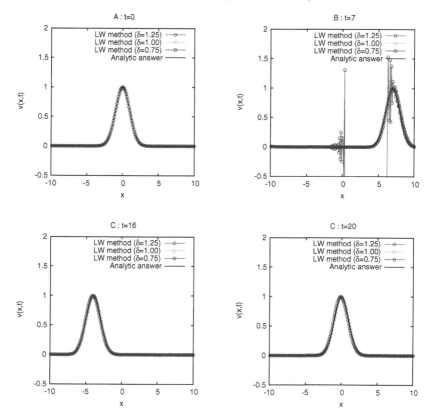

Figure 5.16 Lax–Wendroff method applied to the advection equation.

For the specific case of the advection equation, $F(v) = c\,v$ and $F'(v) = c$. Then Eq. (5.164) reduces to

$$v_{i,j+1} = v_{i,j} - \frac{c\,h}{2\Delta x}\left[v_{i+1,j} - v_{i-1,j}\right] + \frac{c^2 h^2}{2(\Delta x)^2}\left[v_{i+1,j} - 2v_{i,j} + v_{i-1,j}\right].$$

$$(5.165)$$

Figure 5.16 shows the application of this method to the advection equation. Although the method is still unstable for $\gamma > 1$, the damping of the solution is greatly reduced for the case $\gamma = 0.75$ as compared to the Lax method.

5.5.4 The wave equation

Let us now return and adapt the methods which we have developed for the advection equation to the wave equation. In fact we shall write our formulae for a general K-component hyperbolic conservation law:

$$\frac{\partial v^{(k)}}{\partial t} = -\frac{\partial}{\partial x}\left[F^{(k)}(\mathbf{v})\right] \qquad k = 1\ldots K, \qquad (5.166)$$

and then specialise our results to the case of Eq. (5.147). Life is complicated by the fact that we are now required to keep track of an additional index – that labelling the components of $\mathbf{v}$. Just to be clear, the notation $w_{i,j}^{(k)}$ denotes the kth component of a K-vector $\mathbf{w}$ evaluated at the space-time point (x_i, t_j). The Lax method, Eq. (5.154), simply generalises component-wise:

$$v_{i,j+1}^{(k)} = \frac{1}{2}(v_{i+1,j}^{(k)} + v_{i-1,j}^{(k)}) - \frac{h}{2\Delta x}(F_{i+1,j}^{(k)} - F_{i-1,j}^{(k)}). \qquad (5.167)$$

For the wave equation, $K = 2$ with

$$\mathbf{v} = \begin{pmatrix} v^{(1)} \\ v^{(2)} \end{pmatrix}, \qquad (5.168)$$

$$\mathbf{F} = \begin{pmatrix} 0 & -c \\ -c & 0 \end{pmatrix}\begin{pmatrix} v_1 \\ v_2 \end{pmatrix} = \begin{pmatrix} -c\,v^{(2)} \\ -c\,v^{(1)} \end{pmatrix}, \qquad (5.169)$$

so that Eq. (5.167) becomes:

$$v_{i,j+1}^{(1)} = \frac{1}{2}(v_{i+1,j}^{(1)} + v_{i-1,j}^{(1)}) + \frac{h\,c}{2\,\Delta x}\left[v_{i+1,j}^{(2)} - v_{i-1,j}^{(2)}\right], \qquad (5.170)$$

$$v_{i,j+1}^{(2)} = \frac{1}{2}(v_{i+1,j}^{(2)} + v_{i-1,j}^{(2)}) + \frac{h\,c}{2\,\Delta x}\left[v_{i+1,j}^{(1)} - v_{i-1,j}^{(1)}\right].$$

The Lax–Wendroff method for the vectorial case requires considerable care. The complications arise when calculating the analogue of Eq. (5.160), which, after careful application of the multi-variable chain rule, comes out as

$$\left.\frac{\partial^2 v^{(k)}}{\partial t^2}\right|_{i,j} = \sum_{l=1}^{K} \frac{\partial}{\partial x}\left[\frac{\partial F^{(k)}}{\partial v^{(l)}}\frac{\partial F^{(l)}}{\partial x}\right]\bigg|_{i,j}. \qquad (5.171)$$

The resulting Lax–Wendroff method is

$$v_{i,j+1}^{(k)} = v_{i,j}^{(k)} - \frac{h}{2\Delta x}\left[F_{i+1,j}^{(k)} - F_{i-1,j}^{(k)}\right]$$
$$+ \frac{1}{2}\frac{h^2}{\Delta x}\sum_{l=1}^{K}\left[\frac{\partial F^{(k)}}{\partial v^{(l)}}\bigg|_{i+\frac{1}{2},j}\left(\frac{F_{i+1,j}^{(l)} - F_{i,j}^{(l)}}{\Delta x}\right)\right.$$
$$\left. - \frac{\partial F^{(k)}}{\partial v^{(l)}}\bigg|_{i-\frac{1}{2},j}\left(\frac{F_{i,j}^{(l)} - F_{i-1,j}^{(l)}}{\Delta x}\right)\right]. \qquad (5.172)$$

For the wave equation, this rather daunting formula simplifies considerably due to the fact that

$$\frac{\partial F^{(1)}}{\partial v^{(1)}} = 0 \quad \frac{\partial F^{(1)}}{\partial v^{(2)}} = -c,$$

$$\frac{\partial F^{(2)}}{\partial v^{(1)}} = -c \quad \frac{\partial F^{(2)}}{\partial v^{(2)}} = 0,$$

which reduces Eq. (5.172) to

$$v_{i,j+1}^{(1)} = v_{i,j}^{(1)} + \frac{h\,c}{2\,\Delta x}\left[v_{i+1,j}^{(2)} - v_{i-1,j}^{(2)}\right] + \frac{1}{2}\frac{h^2 c^2}{(\Delta x)^2}\left[v_{i+1,j}^{(1)} - 2\,v_{i,j}^{(1)} + v_{i-1,j}^{(1)}\right],$$

$$v_{i,j+1}^{(2)} = v_{i,j}^{(2)} + \frac{h\,c}{2\,\Delta x}\left[v_{i+1,j}^{(1)} - v_{i-1,j}^{(1)}\right] + \frac{1}{2}\frac{h^2 c^2}{(\Delta x)^2}\left[v_{i+1,j}^{(2)} - 2\,v_{i,j}^{(2)} + v_{i-1,j}^{(2)}\right].$$

Notice how the higher order term in h again introduces numerical diffusion.

5.5.5 Boundary conditions for the wave equation

We now consider the wave equation on a finite interval, $[x_L, x_R]$. Periodic
boundary conditions, as always, are very straightforward to incorporate
into the algorithms we have developed. Implementing other boundary
conditions can be tricky in general. Here we shall look explicitly at the
case of Dirichlet conditions.

Consider the wave equation in the form

$$\frac{\partial v^{(1)}}{\partial t} = c\frac{\partial v^{(2)}}{\partial x},\tag{5.173}$$

$$\frac{\partial v^{(2)}}{\partial t} = c\frac{\partial v^{(1)}}{\partial x},\tag{5.174}$$

on the interval $[x_L, x_R]$ with initial conditions:

$$v^{(1)}(x,0) = V_1(x),$$
$$v^{(2)}(x,0) = V_2(x),\tag{5.175}$$

and Dirichlet boundary conditions:

$$v^{(1)}(x_L, t) = D_L(t),$$
$$v^{(1)}(x_R, t) = D_R(t).\tag{5.176}$$

From what we have learned already, our first thought is to introduce
false points, x_{-1} and x_N, and proceed as before. We quickly encounter a
problem, however. Equations (5.176) only specify the behaviour of $v^{(1)}$ at

the false points. What about $v^{(2)}$? Luckily, this problem is rather easily solved: the PDEs themselves can be used to determine the behaviour of $v^{(2)}$ at the boundaries given that we know the behaviour of $v^{(1)}$.

At $x = x_L$, Eq. (5.173) implies

$$\frac{\partial v^{(1)}}{\partial t}(x_L, t) = c\,\frac{\partial v^{(2)}}{\partial x}(x_L, t)$$

$$\Rightarrow \frac{\partial v^{(2)}}{\partial x}(x_L, t) = \frac{1}{c}\frac{dD_L}{dt} \equiv D'_L(t). \qquad (5.177)$$

Likewise, at $x = x_R$:

$$\frac{\partial v^{(1)}}{\partial t}(x_R, t) = c\,\frac{\partial v^{(2)}}{\partial x}(x_R, t)$$

$$\Rightarrow \frac{\partial v^{(2)}}{\partial x}(x_R, t) = \frac{1}{c}\frac{dD_R}{dt} \equiv D'_R(t). \qquad (5.178)$$

From Eqs. (5.177) and (5.178), we conclude that Dirichlet conditions for $v^{(1)}$ imply Neumann conditions for $v^{(2)}$. We can now determine the behaviour of the solution at the false points as we did for the diffusion equation in Section 5.4.2. For $v^{(1)}$ the boundary conditions determine the solution at the false points directly:

$$v^{(1)}_{-1}(t_j) = D_L(t_j),$$
$$v^{(1)}_N(t_j) = D_R(t_j). \qquad (5.179)$$

For $v^{(2)}$ we wish to choose, $v^{(2)}_{-1,j}$ and $v^{(2)}_{N,j}$ so that

$$\frac{\partial v^{(2)}}{\partial x}(x_L, t_j) \approx \frac{v^{(2)}_{1,j} - v^{(2)}_{-1,j}}{2\Delta x} = D'_L(t_j),$$

$$\frac{\partial v^{(2)}}{\partial x}(x_R, t_j) \approx \frac{v^{(2)}_{N,j} - v^{(2)}_{N-2,j}}{2\Delta x} = D'_R(t_j),$$

from which we are led to choose

$$v^{(2)}_{-1,j} = v^{(2)}_{1,j} - 2\,\Delta x\,D'_L(t_j),$$
$$v^{(2)}_{N,j} = v^{(2)}_{N-2,j} + 2\,\Delta x\,D'_R(t_j). \qquad (5.180)$$

Equations (5.179) and (5.180) together with Eq. (5.170) or Eq. (5.173) now determine the finite difference algorithm completely.

Imposing boundary conditions on wave equations usually introduces a new phenomenon into the game: wave reflections. Figure 5.17 illustrates this effect for a numerical solution of the wave equation on the spatial

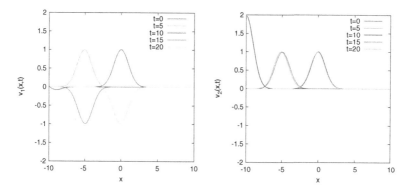

Figure 5.17 A wave reflecting from a boundary. The wave equation was solved with zero Dirichlet conditions at the boundaries and integrated using the Lax method as detailed in the text. The left panel shows the evolution of $v^{(1)}(x,t)$ and the right panel shows the evolution of $v^{(2)}(x,t)$.

domain $[-10:10]$ with initial conditions

$$V_1(x) = e^{-x^2},$$
$$V_2(x) = e^{-x^2},$$

and Dirichlet boundary conditions $v^{(1)}(-10,t) = v^{(1)}(10,t) = 0$. The Lax method was used with $c = 1$, $\Delta x = 0.1$ and $h = 0.1$. Notice how the pulse is reflected from the boundary with a change of sign.

References

[1] Wikipedia. 2008. *Linear multistep method – Wikipedia, The Free Encyclopedia.* http://en.wikipedia.org/w/index.php?title=Linear_multistep_method&oldid=250111301.

[2] Moler, C. 2003. *MATLAB News & Notes May 2003 – Stiff Differential Equations.* http://www.mathworks.com/company/newsletters/news_notes/clevescorner/may03_cleve.html.

[3] Shampine, L.F. and Thompson, S. 2007. Stiff systems. *Scholarpedia*, **2**(3), 2855.

6

Stochastic methods in economics and finance

Vassili N. Kolokoltsov

Abstract

Economic behavior and market evolution present notoriously difficult complex systems, where physical interacting particles become purpose-pursuing interacting agents, thus providing a kind of a bridge between physics and social sciences.

We systematically develop the mathematical content of the basic theory of financial economics that can be presented rigorously using elementary probability and calculus, that is, the notions of discrete and absolutely continuous random variables, their expectation, notions of independence and of the law of large numbers, basic integration – differentiation, ordinary differential equations and (only occasionally) the method of Lagrange multipliers. We do not assume any knowledge of finance, apart from an elementary understanding of the idea of compound interest, which can be of two types: (i) simple compounding with rate r and a fixed period of time means your capital in this period is multiplied by $(1 + r)$; (ii) continuous compounding with rate r means your capital in a period of time of length t is multiplied by e^{rt}.

This chapter is based on several lecture courses for statistics and mathematics students at the University of Warwick and on invited mini-courses presented by the author at various other places. Sections 6.2 and 6.3 are developed from the author's booklet [9]. The chapter is written in a rather concise (but comprehensive) style in attempt to pin down as clear as possible the mathematical relations that govern the laws of financial economics. Numerous heavy volumes are devoted to the detail discussion of the economic content of these mathematical relations, see e.g. [5], [6], [8], [15], [17].

6.1 Utility theory

Here we sketch the basics of utility theory presenting the main definitions, but omitting the proofs of key theorems (which do not contribute
much to their economic content). The full story can be found in [1].

6.1.1 Order, utility and expected utility

We start with an abstract definition forming the basis for the theory.

Let M be a subset of a Euclidean space. A *binary relation* R on M is
defined abstractly just as a subset of the product $M \times M$, i.e. a subset
of the set of pairs (x, y), $x, y \in M$. One writes xRy, if the pair (x, y)
belongs to this subset. With some ambiguity, we shall normally use the
standard symbol $\leq$ for an abstract relation R.

The relation $\leq$ is called *transitive*, if $x \leq y$ and $y \leq z$ imply $x \leq z$,
reflexive, if $x \leq x$ for all x, *symmetric* if $x \leq y \iff y \leq x$, *anti-
symmetric*, if $x \leq y$ and $y \leq x$ imply $x = y$, *complete* if either $x \leq y$
or $y \leq x$ hold for any pair x, y, *continuous*, if the sets $\{x : x \geq y\}$ and
$\{x : x \leq y\}$ are closed.

An *equivalence relation* is a symmetric, reflexive and transitive relation.

A *pre-order* is a reflexive and transitive relation.

If $\leq$ is a pre-order, one defines associated relations $x < y$ meaning
that $x \leq y$, but not $y \geq x$. Also $x \geq y$ means $y \leq x$.

A *partial order* is a reflexive, transitive and anti-symmetric relation.
A complete partial order is called a *complete order*.

Main examples of partial orders in $\mathbf{R}^d$ are the *Pareto order*, where
$x = (x_1, \dots, x_d) \leq (y_1, \dots, y_d)$ means $x_j \leq y_j$ for all j, and *lexicographic
order*, where $x = (x_1, \dots, x_d) \leq (y_1, \dots, y_d)$ means that either $x_1 < y_1$,
or $x_1 = y_1$ and $x_2 < y_2$, or $x_1 = y_1$, $x_2 = y_2$ and $x_3 < y_3$, etc. Clearly
Pareto order is not complete and lexicographic order is not continuous.

Utility functions were introduced in an attempt to measure prefer-
ences (pre-orders) quantitatively. One can say that a *utility function* is
a measure of satisfaction. For any function U on a set M one can define
the relation $x \leq y$ that holds if and only if $U(x) \leq U(y)$ (the latter
relation is the usual order on real numbers). Function U is then called
the *utility function* corresponding to (or defining) this pre-order. For any
U this relation is a complete pre-order. If U is continuous, this pre-order
is continuous. Remarkably enough, the inverse statement is also true, as
the following theorem claims.

Theorem 6.1.1   Main theorem on utility functions. *If a pre-order on* $\mathbf{R}^d$ *is complete and continuous, it can be specified via a continuous (utility) function.*

In economics it is vital to compare uncertain outcomes, leading to the necessity to define pre-orders on random variables. The simplest random variables are those with a finite range, which can be identified with what is usually referred to as simple lotteries. Let Y be a subset of a Euclidean space. A *simple lottery* on Y is a finite convex combination of Dirac's point measures:

$$p = \sum_{y \in \mathrm{supp}(p)} p(y)\delta_y, \qquad \sum_{y \in \mathrm{supp}(p)} p(y) = 1,$$

so that the measure p is identified with a function $p : Y \to [0,1]$ with a finite support $\mathrm{supp}(p)$ (economists' notation $\sum_{y \in \mathrm{supp}(p)} p(y)1_y$). If $\mathrm{supp}(p)$ is one point, p is called a *degenerate lottery*. Countable mixtures of Dirac's point measures, called *discrete lotteries*, are also sometimes used in economics, but much more rarely, and we shall not touch them here.

Let us denote by $\Delta(Y)$ the set of all simple lotteries.

Any utility function v on Y induces a utility on $\Delta(Y)$ called *expected utility*:

$$\mathbf{E}v(p) = \sum_{y \in \mathrm{supp}(p)} p(y)v(y) = \int_Y v(y)p(dy).$$

Clearly, a relation on $\Delta(Y)$ specified by expected utility enjoys the following properties:

(O) ordering: $\leq$ is a complete pre-order on $\Delta(Y)$;
(I) independence: $\lambda > \mu \implies \alpha\lambda + (1-\alpha)\nu > \alpha\mu + (1-\alpha)\nu$
for all $\alpha \in (0,1)$, $\lambda, \mu, \nu \in \Delta(Y)$;
(C) continuity: if $\lambda, \mu, \nu \in \Delta(Y)$ and $\lambda > \mu > \nu$, then $\exists \alpha, \beta \in (0,1)$:

$$\alpha\lambda + (1-\alpha)\nu > \mu > \beta\lambda + (1-\beta)\nu;$$

Theorem 6.1.2   Main theorem on expected utility. *If a relation* $\leq$ *on* $\Delta(Y)$ *satisfies (O), (I), (C), then there exists a function* v *on* Y, *unique up to a linear equivalence (also called ordinal equivalence in this context), such that* $\leq$ *is generated by the expected utility* $\mathbf{E}v$.

6.1.2 Utility on monetary outcomes

Let a real function u on $\mathbf{R}$ or $\mathbf{R}_+$ be twice differentiable.

One says that u satisfies the *principle of non-satiation*, if $u' > 0$; u is *risk averse* (resp. *risk seeking* or *risk indifferent*) if $u'' < 0$ (resp. $u'' > 0$ or $u'' = 0$). The derivative u' designates the rate of growth of utility with the increase of monetary outcomes. Risk aversion means that when your wealth increases, your satisfaction from a given incremental increase decreases (e.g. when you have \$1 or \$1000, your interest in another \$1 is quite different).

Another interpretation: let u be risk averse (that is increasing and concave), $p \in (0, 1)$, and P the lottery winning a or b with probabilities p or $1 - p$ resp., with expectation $\mathbf{E}(P) = pa + (1 - p)b$. Then

$$\mathbf{E}u(P) = pu(a) + (1 - p)u(b) \leq u(pa + (1 - p)b) = u(\mathbf{E}(P)), \quad (6.1)$$

that is, the expected utility of a lottery with the expectation $c = \mathbf{E}(P)$ is less then the utility of the sure $\mathbf{E}(P)$.

Certainty equivalent $s(P)$ of a lottery P is defined via $u(s(P)) = \mathbf{E}u(P)$. By (6.1), $s(P) \leq \mathbf{E}(P)$. See Fig. 6.1.

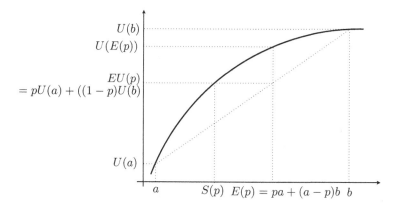

Figure 6.1 Certainty equivalent.

Utility functions can be quantitatively compared by their *absolute* and *relative risk aversions*, *ARA* and *RRA* respectively, which are defined as

$$ARA(u) = -\frac{u''(w)}{u'(w)}, \quad RRA(u) = -w\frac{u''(w)}{u'(w)}.$$

For a given ARA $\rho(x)$, there exists a unique utility with $u(0) = u'(0) =$

0, namely

$$u(x) = \int_0^x \exp\{-\int_0^y \rho(y)dy\}\, dx. \tag{6.2}$$

Basic examples of utility functions are the following. A utility function of form $u(x) = a + bx + cx^2$ is called quadratic (of course). A utility function is called *CARA* (constant ARA) if its ARA is a constant. For a CARA utility, the acceptance of a gamble g does not depend on an initial capital, that is the validity of the inequality $\mathbf{E}u(x + g) > u(x)$ does not depend on the constant x. A utility is called *CRRA* (constant RRA) if its RRA is a constant. A utility is called *HARA* (hyperbolic ARA), if its ARA is inverse to linear, that is of the form $1/(ax+b)$ with constants a, b.

6.1.3 Dominance

Expected utilities allows one to compare random variables. But utility functions are difficult to measure. A question arises whether one can characterize intrinsically the order arising from all utilities satisfying non-satiation principle and/or risk aversion.

For two random variables A and B with distribution functions F_A and F_B, A is said to *stochastically dominate B in the first order* non-strictly (denote $A \geq_1 B$), if for all x

$$F_A(x) \leq F_B(x) \Longleftrightarrow \mathbf{P}(A > x) \geq \mathbf{P}(B > x),$$

and strictly (denote $A >_1 B$) if additionally B does not dominate A, that is there exists x such that $F_A(x) < F_B(x)$.

If $A \geq_1 B$, then there exist two random variables A', B' on one and the same probability space such that A' (resp. B') has the same distribution as A (resp. B) and $A' \geq B'$ point-wise. In fact, one can take $A' = F_A^{-1}$, $B' = F_B^{-1}$ (see Proposition 6.8.1 for the inverse function method employed here).

For two random variables A and B with range $[a, b]$ and distribution functions F_A and F_B, A *stochastically dominates B in the second order* (or in the sense of the *stop-loss stochastic order*) non-strictly (denote $A \geq_2 B$), if for all x

$$\Phi_A(x) = \int_a^x F_A(y)dy \leq \Phi_B(x) = \int_a^x F_B(y)dy,$$

and strictly (denote $A >_2 B$) if additionally B does not dominate A, that is there exists x such that $\Phi_A(x) < \Phi_B(x)$.

It is easy to see that the first- and second-order dominance specify partial orders on the set of random variables.

For a random variable A and a utility function u we shall write shortly $u(A)$ for $\mathbf{E}u(A)$.

Proposition 6.1.3 Let U^1 be the set of all utility functions u on $[a, b]$ with $u' > 0$. Then

$$A \geq_1 B \Longleftrightarrow u(A) \geq u(B) \quad \forall u \in U^1,$$

$$A >_1 B \Longleftrightarrow u(A) > u(B) \quad \forall u \in U^1.$$

Proof From integration by parts

$$u(A) - u(B) = -\int_a^b u'(x)(F_A(x) - F_B(x))dx,$$

implying what was claimed. $\qquad\qquad\square$

Proposition 6.1.4 Let U^2 be the set of all utility functions u on $[a, b]$ with $u' > 0$ and $u'' < 0$. Then

$$A \geq_2 B \Longleftrightarrow u(A) \geq u(B) \quad \forall u \in U^2,$$

$$A >_2 B \Longleftrightarrow u(A) > u(B) \quad \forall u \in U^2.$$

Proof

$$u(A) - u(B) = -\int_a^b u'(x)(F_A(x) - F_B(x))dx$$

$$= -u'(b)(\Phi_A(b) - \Phi_B(b)) + \int_a^b u''(x)(\Phi_A(x) - \Phi_B(x))dx,$$

implying what was claimed. $\qquad\qquad\square$

6.1.4 Utility on $\mathbf{R}^d_+$

Any combination of available goods is referred to as a *consumption bundle*. An *indifference curve* (or *surface*) comprises all bundles of equal utility. A *marginal rate of substitution* is the amount of one good that a consumer is prepared to swap for one extra unit of another good. They are given by slopes of indifferent surfaces (or curves obtained from the intersections of an indifference surface with a two-dimensional section of $\mathbf{R}^d_+$).

A function u on $\mathbf{R}_+^d$ is called *quasi-concave* (resp. *strictly quasi-concave*), if $\{w : u(w) \geq w\}$ is convex for each u (resp. strictly convex).

Two interpretations of quasi-concavity are worthy of mention. Namely, if u is quasi-concave and increasing (in all variables), then:

(a) mixtures of bundles I and II of equal utility are not worse than I or II (see Fig. 6.2);

(b) consumer preference exhibits *diminishing marginal rates of substitution*: if a person is happy to exchange a cherry for an apple, it would take more than one apple to persuade him/her to give another cherry (see Fig. 6.3).

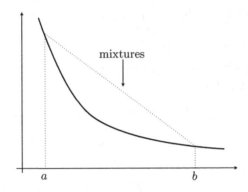

Figure 6.2 Mixtures of consumption bundles.

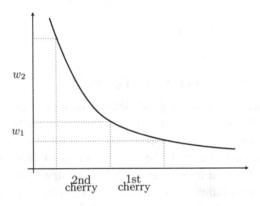

Figure 6.3 Diminishing marginal rate of substitution.

Consumer choice cost-function (or *expenditure function*) for a given set of prices $p \in \mathbf{R}_+^d$ is defined as

$$c(p, w) = \min_q \{(p, q) : u(q) \geq w\}. \tag{6.3}$$

If u is continuous, then c is a nondecreasing homogeneous and concave function of p. The dual function

$$u^*(q) = \max\{w : (p, q) \geq c(p, w) \forall p\}$$

allows us to reconstruct utility from the cost-function, due to the following fundamental result.

Theorem 6.1.5   Shephard's lemma and duality theorem. *If u is strictly quasi-concave, smooth and increasing (in all variables), then there exists a unique q^* minimizing the r.h.s. of* (6.3) *and*

$$q^* = \frac{\partial c}{\partial p}(p, u), \quad u^* = u.$$

6.2 Variance – spread – risk

6.2.1 Variance and correlation

Trying to describe a random variable X by some simple characteristics, one is naturally led to look for its location statistics and its spread, i.e. one looks for a point d, where the spread of the deviation $\mathbf{E}[(X - d)^2]$ is minimal and then assesses this spread.

As by the linearity of the mean

$$\mathbf{E}[(X - d)^2] = \mathbf{E}(X^2) - 2d\mathbf{E}(X) + d^2,$$

one easily finds that the minimum is attained when $d = \mathbf{E}(X)$, yielding a characterization of the expectation as the location statistics. The minimum itself

$$Var(X) = \mathbf{E}[(X - \mathbf{E}(X))^2] = \mathbf{E}(X^2) - [\mathbf{E}(X)]^2 \tag{6.4}$$

is called the *variance* of X and constitutes the second basic characteristics of a random variable describing its spread statistics. In applications, the variance often represents a natural measure of risk, as it specifies the average deviation from the expected level of gain, loss, win, etc. Variance is measured in square units, and the equivalent spread statistics measured in the same units as X is the *standard deviation* of X:

$$\sigma = \sigma_X = \sqrt{Var(X)}.$$

For instance, if $\mathbf{1}_A$ is the indicator function on a probability space, then $\mathbf{E}(\mathbf{1}_A) = \mathbf{P}(A)$ and $\mathbf{1}_A = (\mathbf{1}_A)^2$ so that

$$Var(\mathbf{1}_A) = \mathbf{E}(\mathbf{1}_A) - [\mathbf{E}(\mathbf{1}_A)]^2 = \mathbf{P}(A)[1 - \mathbf{P}(A)].$$

In particular, for a Bernoulli random variable X taking values 1 and 0 with probabilities p and $1 - p$ it implies

$$Var(X) = p(1 - p). \tag{6.5}$$

The simplest measure of the dependence of two random variables X, Y (forming a random vector, i.e with the joint probabilities specified) is supplied by their *covariance*:

$$Cov(X, Y) = \mathbf{E}[(X - \mathbf{E}(X))(Y - \mathbf{E}(Y))],$$

or equivalently by their *correlation coefficient*:

$$\rho = \rho(X, Y) = \frac{Cov(X, Y)}{\sigma_X \sigma_Y}.$$

As easily follows from definition,

$$Cov(X, Y) = \mathbf{E}(XY) - \mathbf{E}(X)\mathbf{E}(Y). \tag{6.6}$$

If $Cov(X, Y) = 0$, one says that the random variable X and Y are *uncorrelated*. If X and Y are independent, they are clearly uncorrelated. Generally speaking, the converse statement does not hold. However, it is possible to show (we will not go into detail here) that it does hold for Gaussian random variables, i.e. two jointly Gaussian random variables are uncorrelated if and only if they are independent. This implies, in particular, that the dependence structure of, say, two standard Gaussian random variables can be practically described by a singe parameter, their correlation. This property makes Gaussian random variable particularly handy when assessing dependence, as is revealed, say, by the method of Gaussian copulas, see Section 6.8.

The linearity of the expectation implies that for a collection of random variables $X_1, \ldots, X_k$ one has

$$Var(\sum_{i=1}^{k} X_i) = \sum_{i=1}^{k} Var(X_i) + \sum_{i \neq j} Cov(X_i, X_j).$$

In particular, if random variables $X_1, \ldots, X_k$ are pairwise uncorrelated, for example if they are independent, the variance of their sum equals the sum of their variances:

$$Var(X_1 + \cdots + X_k) = Var(X_1) + \cdots + Var(X_k). \tag{6.7}$$

This linearity is very useful in practical calculations. For example, if X is a binomial (n, p) random variable, then

$$X = \sum_{i=1}^{n} X_i,$$

where X_i are independent Bernoulli trials, so that by (6.5) and by linearity one gets

$$Var(X) = np(1 - p). \tag{6.8}$$

By the definition of the expectation for continuous random variables X described by a probability density function f, the variance can be calculated as

$$Var(X) = \int [x - \mathbf{E}(X)]^2 f_X(x)\, dx. \tag{6.9}$$

As an example, let us calculate the variance for a normal random variable. Recall first that a *normal or Gaussian* random variable is specified by the probability density functions of the form

$$f(x) = \frac{1}{\sqrt{2\pi}\sigma} \exp\left\{ -\frac{1}{2}\left(\frac{x - \mu}{\sigma}\right)^2 \right\}, \tag{6.10}$$

where $\mu \in \mathbf{R}$ and $\sigma > 0$ are two parameters. These random variables are usually denoted $N(\mu, \sigma^2)$. Random variables $N(0, 1)$ are called *standard normal* and have the density

$$f(x) = \frac{1}{\sqrt{2\pi}} \exp\left\{ -\frac{x^2}{2} \right\}. \tag{6.11}$$

Its distribution function

$$\Phi(x) = \frac{1}{\sqrt{2\pi}} \int_{-\infty}^{x} \exp\left\{ -\frac{y^2}{2} \right\} dy \tag{6.12}$$

can not be expressed in closed form, but is tabulated in standard statistical tables with great precision due to its importance. The properties of general $N(\mu, \sigma^2)$ are easily deduced from the standard normal, due to their simple linear connections. Namely, if X is $N(0, 1)$, then

$$Y = \mu + \sigma X \tag{6.13}$$

is $N(\mu, \sigma^2)$. In fact,

$$F_Y(x) = \mathbf{P}(\mu + \sigma X \leq x) = \mathbf{P}\left(X \leq \frac{x - \mu}{\sigma}\right),$$

so that the distribution function of Y is

$$F_Y(x) = \Phi\left(\frac{x - \mu}{\sigma}\right).$$

Differentiating with respect to x yields for the probability density of Y the expression (6.10) as was claimed.

Let X be $N(0, 1)$. Then by (6.9) and (6.10) (taking into account that the expectation of X vanishes and using integration by parts)

$$Var(X) = \frac{1}{\sqrt{2\pi}} \int_{-\infty}^{\infty} x^2 \exp\left\{-\frac{1}{2}x^2\right\} dx$$

$$= -\frac{1}{\sqrt{2\pi}} x \exp\left\{-\frac{1}{2}x^2\right\}\Big|_{-\infty}^{\infty} + \frac{1}{\sqrt{2\pi}} \int_{-\infty}^{\infty} \exp\left\{-\frac{1}{2}x^2\right\} dx = 1,$$

as the second term is the integral of a probability density function. It follows from (6.13) that the variance of a $N(\mu, \sigma^2)$ normal random variable equals σ^2 and its expectation μ.

6.2.2 Volatility and correlation estimators

A standard problem in applications of probability is estimating the mean, variance and correlations of random variables on the basis of their observed realizations during a period of time. Such estimates are routinely performed, say, by traders, for assessing the volatility of the stock market or a particular common stock. In its simplest mathematical formulation the problem is to estimate the mean and variance of a random variable X, when a realization of a sequence $X_1, \ldots, X_n$ of independent random variables distributed like X was observed. It is of course natural to estimate the expectation μ of X by its empirical mean

$$\hat{\mu}_n = \frac{1}{n} \sum_{i=1}^{n} X_i.$$

This estimate is *unbiased* in the sense that $\mathbf{E}(\hat{\mu}_n) = \mu$ and *asymptotically effective* in the sense that $\hat{\mu}_n$ converges to μ by the law of large numbers. As the variance σ^2 is the expectation of $(X - \mu)^2$, the above reasoning suggests also the estimate for the variance in the form

$$\hat{\sigma}_1^2 = \frac{1}{n} \sum_{i=1}^{n} (X_i - \mu)^2,$$

which is again unbiased and asymptotically effective. The problem with this estimate lies in utilizing the unknown μ. To remedy this shortcoming, it is natural to plug in the above given estimate instead of μ leading to the estimate

$$\hat{\sigma}_2^2 = \frac{1}{n}\sum_{i=1}^{n}(X_i - \hat{\mu}_n)^2 = \frac{1}{n}\sum_{i=1}^{n}X_i^2 - \hat{\mu}_n^2.$$

But here a surprise is awaiting. This estimate is no longer unbiased. In fact,

$$\mathbf{E}(\hat{\mu}_n^2) = \frac{1}{n}(\mathbf{E}(X^2) + (n-1)[\mathbf{E}(X)]^2) = \frac{1}{n}\sigma^2 + \mu^2,$$

implying

$$\mathbf{E}(\hat{\sigma}_2^2) = \sigma^2 + \mu^2 - (\frac{1}{n}\sigma^2 + \mu^2) = \frac{n-1}{n}\sigma^2.$$

So, to have unbiased estimate one has to take, instead of $\hat{\sigma}_2^2$, the estimate

$$\hat{\sigma}_3^2 = \frac{1}{n-1}\sum_{i=1}^{n}(X_i - \hat{\mu}_n)^2.$$

Of course, for large n the difference between $\hat{\sigma}_2^2$ and $\hat{\sigma}_3^2$ disappears, and both these estimates are asymptotically effective.

Similarly, suppose we observe two sequences of independent identically distributed (i.i.d.) random variables $X_1, X_2, \ldots, Y_1, Y_2, \ldots$ distributed like X and Y respectively (let us stress that actually we observe the realizations of i.i.d. random vectors $(X_1, Y_1), (X_2, Y_2), \ldots$). An unbiased and asymptotically effective estimate for the covariance can be constructed as

$$\hat{Cov}(X,Y) = \frac{1}{n-1}\sum_{i=1}^{n}(X_i - \hat{\mu}_n)(Y_i - \hat{\nu}_n),$$

where $\hat{\nu}_n = (Y_1 + \cdots + Y_n)/n$.

6.2.3 Waiting time paradox

As an illustration of an application of variance, let us discuss the so-called *waiting time paradox*.

One could be very annoyed by regularly waiting an hour for a bus at a bus stop, where buses are scheduled to run at 20-minute intervals. However, only if the buses run precisely at 20-minute intervals, your average waiting time (when you arrive at the stop at a random moment) will be 10 minutes. Of course this effect is relevant not only for buses, but for a variety of situations, when waiting times and queues are to be handled (i.e. supermarket cashpoints, internet sites, mobile phone networks, etc). The full development belongs to a highly applied domain of probability, called queueing theory. Here we shall only sketch a (not quite rigorous) argument leading to the calculation of the average waiting times.

To simplify the matter, let us assume that possible intervals between busses take only finite number of values, say $T_1, \ldots, T_k$, with certain probabilities $p_1, \ldots, p_k$. The average interval (usually posted on a time-table) is therefore

$$\mathbf{E}(T) = \sum_{j=1}^{k} p_j T_j.$$

We are interested in the average waiting time. Suppose first that all possible intervals follow each other periodically in a strictly prescribed order, say $T_1, T_2, \ldots, T_k, T_1, T_2, \ldots$ What will be your average waiting time, if you arrive at the stop at a random time t uniformly distributed on the interval $[0, T_1 + \cdots + T_k]$? If $T_1 + \cdots + T_{j-1} < t \le T_1 + \cdots + T_j$ (i.e. your arrive at the jth period between buses), then you will wait for a time $W = T_1 + \cdots + T_j - t$. Taking expectation with respect to the uniform distribution of t yields

$$\mathbf{E}_{T_1, \ldots, T_k}(W)$$

$$= \frac{1}{T_1 + \cdots + T_k} \left(\int_0^{T_1} (T_1 - t)\, dt + \int_{T_1}^{T_1 + T_2} (T_1 + T_2 - t)\, dt + \cdots \right.$$

$$\left. + \int_{T_1 + \cdots + T_{k-1}}^{T_1 + \cdots + T_k} (T_1 + \cdots + T_k - t)\, dt \right)$$

$$= \frac{1}{T_1 + \cdots + T_k} \left(\int_0^{T_1} t\, dt + \int_0^{T_2} t\, dt + \cdots + \int_0^{T_k} t\, dt \right)$$

$$= \frac{T_1^2 + \cdots + T_k^2}{2(T_1 + \cdots + T_k)}.$$

Similarly, if during a period of time there were m_1 intervals of length T_1, m_2 intervals of length $T_2, \ldots, m_k$ intervals of length T_k (in any order), and you arrive uniformly randomly on this period, then your average waiting time will be

$$\mathbf{E}_{T_1, \ldots, T_k; m_1, \ldots, m_k}(W) = \frac{m_1 T_1^2 + \cdots + m_k T_k^2}{2(m_1 T_1 + \cdots + m_k T_k)} = \frac{q_1 T_1^2 + \cdots + q_k T_k^2}{2(q_1 T_1 + \cdots + q_k T_k)},$$

where $q_j = k_j / (k_1 + \cdots + k_m)$ denotes the frequencies of the appearance of the intervals T_j. But frequencies are equal approximately to probabilities (and approach them as the number of trials go to infinity), so that, approximately, if the intervals T_j occur with probabilities p_j, the above formula becomes

$$\mathbf{E}(W) = \frac{q_1 T_1^2 + \cdots + q_k T_k^2}{2(q_1 T_1 + \cdots + q_k T_k)} = \frac{\mathbf{E}(T^2)}{2\mathbf{E}(T)}, \qquad (6.14)$$

which can be equivalently rewritten as

$$\mathbf{E}(W) = \frac{1}{2}\left[\mathbf{E}(T) + \frac{Var(T)}{\mathbf{E}(T)}\right]. \qquad (6.15)$$

Hence, as we have expected, for a given average interval time $\mathbf{E}(T)$, the average waiting time can be be arbitrarily large depending on the variance of the interval lengths.

Similar arguments can be used in assessing traffic flows. For example, suppose n cars are driven along the race track, formed as a circumference of radius 1 km, with speeds $v_1, \ldots, v_n$ kilometers per hour, and a speed camera is placed at some point that registers the speeds of passing cars and then calculates the average (arithmetic mean) speed from all the observed ones. Would the average be $(v_1 + \cdots + v_n)/n$? Of course not! In fact, during a time T, the cars will cover $v_1 T, \ldots, v_n T$ circumferences respectively, so that the camera will register $(v_1 + \cdots + v_n)T$ cars with the average speed being

$$\frac{v_1^2 + \cdots + v_n^2}{v_1 + \cdots + v_n}.$$

If the speed of a car is a random variable with a given distribution, then this would turn to the expression $\mathbf{E}(V^2)/\mathbf{E}(V)$, which is similar to (6.14).

6.2.4 Hedging via futures

Futures and forwards represent contracts to buy or sell a commodity or an asset by a fixed price on a prescribed date in the future (the distinction between future and forwards will not be of any relevance to our elementary analysis). Futures markets can be used to hedge the risk, i.e. to neutralize it as far as possible.

Our plan here will be as follows: (1) we shall show how this kind of hedging works on a simple numerical example; (2) deduce the main formula for the optimal hedge ratio; (3) introduce the main idea (arbitrage) underlying the pricing of the futures. Thus we discuss on a simplest possible example the use of variance as a measure of risk and the basic idea of arbitrage for pricing financial securities. Both these ideas are central for financial economics, and will pop out further in many places.

Assume that on the 1st of April an oil producer negotiated a contract to sell 1 million barrels of crude oil on the 1st of August by the market price that will form on this latter date. The point is that this market price is unknown on the 1st of April. Suppose that on the 1st of April the crude oil futures price (on a certain Exchange) for August delivery is $19 per barrel and each futures contract is for the delivery of 1000 barrels. The oil producer can hedge its risk by *shorting* 1000 futures contracts, i.e. by agreeing to sell $1000 \times 1000 = 10^6$ barrels on the 1st of August by the price of $19 per barrel. Let us see what can then happen. Suppose the price for crude oil will go down and on the 1st of August become, say, $18 per barrel. Then the company will realize $18 million from its initial contract. On the other hand, the right to sell something by $19 per barrel when the actual price is $18 per barrel means effectively a gain of $1 per barrel, i.e. the company will realize from its futures contract the sum of $1 million. The total gain then will equal $18 + 1 = \$19$ million. Suppose now the opposite scenario takes place, namely, the price for crude oil goes up and on the 1st of August becomes, say, $21 per barrel. Then the company will realize $21 million from its initial contract. On the other hand, the obligation to sell something by $19 per barrel when the actual price is $21 per barrel means effectively the loss of $2 per barrel, i.e. the company will loose from its futures contract the sum of $2 million. The total gain then will equal $21 - 2 = \$19$ million. In both cases the total gain of the company is the one obtained by selling its oil according to the futures price for the August delivery. Thus the risk is totally eliminated.

Such a perfect hedging is possible because the commodity underlying a futures contract is the same as the commodity whose price is being hedged. However, this is not always possible, so that one can only hedge the risk using futures on a related commodity or an asset (say to hedge a contract on a certain product of crude oil by the futures on the oil itself). To assess (and to minimize) the risk correlations between these commodity prices should be then taken into account.

Suppose N_A units of assets are to be sold at time t_2 and a company is hedging its risk at time $t_1 < t_2$ by shorting futures contract on N_F units of a similar (but not identical asset). The *hedge ratio* is defined as

$$h = N_F/N_A.$$

Let S_1 and F_1 be the (known at time t_1) asset and futures prices at the initial time t_1 (let us stress that F_1 is the price, at which one can agree at time t_1 to deliver a unit of the related asset on the time t_2), and let S_2 and F_2 be the (unknown at time t_1) asset and futures prices at the time t_2 (F_2 is the price, at which one can agree at time t_2 to deliver a unit of the underlying asset at this time t_2, so it is basically equal to the asset price at time t_2). The gain of the company at time t_2 becomes

$$Y = S_2 N_A - (F_2 - F_1) N_F = [S_2 - h(F_2 - F_1)] N_A = [S_1 + (\Delta S - h \Delta F)] N_A,$$

where

$$\Delta S = S_2 - S_1, \quad \Delta F = F_2 - F_1.$$

We are aiming to minimize the risk, i.e. the spread of this random variable around its average. In other words we are aiming to minimize the variance $Var(Y)$. As S_1 and N_A are given, this is equivalent to minimizing

$$Var(\Delta S - h \Delta F).$$

Let σ_S, σ_F and ρ denote respectively the variance of ΔS, the variance of ΔF and the correlation between these random variables. Then

$$Var(\Delta S - h \Delta F) = (\sigma_S^2 + h^2 \sigma_F^2 - 2h\rho\sigma_S\sigma_F).$$

Minimizing this quadratic expression with respect to h one obtains (check!) that the minimum is attained on the value

$$h^\star = \rho \frac{\sigma_S}{\sigma_F}, \tag{6.16}$$

which is the basic formula for the it optimal hedge ratio.

In practice, the coefficients σ_S, σ_F and ρ are calculated from the historical behavior of the prices using the standard statistical estimators (see Section 6.2.2).

Finally, let us give a rough idea on how futures contracts are priced. Let S_0 denote the price of the underlying asset at time t_0 and let r denotes the risk free rates (continuously compound), with which one can borrow and/or lend money (in practice rates r can be variable, but we only consider a simple situation with fixed rates). The futures price F to deliver this asset after a time T (called *time to maturity*) from the initial time t_0 should be equal to

$$F = S_0 e^{rT}. \tag{6.17}$$

Otherwise, *arbitrage opportunities* arise leading to a possibility of a free income and thus driving the price back to this level. In fact, suppose $F > S_0 e^{rT}$ (opposite situation is considered similarly). Then one can buy the asset by the price S_0 and short a future contract on it (i.e. enter into the agreement to sell the asset at time $t_0 + T$). Realizing this contract at time $t_0 + T$ would yield the sum F, which, when discounted to the present time, equals Fe^{-rT}. Since this is greater than S_0, one gets a free income.

6.2.5 Other measures of risk

Variance is definitely the simplest meaningful measure of risk. It is however non-adequate from several points of view. For instance, by increasing your possible winning, your risk can not be increased, but the variance can well do so. Looking for risks measures and analyzing their applicability remains an active area of research both in theory and in practice of finance.

We shortly mention here some basic alternatives. Apart from variance, one uses *semi-variance*

$$\int_{-\infty}^{\mu} (\mu - x)^2 f(x)dx = \mathbf{E}[(X - \mu)^2 \mathbf{1}_{X \leq \mu}],$$

where $\mu = \mathbf{E}X$, *shortfall probability* $\mathbf{P}(X \leq L)$ (that depends on an arbitrary parameter L), and possibly the most popular now is the Value at Risk, $VaR = VaR(X) = VaR_p(X)$, defined via the equation

$$\mathbf{P}(X \leq -VaR) = p.$$

Thus losses exceeding VaR (the event $(X \leq -VaR)$) can happen only with probability p. Often in banking practice $p = 0.01$ or 1%. As the disadvantages of VaR one can mention that (i) the event $(X \leq -VaR)$ (and hence VaR itself) does not capture how bad things can really be, that is how large are the losses $X \leq -VaR$; (ii) it depends on an arbitrary parameter p, which is effectively taken from nowhere, and so it can not be considered as an objective measure of risk.

Let us introduce some more recent trends in the analysis of risk. Let us say that a lottery is a gamble if it has positive expectation, but negative outcomes have nonzero probability. By a strategy let us mean a rule that, capital given, accepts or rejects a gamble with a distribution given. Any mapping Q from a set of gambles to $\mathbf{R}_+$ specifies a strategy S_Q: g is rejected at wealth w if $w < Q(g)$ and is accepted otherwise (i.e. $Q(g)$ is the minimal wealth at which g is accepted). Let us call strategies of type S_Q simple. Let us say that a strategy guarantees no-bankruptcy if for any sequence $G = \{g_1, g_2, \ldots\}$ of gambles (that the strategy would reject or accept according to its rule and current capital w_t) and any initial wealth w_0, one has $\mathbf{P}(\lim_{t\to\infty} w_t = 0) = 0$.

Theorem 6.2.1 *(Foster–Hart, see [7]). There exists a unique mapping R from gambles to $\mathbf{R}_+$ such that the simple strategy S_Q guarantees no-bankruptcy $\iff Q(g) \geq R(g)\forall g$. Moreover, $R(g)$ solves the equation*

$$\mathbf{E}\log\left(1 + \frac{g}{R(g)}\right) = 0.$$

6.3 Optimal betting (Kelly's system) and money management

Suppose you have an edge in betting on a series of similar trials, i.e. the average gain in each bet is positive (say, as you might expect by trading on Forex using some advanced strategy). Then you naturally expect to win in a long run. However, if you are going to put all your bankroll on each bet, you would definitely loose instead. An obvious idea is therefore to bet on a fraction of your current capital. What is the optimal fraction?

To answer this question assume that you are betting on a series of Bernoulli trials with the probability of success p, when you get m times the amount you bet. And otherwise, with probability $1 - p$, you loose your bet. Consequently, if you invest one dollar, the expectation of your gain is mp dollars. This, assuming that you have an edge, is equivalent to assuming $mp > 1$, as we shall do now.

Let V_0 be your initial capital, and your strategy is to invest in each bet a fixed fraction α, $0 < \alpha < 1$, of your current bankroll. Let X_k denote the random variable that equals m, if the kth bet is winning, and equals zero otherwise, so that all X_k are independent and identically distributed. Hence after the first bet your capital becomes $V_1 = (1 - \alpha + \alpha X_1)V_0$, after the second bet it will be

$$V_2 = (1 - \alpha + \alpha X_2)(1 - \alpha + \alpha X_1)V_0,$$

and for any n your capital at time n will become

$$V_n = (1 - \alpha + \alpha X_n)\cdots(1 - \alpha + \alpha X_1)V_0.$$

We are aiming to maximize the ratio V_n/V_0, or equivalently its logarithm

$$\ln\frac{V_n}{V_0} = \ln(1 - \alpha + \alpha X_n) + \cdots + \ln(1 - \alpha + \alpha X_1).$$

Here the law of large numbers comes into play. Namely, according to this law, the average of the winning rates per bet

$$G_n = \frac{1}{n} \ln \frac{V_n}{V_0} = \frac{1}{n}[\ln(1 - \alpha + \alpha X_n) + \cdots + \ln(1 - \alpha + \alpha X_1)]$$

converges to the expectation

$$\phi(\alpha) = \mathbf{E} \ln(1 - \alpha + \alpha X_1)$$
$$= p \ln(1 - \alpha + \alpha m) + (1 - p) \ln(1 - \alpha).$$

Therefore, to maximize the gain in a long run, one has to look for α that maximizes the function $\phi(\alpha)$. But this is an easy exercise in calculus yielding (check it!) that the maximum is attained at

$$\alpha^\star = \frac{pm - 1}{m - 1}. \tag{6.18}$$

This is the final formula we aimed at, known as *Kelly's betting system*. This is not the full story, however, as we did not take into account the risk (we only used averages). A careful analysis (that we are not going into) allows one to conclude that by choosing the betting fraction α slightly smaller than $\alpha^\star$ one can essentially reduce the risk with only a slight decrease of the expected gain. We refer to book [12] for a recent review of the theory and practice of Kelly's betting system.

Kelly's system reveals a distinguished role played by the logarithmic utility function in the theory of betting.

As an application, let us consider now an aspect of financial trading, namely *money management*, which by many active and successful traders is considered to be the most crucial one.

As a first (possibly rather artificial) example (taken from [16]), consider trading on a stock that each week goes up 80% or down 60% with equal probabilities 1/2. You clearly have an edge in this game, as the expectation of your gain is

$$\frac{1}{2}80\% - \frac{1}{2}60\% = 10\%.$$

On the other hand, suppose you invest (and reinvest) your capital for many weeks, say n times. Each time your capital is multiplied either by 1.8 (success) or by 0.4 (failure) with equal probabilities. During a large period of time, you can expect the number of successes and failures to be approximately equal (the law of large numbers!). Thus your initial capital V_0 would become

$$1.8^{n/2} 0.4^{n/2} V_0 = (0.72)^{n/2} V_0,$$

which quickly tends to zero as $n \to \infty$. So, do you really have an edge in this game?

The explanation to this paradoxical situation should be seen from the above discussion of Kelly's system. The point is that you should not invest all your capital at once (money management!).

To properly sort out the situation it is convenient to work in a more general setting. Namely, assume that the distribution of the return is given by a positive random variable R (i.e. in one period of investing your capital is multiplied by R) with a given distribution. Following the line of reasoning of the previous section, we can conclude that if you always invest the fraction α of your capital, then the average rate of return per investment would converge to

$$g(\alpha) = \mathbf{E} \ln(1 - \alpha + \alpha R).$$

We are interested in the maximum of this expression over all $\alpha \in [0, 1]$.

Proposition 6.3.1 Assume $\mathbf{E}(R) > 1$ (i.e. you have an edge in the game). Then (i) if $\mathbf{E}(1/R) \leq 1$ (your edge is overwhelming), then $\alpha = 1$ is optimal and your maximum average return per investment equals $g(1) = \mathbf{E} \ln(R)$; and (ii) if $\mathbf{E}(1/R) > 1$, then there exists a unique $\hat{\alpha} \in (0, 1)$ maximizing $g(\alpha)$, which is found from the equation

$$g'(\hat{\alpha}) = \mathbf{E} \frac{R - 1}{1 + (R - 1)\hat{\alpha}} = 0. \tag{6.19}$$

Proof Since the second derivative

$$g''(\alpha) = -\mathbf{E} \left(\frac{R - 1}{1 + (R - 1)\hat{\alpha}} \right)^2$$

is negative, the function g is concave (i.e. its derivative is decreasing), and there can be at most one solution of equation (6.19) that necessarily would specify a maximum value of g. The assumption $\mathbf{E}(R) > 1$ is equivalent to saying that $g'(0) > 0$. In case (i) one has $g'(1) \geq 0$, so that the solution $\hat{\alpha}$ can lie only to the right of $\alpha = 1$ and hence the maximum of $g(\alpha)$ is attained at $\alpha = 1$. In case (ii), $g'(1) < 0$, implying the existence of the unique solution to (6.19) in $(0, 1)$. $\qquad \square$

Proposition 6.3.2 If R takes only two values m_1 and m_2 with equal probabilities (in the example at the beginning, $m_1 = 1.8$, $m_2 = 0.4$), the condition $\mathbf{E}(1/R) > 1$ rewrites as $m_1 + m_2 > 2m_1 m_2$, and if it holds, the optimal $\hat{\alpha}$ equals

$$\hat{\alpha} = \frac{1}{2(1 - m_1)} + \frac{1}{2(1 - m_2)}.$$

Proof Is left as an exercise.   □

6.4 Portfolio, CAPM and factor models

6.4.1 Portfolio optimization

We introduce the Nobel price winning theory of the Markowitz mean-variance portfolio optimization.

Suppose $S_0^1, \ldots, S_0^n$ are the initial prices of n securities, and let $S_T^1, \ldots,$ S_T^n denote their prices at time T (unknown at the initial time). Let the returns be denoted by

$$1 + r_i = 1 + r_i(T) = S_T^i / S_0^i, \quad r = (r_1, \ldots, r_n),$$

where r_i are *rates of return* (or *returns* for short). The main input data for the analysis, assumed to be known, are the expectations $\mathbf{E}(r_i) = \bar{r}_i$ and the covariance matrix Σ with entries $Cov(r_i, r_j) = \sigma_{ij}$. The initial capital is $y = y_0$.

The control parameters available to an investor are the numbers $(\phi_1, \ldots, \phi_n)$ (not necessarily integers: liquidity!) of securities of each type satisfying the budget constraint:

$$\sum_{i=1}^{n} \phi_i S_0^i = y.$$

In terms of the *portfolio vector* $x = (x_1, \ldots, x_n)$, $x_i = \phi_i S_0^i / y$, this constraint rewrites as $\sum_{i=1}^{n} x_i = 1$.

For a chosen portfolio vector, the wealth at time T becomes $y(T) = \sum_{i=1}^{n} \phi_i S_T^i$ with return

$$1 + r(x) = \frac{y(T)}{y} = \frac{1}{y} \sum_{i=1}^{n} \phi_i S_0^i (1 + r_i) = 1 + \sum_{i=1}^{n} x_i r_i,$$

$$\mathbf{E}(r(x)) = \sum_{i=1}^{n} x_i \bar{r}_i = (x, \bar{r}),$$

$$Var(r(x)) = \sigma^2(r(x)) = \sum_{i,j=1}^{n} x_i x_j \sigma_{ij} = (x, \Sigma x).$$

The two main problems suggested by Markowitz to describe an investor aiming at increasing the expected return and at the same time decreasing the risk measured in terms of the variance, can be formulated as follows: (i) find $\min_x \sigma^2(r(x))$ with $\mathbf{E}(r(x))$ given or bounded below by a given constant R; (ii) find $\max_x \mathbf{E}(r(x))$ with $\sigma^2(r(x))$ given.

As seen from the analysis below, these problems are in some sense equivalent. Let us analyze the first one using the method of Lagrange multipliers. The Lagrange function can be taken as (the coefficient $\frac{1}{2}$ is inserted for convenience):

$$L = \frac{1}{2}(x, \Sigma x) - \lambda((x, \bar{r}) - R) - \mu(\sum_{i=1}^{n} x_i - 1).$$

The equations for critical points

$$\frac{\partial L}{\partial x} = 0, \quad \frac{\partial L}{\partial \lambda} = 0, \quad \frac{\partial L}{\partial \mu} = 0$$

yield the system (**1** is the vector with all coordinates one) of equations

$$\Sigma x = \lambda \bar{r} + \mu \mathbf{1}, \quad (x, \bar{r}) = R, \quad (x, \mathbf{1}) = 1. \tag{6.20}$$

From the first equation

$$x = \Sigma^{-1}(\lambda \bar{r} + \mu \mathbf{1}). \tag{6.21}$$

The last two equations in (6.21) yield the linear system

$$\begin{cases} A\mu + B\lambda = 1, \\ B\mu + C\lambda = R, \end{cases} \tag{6.22}$$

with constant coefficients

$$A = (\mathbf{1}, \Sigma^{-1}\mathbf{1}), \quad B = (\bar{r}, \Sigma^{-1}\mathbf{1}), \quad C = (\bar{r}, \Sigma^{-1}\bar{r}). \tag{6.23}$$

Hence

$$\lambda = \lambda(R) = \frac{AR - B}{AC - B^2}, \quad \mu = \mu(R) = \frac{C - BR}{AC - B^2}. \tag{6.24}$$

By the Cauchy inequality (applied to quadratic form $(x, \Sigma^{-1}x)$), $AC \geq B^2$, and $AC = B^2 \Longleftrightarrow \bar{r}$ is proportional to **1**. Denote $x^* = x^*(R)$ the corresponding vector (6.21). Then $\Sigma x^* = \lambda \bar{r} + \mu$ with λ, μ from (6.24). For $\sigma = \sigma(r(x^*(R)))$ we get

$$\sigma^2 = Var(r(x^*)) = (x^*, \Sigma x^*) = \lambda(x^*, \bar{r}) + \mu(x^*, \mathbf{1})$$

$$= \lambda R + \mu = \frac{AR^2 - 2BR + C}{AC - B^2} = \frac{A(R - B/A)^2}{AC - B^2} + \frac{1}{A}.$$

Hence $\sigma_m^2 = \min_R Var(r(x^*)) = 1/A$, the minimum is attained at $R_m = B/A$ with $\lambda_{\min} = 0$, and the last equation rewrites as

$$\frac{\sigma^2}{\sigma_m^2} - \frac{A}{AC - B^2} \frac{(R - R_m)^2}{\sigma_m^2} = 1. \qquad (6.25)$$

These optimal returns and their deviations R, σ lie on a hyperbola. The upper part of this hyperbola is called (for obvious reasons) the *efficient frontier* (see Fig. 6.4).

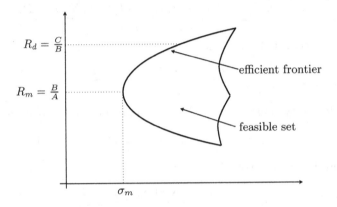

Figure 6.4 Efficient frontier and two funds.

Two feasible sets can be naturally considered corresponding to two reasonable posits of the problem: one with shorting not permitted (that is, all x_i non-negative), and another with shorting allowed (without the above constraint).

Two special points on the hyperbola (6.25) are worth singling out, one with vanishing λ and one with vanishing μ, the corresponding x^* will be denoted x_g and x_d:

(1) $\lambda = 0 \iff R = R_m = \dfrac{B}{A} \implies \mu = \dfrac{1}{A} = \sigma_m^2, \quad x^* = x_g = \dfrac{1}{A}\Sigma^{-1}\mathbf{1}$,

(2) $\mu = 0 \iff R = \dfrac{C}{B} \implies \lambda = \dfrac{1}{B}, \quad x^* = x_d = \dfrac{1}{B}\Sigma^{-1}\bar{r}$.

Notice $\frac{C}{B} \geq \frac{B}{A} = R_m$.

For any R we have

$$x^* = \lambda B x_d + \mu A x_g,$$

leading to the following crucial result:

Theorem 6.4.1   The two-fund theorem. *There exist two efficient portfolios, namely x_d and x_g, such that any efficient portfolio is their linear combination.*

6.4.2 Portfolio optimization with a risk-free asset

Additionally to the above setting, suppose there exists a risk-free asset with the rate of growth r_f and some initial price S_0^0. Let $x = (x_1, \ldots, x_n)$ still denote the part held on risky assets. Then $\sum_{i=1}^n x_i \le 1$ and $x_0 = 1 - (x, \mathbf{1}) = \phi_0 S_0^0 / y_0$ is the part held on the risk-free asset.

The main problem can now be formulated as the problem of finding the minimum of $Var(r(x))/2$ subject to $(x, \bar{r}) + (1 - (x, \mathbf{1}))r_f = R$. The Lagrange function is

$$L = \frac{1}{2}(x, \Sigma x) - \lambda[(x, \bar{r}) + (1 - (x, \mathbf{1}))r_f - R].$$

The optimality conditions are

$$\Sigma x = \lambda(\bar{r} - r_f \mathbf{1}), \quad (x, \bar{r} - r_f \mathbf{1}) = R - r_f. \tag{6.26}$$

Thus, for a solution pair (x^*, λ)

$$x^* = \lambda \Sigma^{-1}(\bar{r} - r_f \mathbf{1}), \tag{6.27}$$

$$\lambda = \frac{R - r_f}{(\bar{r} - r_f \mathbf{1}, \Sigma^{-1}(\bar{r} - r_f \mathbf{1}))} = \frac{R - r_f}{C - 2r_f B + r_f^2 A}. \tag{6.28}$$

Then

$$x^* = \frac{R - r_f}{C - 2r_f B + r_f^2 A} \Sigma^{-1}(\bar{r} - r_f \mathbf{1}), \tag{6.29}$$

and

$$\sigma^2 = \sigma^2(r(x^*)) = (x^*, \Sigma x^*)$$
$$= \lambda^2(\bar{r} - r_f \mathbf{1}, \Sigma^{-1}(\bar{r} - r_f \mathbf{1})) = \frac{(R - r_f)^2}{C - 2r_f B + r_f^2 A}, \tag{6.30}$$

or

$$R = r_f \pm \sigma \sqrt{C - 2r_f B + r_f^2 A}, \tag{6.31}$$

which effectively decomposes into two linear functions. The upper part is called the *efficient frontier* or *efficient line* and has the slope

$$\tan \theta = \sqrt{C - 2r_f B + r_f^2 A}. \tag{6.32}$$

The overall result of the analysis is quite impressive: the totality of data (of order n^2 for n assets) is reduced to one line!

The *tangent portfolio* is defined as the portfolio on the efficient line such that $(x_T, \mathbf{1}) = 1$, that is, the unique efficient portfolio without risk-free investment. Denote $r_T = (x_T, r)$, $\bar{r}_T = (x_T, \bar{r})$ its return and σ_T its deviation. The efficient line (see Fig. 6.5) on (R, σ) plane is then

$$R = r_f + \frac{\bar{r}_T - r_f}{\sigma_T}\sigma.$$

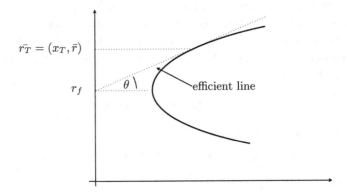

Figure 6.5 Efficient line and tangent portfolio.

For the tangent portfolio we have

$$(x_T, \mathbf{1}) = 1 = \lambda_T[(\mathbf{1}, \Sigma^{-1}\bar{r}) - r_f(\mathbf{1}, \Sigma^{-1}\mathbf{1})],$$

implying

$$\lambda_T = \frac{1}{B - r_f A}, \quad x_T = \frac{1}{B - r_f A}\Sigma^{-1}(\bar{r} - r_f\mathbf{1}), \quad (x_T, \bar{r}) = \frac{C - r_f B}{B - r_f A}. \tag{6.33}$$

Hence for any R

$$x^* = \lambda\Sigma^{-1}(\bar{r} - r_f\mathbf{1}) = \lambda(B - r_f A)x_T, \tag{6.34}$$

leading to the following crucial result:

Theorem 6.4.2 The one-fund theorem. *There exists a portfolio of risky assets, namely x_T: any efficient portfolio can be constructed as a combination of the fund x_T and some risk-free assets.*

Let us find the standard deviation of the tangent portfolio x_T. From (6.30) and the first equation of (6.33),

$$\sigma_T^2 = \lambda_T^2(\bar{r} - r_f 1, \Sigma^{-1}(\bar{r} - r_f 1)) = \frac{C - 2r_f B + r_f^2 A}{(B - r_f A)^2}, \qquad (6.35)$$

and using the last equation in (6.33),

$$\sigma_T^2 = \frac{1}{B - r_f A}\left[(x_T, \bar{r}) + \frac{r_f(r_f A - B)}{B - r_f A}\right] = \frac{\bar{r}_T - r_f}{B - r_f A} = \lambda_T(\bar{r}_T - r_f). \qquad (6.36)$$

Let us now use (6.27) in the opposite direction:

$$\bar{r} - r_f 1 = \frac{1}{\lambda_T}\Sigma x_T. \qquad (6.37)$$

By the definition of Σ, $(\Sigma x)_i = Cov(r_i, (r, x))$ for any x. Hence, by (6.36) and (6.37),

$$\bar{r} - r_f 1 = \frac{Cov(r, r_T)}{\sigma_T^2}(\bar{r}_T - r_f). \qquad (6.38)$$

Thus we have proved the following:

Theorem 6.4.3 *The average returns of the initial assets are expressed in terms of the tangent portfolio by formula (6.38).*

We say that an investor with portfolio x^* *lends money to the market* if the amount put on the risk-free asset is positive, that is $(x^*, 1) < 1$, or equivalently (using (6.34))

$$(x^*, 1) = \lambda(B - r_f A)(x_T, 1) = \lambda(B - r_f A) < 1,$$

which in turn is equivalent to $\lambda < \lambda_T$ or

$$(x^*, \bar{r}) = \lambda(B - r_f A)(x_T, \bar{r}) < (x_T, \bar{r}) = \bar{r}_T.$$

Otherwise he/she borrows money.

6.4.3 Capital Asset Pricing Model (CAPM)

The main conjecture of the *Capital Asset Pricing Model* (CAPM) can be formulated by stating that the totality of all assets on the market is an efficient portfolio, and hence the tangent portfolio (as it does not include risk-free assets, by definition). This universal tangent portfolio is then referred to as the *market portfolio*. An asset's *weight* in this portfolio is then naturally defined as the proportion of the asset's capital value to the total market capital value. Briefly, the CAPM states that *the market portfolio is efficient.*

Thus the tangent portfolio x_T is now the collection of all assets of the market x_M. The efficient line is then called the *capital market line*:

$$R = r_f + \frac{\bar{r}_M - r_f}{\sigma_M}\sigma. \qquad (6.39)$$

Its slope

$$\lambda_M = \frac{\bar{r}_M - r_f}{\sigma_M}$$

is called the *market price of risk.*

The fundamental equation (6.39) can be formulated as the following 'equation in words':

Expected return = Price of time + (Price of risk) × (Amount of risk).

Practically, for choosing an appropriate investment on the efficient line, one uses an individual utility function that somehow measures an investor's aversion to risk. The simplest class of utility functions is of course quadratic, leading to the following optimization problem: for the utility function $R - c\sigma^2$ find the optimal point (maximizing the utility) on the capital line. Here c is a constant specifying the risk aversion of an investor. The inverse number, $1/c$, is referred to as *risk tolerance.*

The main analytic result of CAPM represents a direct consequence of Theorem 6.4.3, which can be formulated as follows.

Theorem 6.4.4   CAPM: main theorem. *If the market portfolio M is efficient (i.e. the CAPM conjecture holds), the expected return of any asset satisfies*

$$\bar{r}_i - r_f = \beta_i(\bar{r}_M - r_f), \quad \beta_i = \frac{\sigma_{iM}}{\sigma_M^2}, \quad \sigma_{iM} = Cov(r_i, r_M). \qquad (6.40)$$

Function (6.40), as a function of β_i, or the corresponding line in the (R, β)-plane, is called the *security market line* (SML). Alternatively, one represents SML as a line in the (R, σ_{iM})-plane:

$$\bar{r}_i = r_f + \frac{\lambda_M}{\sigma_M}\sigma_{iM}.$$

Remark 6.4.5   Some authors refer to $\lambda'_M = (\bar{r}_M - r_f)/\sigma_M^2$ as the market price of risk, so that CAPM in the (R, σ_{iM})-plane becomes $\bar{r}_i = r_f + \lambda'_M\sigma_{iM}$.

As a simple numeric example, let us assume that $r_f = 8\%, \bar{r}_M = $

$12\%, \sigma_M = 15\%$, and an asset has covariance 0.045 with the market. Under CAPM, we can find its average return $\bar{r}$ as follows:

$$\beta = \frac{0.045}{(0.15)^2} = 2.0; \quad \bar{r} = 0.08 + 2.0(0.12 - 0.08) = 0.16.$$

Remark 6.4.6 β is a linear function on assets, i.e. for $r = \sum_{i=1}^{k} w_i r_i$, $Cov(r, r_M) = \sum_{i=1}^{k} w_i Cov(r_i, r_M)$ and $\beta = \sum_{i=1}^{k} w_i \beta_i$.

The theory above was mostly concerned with the averages $\bar{r}_i$. What about r_i? Clearly one can write

$$r_i = r_f + \beta_i(r_M - r_f) + \epsilon_i, \quad \mathbf{E}(\epsilon_i) = 0.$$

Then, remarkably enough,

$$cov(\epsilon_i, r_M) = -\beta_i \sigma_M^2 + \sigma_{iM} = 0,$$

so that ϵ_i and r_M are uncorrelated, and hence

$$\sigma_i^2 = \beta_i^2 \sigma_M^2 + Var(\epsilon_i).$$

The first term here is called the *systematic risk*, the second is the *idiosyncratic or specific risk*.

6.4.4 Performance evaluation

Usually the estimates $\tilde{\bar{r}}$ and $\tilde{\bar{r}}_M$ for $\bar{r}$ and $\bar{r}_M$ respectively are calculated as empirical averages over past years for a company and the market (using, say, the *S&P* index for the latter). Similarly, β is usually meant to be the empirical β. Then writing

$$\tilde{\bar{r}}_i - r_f = J_i + \beta_i(\tilde{\bar{r}}_M - r_f), \tag{6.41}$$

one calls J_i the *Jensen index*. It measures how much a firm deviates from the theoretical value 0. $J_i > 0$ (resp. $J_i < 0$) means better (resp. worse) performance than average. However, one has to check whether this deviation is not due entirely to the natural fluctuations of $\tilde{\bar{r}}_i - \bar{r}_i$.

Alternatively, writing

$$\tilde{\bar{r}}_i - r_f = S_i \sigma_{iM}, \tag{6.42}$$

one calls S_i the *Sharpe index*. Again it measures how much the efficiency deviates from the theoretical value λ_M / σ_M.

6.4.5 CAPM as pricing formula

Let P denote the present price of certain product, and Q its future price, with average $\bar{Q}$. By CAPM:

$$\frac{\bar{Q} - P}{P} = r_f + \beta(\bar{r}_M - r_f).$$

This implies the *pricing form of the CAPM*:

$$P = \frac{\bar{Q}}{1 + r_f + \beta(\bar{r}_M - r_f)}. \tag{6.43}$$

Here $r_f + \beta(\bar{r}_M - r_f)$ is referred to as the *risk-adjusted interest rate*.

However, this formula ignores the dependence of β on P:

$$\beta = \frac{Cov(Q/P - 1, r_M)}{\sigma_M^2} = \frac{Cov(Q, r_M)}{P\sigma_M^2}.$$

Substituting this in (6.43) and rearranging yields the *certainty equivalent pricing formula*:

$$P = \frac{1}{1 + r_f}\left[\bar{Q} - \frac{Cov(Q, r_M)(\bar{r}_M - r_f)}{\sigma_M^2}\right]. \tag{6.44}$$

For obvious reasons, the term in brackets is called the *certainty equivalent* of Q.

This theory can be used to evaluate projects via the *net present value* (NPV).

$$NPV = -P + \frac{1}{1 + r_f}\left[\bar{Q} - \frac{Cov(Q, r_M)(\bar{r}_M - r_f)}{\sigma_M^2}\right]. \tag{6.45}$$

Roughly speaking, a good project should have large positive NPV.

6.4.6 Multi-factor models (arbitrage pricing)

By a *multi-factor model* one means a model for the rates of return r_i of various firms of the form

$$r_i = a_i + \sum_{i=1}^{k} b_{ij} f_j + e_i, \quad i = 1, \ldots, m,$$

where $\{f_i\}$ are common economic factors (practically, a number of factors k should be small compared to m), and e_i individual noises (independent, and with zero expectations). In the basic case of one factor, r_M, the model becomes

$$r_i - r_f = \alpha_i + \beta_i(r_M - r_f) + e_i.$$

Then

$$Cov(r_i - r_f, r_M) = \beta_i \sigma_M^2 \implies \beta_i = \frac{Cov(r_i, r_M)}{\sigma_M^2},$$

as in CAPM, but additionally with α_i. Interpretation: $\alpha_i > 0$ (resp. $\alpha_i < 0$) means better (resp. worse) performance than average, hence stocks are underpriced (resp. overpriced).

Let us use a simple one-factor model to illustrate again the basic idea of *arbitrage pricing*, which was introduced earlier for futures. As we shall see, the idea of arbitrage (better to say, the requirement of its absence) poses strong restrictions to the model, reducing essentially the number of possible parameters. Thus, suppose the return on some class of assets is given by the one-factor model

$$r_i = a_i + b_i f, \quad i = 1, \dots, m,$$

where a_i, b_i are constants and f a random variable (single factor). Assume all b_i are different.

A portfolio mixing two assets i and j yields a return

$$r = w a_i + (1 - w) a_j + [w b_i + (1 - w) b_j] f.$$

Choosing $w = b_j / (b_j - b_i)$ makes the coefficient at f zero and the corresponding asset risk free. Hence, the absence of arbitrage opportunities implies that its return should be the same number r_f for all pairs (and equal to the underlying basic rate of return of a risk-free asset, if this is available), i.e.

$$\frac{a_i b_j}{b_j - b_i} + \frac{a_j b_i}{b_i - b_j} = r_f.$$

Rearranging yields

$$\frac{a_j - r_f}{b_j} = \frac{a_i - r_f}{b_i}.$$

As this ratio should be the same for all i, it is a constant, say c. Consequently,

$$a_i = r_f + b_i c, \quad i = 1, \dots, m.$$

It is thus shown that a_i should be given by certain fixed (independent of i) linear functions of b_i.

6.4.7 The Wilkie model

The Wilkie model (in the simplest form) describes the evolution of the *force of inflation* $I(t)$ as an AR(1) model (auto-regression of length one):

$$I(t) = a + bI(t-1) + cZ(t),$$

where a, b, c are constants and $Z(t)$, $t = 0, 1, \ldots$, is a sequence of independent standard normal random variables. The long-term *average rate of inflation* is the fixed point of the corresponding equation for averages: $I_\infty = a + bI_\infty$.

$I(t)$ is expressed via the *Consumer price index* Q (called also *PRI index* or *inflation index*) as

$$I(t) = \log \frac{Q(t)}{Q(t-1)}.$$

The *inflation* over year t equals $Q(t)/Q(t-1)$.

6.5 Derivative securities in discrete time: arbitrage pricing and hedging

6.5.1 Efficient market hypothesis

Efficient market hypothesis is known in three basic forms:

- *Strong*: market prices incorporate all information, available both publicly and to insiders.
- *Semi-strong*: market prices incorporate all publicly available information.
- *Weak*: market prices incorporate all information contained in the price history of that investment, that is, prices represent a Markov process. In particular, this assumption makes useless almost all of the so-called technical analysis.

Whether this hypothesis holds in any of these forms is still an open problem, widely discussed in the abundant economic literature on all levels, see e.g. [5], [6], [15], [17] and references therein. We only make it clear here that most of the models used in financial mathematics for asset pricing (and only those will be discussed in our presentation) do enjoy the Markov property, meaning that the future depends on the past only via the present state.

6.5.2 Binomial model: basic assumptions

General underlying assumptions of the binomial models of asset pricing that are shared with many other basic models (including the most famous Black–Sholes one) can be listed as follows:

- Shares can be subdivided in arbitrary fractions for sale or purchase.
- The interest rate is constant and the same for investing or borrowing. No limits are assumed for borrowing.
- Purchase price equals selling price, that is, no bid-ask spread is taken into account.
- Transaction costs are assumed to be negligibly small.

On the other hand, more specific assumptions can be roughly described as follows: time is discrete, there is only one stock on the market with only two values of possible relative price changes for each period. Before plunging into the proper mathematical formulation, we shall recall the notion of conditional expectation in the context of the model of Bernoulli trials.

6.5.3 Conditional expectation and martingales on Bernoulli spaces

Recall that, for a probability law $\mathbf{P}$ on a finite sample space Ω, the conditional expectation of an event A given that the event B occurred is defined as $\mathbf{P}(A|B) = \mathbf{P}(A \cap B)/\mathbf{P}(B)$, $A, B \subset \Omega$.

Notice that $\mathbf{P}(A|B)$, as a function of $A \subset B$, is a probability law on B.

The *conditional expectation* $\mathbf{E}(f(\omega)|B)$ of a random variable f conditioned on the event B is defined as the expectation of f with respect to the law $\mathbf{P}(.|B)$:

$$\mathbf{E}(f(\omega)|B) = \sum_{\omega \in B} f(\omega) \frac{\mathbf{P}(\omega)}{\mathbf{P}(B)}.$$

For Bernoulli experiments the probability space is defined as $\Omega_N = \{1, 0\}^N$ or $\Omega_N = \{u, d\}^N$ (u and d stand for up and down moves of stock prices), etc, that is, it is the set of sequences of N symbols taking two possible values ($1, 0$, or u, d, etc), the probabilities of these two symbols being p and $q = 1 - p$, independently for each experiment. That is, the corresponding probability space $\Omega_N(p)$ assigns probability

$$p^{\#u(\omega_1,\ldots,\omega_N)} q^{\#d(\omega_1,\ldots,\omega_N)} \tag{6.46}$$

to an elementary event $\omega = (\omega_1, \ldots, \omega_N)$, where the symbol $\#$ means 'the number of', so that, say, $\#u(\omega_1, \ldots, \omega_N)$ is the number of u that are present in the sequence $\omega_1, \ldots, \omega_N$.

One often uses special notation:

$$\mathbf{E}_n[X](\omega_1, \ldots, \omega_n) = \mathbf{E}(X|\omega_1, \ldots, \omega_n), \quad \omega_i = u, d,$$

for conditioning with respect to the event $(\omega_1, \ldots, \omega_n)$ that specifies first n symbols in the sequence. Hence

$$\mathbf{E}_{N-1}[X](\omega_1, \ldots, \omega_{N-1}) = pX(\omega_1, \ldots, \omega_{N-1}, u) + qX(\omega_1, \ldots, \omega_{N-1}, d),$$
$$\mathbf{E}_{N-2}[X](\omega_1, \ldots, \omega_{N-2}) = p^2 X(\omega_1, \ldots, \omega_{N-2}, u, u)$$
$$+ q^2 X(\omega_1, \ldots, \omega_{N-2}, d, d)$$
$$+ pq(X(\omega_1, \ldots, \omega_{N-2}, u, d)$$
$$+ X(\omega_1, \ldots, \omega_{N-2}, d, u)),$$

and generally

$$\mathbf{E}_n[X](\omega_1, \ldots, \omega_n)$$
$$= \sum_{\omega_{n+1}, \ldots, \omega_N} p^{\#u(\omega_{n+1}, \ldots, \omega_N)} q^{\#d(\omega_{n+1}, \ldots, \omega_N)} X(\omega_1, \ldots, \omega_n, \omega_{n+1}, \ldots, \omega_N).$$

In particular, $\mathbf{E}_0[X] = \mathbf{E}(X)$, $\mathbf{E}_N[X] = X$.

Basic properties, which are easy to check, but are constantly used in calculations, can be listed as follows:

- *Chain rule*: $\mathbf{E}_n[\mathbf{E}_m[X]] = \mathbf{E}_n[X]$, $n \le m \le N$;
- *Independence*: $\mathbf{E}_n[X] = \mathbf{E}[X]$, if X depends only on $\omega_{n+1}, \ldots, \omega_N$;
- *'Taking out what is known' rule*: $\mathbf{E}_n[XY] = X\mathbf{E}[Y]$, if X depends on $\omega_1, \ldots, \omega_n$.

Random sequence $M_0, M_1, \ldots$ on $\Omega = \{u, d\}^N$ is called *adapted*, if M_n only depends on $\omega_1, \ldots, \omega_n$ (in particular, M_0 is a constant). An adapted sequence $M_0, M_1, \ldots$ is called a *martingale*, if $M_n = \mathbf{E}_n(M_{n+1})$ for all n, and sub(super)martingale if $M_n \le \mathbf{E}_n(M_{n+1})$ (resp. $M_n \ge \mathbf{E}_n(M_{n+1})$).

For a martingale, $\mathbf{E}M_n = \mathbf{E}\mathbf{E}_n(M_{n+1}) = \mathbf{E}(M_{n+1})$ implying $\mathbf{E}M_n = M_0$ for all n.

6.5.4 Binomial model: definition and risk-neutral laws

A *binomial model* of asset pricing is specified by two positive parameters $d < u$. It is then assumed that, given the price $S_n > 0$ of a stock at time n, the price S_{n+1} at the next moment is either uS_n or dS_n with some probability p (which may be not specified explicitly). Given S_0, a probability p of the event u (with $q = 1 - p$ denoting the probability of the complementary event d) defines a random process $\{S_n\}$ on $\Omega_N(p)$.

Proposition 6.5.1 Given r (interpreted as risk-free rates) such that $d < 1 + r < u$, there exists a unique $\tilde{p} \in (0, 1)$ called the *risk-neutral probability law* such that the discounted stock price process $S_n/(1+r)^n$ is a martingale.

Proof The martingale condition reads as

$$\frac{S_n}{(1+r)^n} = \tilde{\mathbf{E}}_n \frac{S_{n+1}}{(1+r)^{n+1}},$$

where $\tilde{\mathbf{E}}$ denotes the expectation on $\Omega_N(\tilde{p})$, or equivalently

$$\tilde{\mathbf{E}}_n S_{n+1} = (1+r)S_n \Longleftrightarrow \tilde{p}u + (1 - \tilde{p})d = 1 + r,$$

with the only solution

$$\tilde{p} = \frac{1 + r - d}{u - d}, \quad \tilde{q} = 1 - \tilde{p} = \frac{u - 1 - r}{u - d}.$$

$\square$

If X_n is the capital of an investor at time n and she chooses to have (or to buy) Δ_n shares, putting the rest on the money market, the capital at the next moment clearly becomes

$$X_{n+1} = \Delta_n S_{n+1} + (1 + r)(X_n - \Delta_n S_n). \tag{6.47}$$

For a given adapted process $\{\Delta_n\}$, $n = 0, 1, \dots$ (called a portfolio process) and an initial (non-random) capital X_0, the corresponding *wealth process* is defined via recursive equation (6.47).

Proposition 6.5.2 Given X_0 and an adapted portfolio process $\{\Delta_n\}$, $n = 0, 1, \dots$, the discounted wealth process $X_n/(1+r)^n$ is a martingale under the risk-neutral law.

Proof

$$\tilde{\mathbf{E}}_n \frac{X_{n+1}}{(1+r)^{n+1}} = \Delta_n \tilde{\mathbf{E}}_n \frac{S_{n+1}}{(1+r)^{n+1}} + \frac{1}{(1+r)^n}(X_n - \Delta_n S_n) = \frac{X_n}{(1+r)^n},$$

where the 'Taking out what is known' rule was used for the first equation and Proposition 6.5.1 for the second. $\square$

Let us stress that in the binomial model simple compounding is used for each period of time. The amount of money X becomes $(1 + r)X$ in the next period.

6.5.5 Binomial model: replication and risk-neutral evaluation of contingent claims

European contingent claims or *derivative securities* or *options* can be identified with a random variable V_N (final payoffs) on $\Omega_N(p)$. An investor who sold such a claim at the initial time has to pay V_N at the expiration time N.

Main examples represent a standard *European call* and a standard *European put* specified by payoffs $V_N = (S_N - K)^+$ and $V_N = (K - S_N)^+$ respectively, where a constant K is called the *strike price*.

This payoff arises from the following interpretation (or original definition): *European call* (respectively a European put) confers on its owner the right (not the obligation) to buy (respectively to sell) one share of the stock at time N for the strike price K.

The following is the main result on the possibility to replicate any contingent claim, specified by an arbitrary random variable V_N, by a certain portfolio process.

Theorem 6.5.3 *For any random variable V_N on $\Omega_N(p)$ there exists a unique adapted portfolio process $\{\Delta_n\}$, $n = 0, 1, \ldots N$, and a unique number X_0 such that for the corresponding portfolio process X_n we have $X_N = V_N$ almost surely, that is for all events in $\Omega_N(p)$.*

Proof This is done by backwards induction. Namely, the last step equation

$$V_N = X_N = \Delta_{N-1}S_N + (1+r)(X_{N-1} - \Delta_{N-1}S_{N-1}),$$

for $V_{N-1} = X_{N-1}, \Delta_{N-1}$, rewrites as a system of two equations:

$$V_N(\omega_1, \ldots, \omega_{N-1}, u)$$
$$= \Delta_{N-1}(\omega_1, \ldots, \omega_{N-1})S_N(\omega_1, \ldots, \omega_{N-1}, u) + (1+r)(V_{N-1}$$
$$\times (\omega_1, \ldots, \omega_{N-1}) - \Delta_{N-1}(\omega_1, \ldots, \omega_{N-1})S_{N-1}(\omega_1, \ldots, \omega_{N-1})),$$
$$V_N(\omega_1, \ldots, \omega_{N-1}, d)$$
$$= \Delta_{N-1}(\omega_1, \ldots, \omega_{N-1})S_N((\omega_1, \ldots, \omega_{N-1}, d) + (1+r)(V_{N-1}$$
$$\times (\omega_1, \ldots, \omega_{N-1}) - \Delta_{N-1}(\omega_1, \ldots, \omega_{N-1})S_{N-1}(\omega_1, \ldots, \omega_{N-1})),$$

with the unique solution

$$\Delta_{N-1}(\omega_1, \ldots, \omega_{N-1}) = \frac{V_N(\omega_1, \ldots, \omega_{N-1}, u) - V_N(\omega_1, \ldots, \omega_{N-1}, d)}{S_N(\omega_1, \ldots, \omega_{N-1}, u) - S_N(\omega_1, \ldots, \omega_{N-1}, d)},$$

$$V_{N-1}(\omega_1, \ldots, \omega_{N-1}) = \frac{1}{1+r}[\tilde{p}V_N(\omega_1, \ldots, \omega_{N-1}, u)$$
$$+ \tilde{q}V_N(\omega_1, \ldots, \omega_{N-1}, d)],$$

where $\tilde{p}, \tilde{q}$ are risk-neutral probabilities. Similarly for any $n < N$ we obtain

$$\Delta_n(\omega_1, \ldots, \omega_n) = \frac{V_{n+1}(\omega_1, \ldots, \omega_n, u) - V_{n+1}(\omega_1, \ldots, \omega_n, d)}{S_{n+1}(\omega_1, \ldots, \omega_n, u) - S_{n+1}(\omega_1, \ldots, \omega_n, d)}, \quad (6.48)$$

$$V_n(\omega_1, \ldots, \omega_n) = \frac{1}{1+r}[\tilde{p}V_{n+1}(\omega_1, \ldots, \omega_n, u) + \tilde{q}V_{n+1}(\omega_1, \ldots, \omega_n, d)].$$
$$(6.49)$$

The last equation rewrites as

$$V_n(\omega_1, \ldots, \omega_n) = \frac{1}{1+r}\tilde{\mathbf{E}}_n[V_{n+1}(\omega_1, \ldots, \omega_n, \omega_{n+1})]. \quad (6.50)$$

These equations define a unique sequence of adapted pairs $V_n = (X_n, \Delta_n)$. $\qquad \square$

It is natural to call the corresponding initial value $X_0 = V_0$ the *fair price* or *hedging price* of the *contingent claim* (or of the *derivative security*) specified by the random variable V_N, because, having such a capital at the initial moment of time, an investor will be able to fulfill the obligation (to pay out the amount V_N) at the final time N obtaining zero surplus, and moreover, it is the minimal amount with which this is possible.

Theorem 6.5.4 *At any time $n < N$, the fair or hedging price of the contingent claim specified by a random variable V_N is given by the following risk-neutral pricing formula:*

$$V_n = \tilde{\mathbf{E}}_n \frac{V_N}{(1+r)^{N-n}}, \quad (6.51)$$

and in particular, the initial price is

$$V_0 = \tilde{\mathbf{E}} \frac{V_N}{(1+r)^N}. \quad (6.52)$$

Proof This follows from the possibility to replicate V_N by a wealth process (Theorem 6.5.3) and the fact that any wealth process is a martingale under a risk-neutral law (Proposition 6.5.2). $\qquad \square$

Corollary 6.5.5 *If V_N depends only on S_N, the stock price at time N, the initial hedging price can be calculated via the following CRR (Cox–Ross–Rubinstein) formula:*

$$V_0 = \frac{1}{(1+r)^N} \sum_{n=0}^{N} \binom{N}{n} \tilde{p}^n \tilde{q}^{N-n} V_N(u^n d^{N-n}). \quad (6.53)$$

Moreover, in this case V_n and Δ_n depend only on S_n and the recursive formulas (6.48), (6.49) simplify to

$$\Delta_n(S_n) = \frac{V_{n+1}(S_n u) - V_{n+1}(S_n d)}{S_n(u - d)}, \tag{6.54}$$

$$V_n(S_n) = \frac{1}{1 + r}[\tilde{p}V_{n+1}(S_n u) + \tilde{q}V_{n+1}(S_n d)]. \tag{6.55}$$

Proof This is straightforward from (6.52) and (6.46). $\square$

Corollary 6.5.6 *The binomial model admits no arbitrage, that is, there does not exist a portfolio process $\{\Delta_n\}$ such that the corresponding wealth process $\{X_n\}$ with the initial $X_0 = 0$ has two properties: $X_N \geq 0$ everywhere on Ω_N and $X_N > 0$ for some $\omega \in \Omega_N$.*

Since for any $p > 0$ all events in $\Omega_N(p)$ have positive probabilities, the following formulation is equivalent.

Corollary 6.5.7 *For any $p > 0$, there does not exist a portfolio process $\{\Delta_n\}$ such that the corresponding wealth process $\{X_n\}$ with the initial $X_0 = 0$ has two properties: $X_N \geq 0$ almost surely in $\Omega_N(p)$ and $\mathbf{P}(X_N > 0) > 0$.*

Proof For any wealth process $\{X_n\}$, $X_0 = 0$ implies $\tilde{\mathbf{E}}X_N = 0$ by (6.52). Hence $\tilde{\mathbf{P}}(X_N > 0) > 0$ necessarily implies $\tilde{\mathbf{P}}(X_N < 0) > 0$, which contradicts to $X_N \geq 0$ everywhere. $\square$

Remark 6.5.8 An important modification of the standard model is the model with stochastic volatility and/or interest rates. In this model, S_{n+1} is obtained by multiplying S_n by either $u(\omega_1, \ldots, \omega_n)$ or $d(\omega_1, \ldots, \omega_n)$ and interest rate in period from n to $n+1$ is $r(\omega_1, \ldots, \omega_n)$, where d, u, r are given functions (random variables) on Ω_n. All formulas are similar, but risk-neutral probabilities become time dependent (and random).

6.5.6 Put-call parity

Consider two portfolios (r interest rates, continuously compounded, T maturity, K strike price) depending on one stock with prices S_0 and S_T initially and at maturity respectively):

A: one European call + cash Ke^{-rt},
B: one European put + one share.

Both are worth

$$\max(S_T, K) = (S_T - K)^+ + K = (K - S_T)^+ + S_T$$

at expiration. Hence, by the arbitrage pricing, their prices today should also coincide leading to the following *put-call parity* relation:

$$C_T + Ke^{-rT} = P_T + S_0.$$

If B is overpriced relative to A, the correct arbitrage strategy is to buy the security in portfolio A and short the securities in portfolio B.

Equivalent formulation: if F_T is the price of a forward contract to buy a share for K, then

$$C_T = F_T + P_T,$$

as at maturity

$$(S_T - K)^+ = (K - S_T)^+ + (S_T - K).$$

Consider a numeric example (taken from [8]). Suppose $S_0 = 31, K = 30, r = 10\%$ p.a. (continuously compound), T is 3 months.

The first scenario: $C_T = 3, P_T = 2.25$. Then the prices of the portfolios are:

A: $C_T + Ke^{-rT} = 3 + 30e^{-0.1\times3/12} = 32.26,$
B: $P_T + S_0 = 2.25 + 31 = 33.25 > 32.26.$

Arbitrage strategy: buy call and short put and stock generating a positive initial (up-front) flow

$$-3 + 2.25 + 31 = 30.25,$$

which grows to $30.25 \times e^{0.1\times0.25} = 31.02$ in three months. What can happen at time T?

If $S_T > 30$, you exercise the call (put is useless), and if $S_T < 30$, a put holder exercises put (your call is useless) so that in both cases you ends up buying one share for 30 (closing your positions on put and call) which is used to close the short position in stock. The net profit is $31.02 - 30.00 = 1.02$.

The second scenario: $C_T = 3, P_T = 1$. Then the price of A is the same 32.26, but the price of B becomes $1 + 31 = 32 < 32.26$, so that now A is overpriced.

Arbitrage strategy: short call and buy put and stock with initial investment

$$31 + 1 - 3 = 29,$$

which grows to $29.25 \times e^{0.1 \times 0.25} = 29.73$ in three months.

As above: if $S_T > 30$, the call is exercised (put is useless), and if $S_T < 30$, you exercise your put (call is useless). In both cases you sell one share for 30 closing all your positions. The net profit is $30 - 29.73 = 0.27$.

6.5.7 Put-call parity in capital structure (Merton's model for credit risk)

Assume that the assets of a company are financed by zero-coupon bonds and equity. The bonds mature in T years, when payment K is required. No dividends are paid earlier. Let A_0 and A_T denote the value of assets at time 0 and T respectively.

At maturity, if $A_T > K$, the equity holders choose to repay bondholders, otherwise they declare bankruptcy (bondholders become the owners). Thus the value of equity is $(A_T - K)^+ = C_T$ (which looks like European call option premium) and the value of dept (that bondholders receive) is $\min(A_T, K) = K - (K - A_T)^+ = K - P_T$ (K minus the European put premium). The present value is the sum of the values of equity and dept:

$$A_0 = C_T + [PV(K) - P_T],$$

where $PV(K) = e^{-rT}K$ is the present value of K (r being interest rate), or

$$C_T + PV(K) = P_T + A_0,$$

which is the *put-call parity* for options on the assets of the company.

6.5.8 Exotic options

A specific feature of the standard European put and call options is revealed in the dependence of the final premium on the final value of the stock price only.

In practice one can find lots of options with the final payoffs depending on the whole history of stock prices. These options with path dependent final payoffs are often referred to as *exotic options*. Calculations of the risk-neutral prices and replicating portfolios can still be performed by formulas (6.48) and (6.52). However, simplified algorithms based on (6.54), (6.63) are not applicable anymore, making calculations more elaborate. There is an extensive literature developing effective calculation algorithms for particular classes of these options.

Let us consider two examples of exotic options. As above, we denote by S_k, $k = 0, 1, \ldots$, the history of underlying stock prices.

Let $Y_n = \sum_{k=0}^{n} S_k$. The *Asian option* expiring at time N with the strike price K is defined via the final payoff

$$V_N = (\frac{1}{N+1} Y_N - K)^+.$$

Let $v_k(s, y)$ denote the price of this option at time k and $\delta_n(s, y)$ the number of shares of stock that should be held by the replicating portfolio at time k, if $S_k = s$, $Y_k = y$. In particular, $v_N(s, y) = (-K + y/(N+1))^+$. Adjusting the general formulas from Theorem 6.5.3 leads to the following recursive formulas:

$$v_k(y, s) = \frac{1}{r+1} [\tilde{p} v_{k+1}(y + us, us) + \tilde{q} v_{k+1}(y + ds, ds)],$$

$$\delta_k(y, s) = \frac{v_{k+1}(y + us, us) - v_{k+1}(y + ds, ds)}{s(u - d)}.$$

Now let time $m \in [1, N-1]$ be given. The *Chooser option* or '*As you like it*' *option* confers the right to receive either call or put at time m, both having maturity N and strike price K.

To price this option, we deduce from the put-call parity that

$$\mathbf{E}_m \frac{(S_N - K)^+}{(1+r)^{N-m}} = \mathbf{E}_m \frac{(K - S_N)^+}{(1+r)^{N-m}} + S_m - \frac{K}{(1+r)^{N-m}}$$

(risk-neutral expectations are assumed), and hence

$$\max \left(\mathbf{E}_m \frac{(S_N - K)^+}{(1+r)^{N-m}}, \mathbf{E}_m \frac{(K - S_N)^+}{(1+r)^{N-m}} \right)$$

$$= \mathbf{E}_m \frac{(K - S_N)^+}{(1+r)^{N-m}} + \left(S_m - \frac{K}{(1+r)^{N-m}} \right)^+.$$

Consequently, the time-zero price of the chooser is the sum of the time-zero price of the put with expiration N and strike K and of the call with expiration m and strike $K(1+r)^{-(N-m)}$.

6.5.9 Options on a dividend-paying stock

In a dividend-paying model, adapted random variables A_n with values in $(0, 1)$ are given, specifying the payment $A_n Y_n S_{n-1}$ at time n, where Y_n equals u or d that occurred at time n. After the dividend is paid, the stock price at time n becomes $S_n = (1 - A_n)Y_nS_{n-1}$. The wealth equation becomes

$$X_{n+1} = \Delta_n S_{n+1} + (1+r)(X_n - \Delta_n S_n) + \Delta_n A_{n+1} Y_{n+1} S_n$$
$$= \Delta_n Y_{n+1} S_n + (1+r)(X_n - \Delta_n S_n).$$

It turns out that the discounted wealth process is still a martingale under the risk-neutral measure (which are the same as in the model without dividends), that is,

$$\mathbf{E}_m \frac{X_n}{(1+r)^n} = \frac{X_m}{(1+r)^m}, \quad n \geq m,$$

implying that the risk-neutral pricing formula

$$V_n = \mathbf{E}_n \frac{V_N}{(1+r)^{N-n}}$$

for option pricing still applies. In fact, it is sufficient to show the first formula for $n = m + 1$, and we have

$$\mathbf{E}_m \frac{X_{m+1}}{(1+r)^{m+1}} = \frac{\Delta_m S_m}{(1+r)^{m+1}} \mathbf{E} Y_{m+1} + \frac{X_m - \Delta_m S_m}{(1+r)^m}$$
$$= \frac{\Delta_m S_m}{(1+r)^m} + \frac{X_m - \Delta_m S_m}{(1+r)^m} = \frac{X_m}{(1+r)^m},$$

as required.

Though, generally speaking, the discounted stock price is not a martingale anymore, it is straightforward to see that if $A_n = a$ is a constant, then

$$\frac{S_n}{(1-a)^n(1+r)^n}$$

is a martingale.

6.5.10 Setting up (or fitting to the data) a binomial model

Clearly the binomial model, being very simple, is at the same time rather artificial. Hence a problem arises to build a reasonable binomial model based on market data.

Given a stock with volatility (of its logarithm) σ and the rate ν (continuously compound) of expected growth, let us find appropriate u, d, p

depending on a chosen period of time δ (important: use the same units of time for all data). That is,

$$Var\left(\log\frac{S_\delta}{S_0}\right) = \sigma^2\delta,$$

$$\mathbf{E}\left(\log\frac{S_\delta}{S_0}\right) = \nu\delta.$$

Using that the variance of a Bernoulli random variable with two values A and B is $p(1-p)(A-B)^2$ yields

$$\begin{cases} p(1-p)(\log u - \log d)^2 = \sigma^2\delta, \\ p\log u + q\log d = \nu\delta. \end{cases} \tag{6.56}$$

These are two equations for three unknowns.

In practice one use the following well-established constraint to fix the uncertainty: $d = 1/u$. This yields

$$\begin{cases} 4p(1-p)(\log u)^2 = \sigma^2\delta, \\ (2p-1)\log u = \nu\delta. \end{cases} \tag{6.57}$$

Squaring the second equation and adding the first yields

$$(\log u)^2 = \sigma^2\delta + \nu^2\delta^2.$$

This implies, for small δ, that approximately $\log u = \sigma\sqrt{\delta}$, yielding

$$u = e^{\sigma\sqrt{\delta}}, \quad d = e^{-\sigma\sqrt{\delta}}. \tag{6.58}$$

Substituting to the second equation of (6.57) yields

$$p = \frac{1}{2} + \frac{1}{2}\frac{\nu}{\sigma}\sqrt{\delta}. \tag{6.59}$$

Alternative fitting (seemingly more used) can be performed as follows: instead of expected growth of the logarithm of the price jump, one fits the growth of the jump itself:

$$\begin{cases} p(1-p)(\log u - \log d)^2 = \sigma^2\delta, \\ pu + (1-p)d = e^{\nu\delta} = 1+r, \end{cases} \tag{6.60}$$

where r is a usual growth rate for a binomial model (simply compound). Assuming u and d are close to 1, so that approximately

$$u = 1 + \log u, \quad d = 1 + \log d,$$

reduces (6.60) to (6.56) leading to the same approximation (6.58) for u, d, but now the second equation of (6.60) yields

$$p = \frac{1+r-d}{u-d} = \frac{e^{\nu\delta}-d}{u-d}. \tag{6.61}$$

These are actual (*real world*) probabilities as long as r denotes the expected rates specific to the stock. If r here were the risk-neutral rate (on a bank account), formula (6.61) would give the usual risk-neutral probability $\tilde{p}$.

If u, d are given by (6.58) and δ is small, (6.61) reduces to (6.59).

6.5.11 Deflator: connecting actual and risk-neutral probability

For our usual probability space $\Omega_N = \{\omega = (\omega_1, \ldots, \omega_N)\}$, with ω_i being u or d, let $\mathbf{P}$ and $\tilde{\mathbf{P}}$ be actual and risk-neutral probabilities:

$$\mathbf{P}(\omega) = p^{\#u(\omega)} q^{\#d(\omega)}, \quad \tilde{\mathbf{P}}(\omega) = \tilde{p}^{\#u(\omega)} \tilde{q}^{\#d(\omega)}, \tag{6.62}$$

where $\#u(\omega)$ (resp. $\#d(\omega)$) denotes the number of u (resp. d) in the string ω, $p \in (0,1)$ and $\tilde{p} \in (0,1)$ denote respectively the actual and the risk-neutral probability of u in a single period, $q = 1 - p$, $\tilde{q} = 1 - \tilde{p}$.

Denote by $\mathbf{E}$ and $\tilde{\mathbf{E}}$ the corresponding expectations and by $Z(\omega) = \tilde{\mathbf{P}}(\omega)/\mathbf{P}(\omega)$ the (Radon–Nikodym) density. Clearly

$$\mathbf{E}Z = \sum_{\omega \in \Omega} \frac{\tilde{\mathbf{P}}(\omega)}{\mathbf{P}(\omega)} \mathbf{P}(\omega) = 1,$$

and, for any random variable Y,

$$\tilde{\mathbf{E}}(Y) = \sum_{\omega \in \Omega} \frac{\tilde{\mathbf{P}}(\omega)}{\mathbf{P}(\omega)} \mathbf{P}(\omega) Y(\omega) = \mathbf{E}(ZY).$$

The *deflator* (or the *state price density* or the *state price deflator*) in the N period model is defined by the formula

$$A_N(\omega) = \frac{Z(\omega)}{(1+r)^N},$$

so that by (6.62),

$$A_N(\omega) = \frac{1}{(1+r)^N} \left(\frac{\tilde{p}}{p}\right)^{\#u(\omega)} \left(\frac{\tilde{q}}{q}\right)^{\#d(\omega)}.$$

Thus the risk-neutral evaluation formula for an option specified by a final payoff V_N rewrites as

$$\tilde{E}\frac{V_N}{(1+r)^N} = \mathbf{E}(A_N V_N).$$

In particular, if V_N depends only on the price of the stock at time N, it implies the following version of CRR formula in the real-world probabilities:

$$C_N = \sum_{n=0}^{N} \binom{N}{n} A_N p^n q^{N-n} V_N(u^n d^{N-n}).$$

6.5.12 Trading with options

By a *spread* in a trading jargon one means a collection of positions on stocks and options taken simultaneously.

The spreads can be *vertical*, if all positions have the same expiration date, *horizontal*, if only options with the same strike are included, or *diagonal*, if both strike and maturity are varying.

It is useful to have in mind various trading wordings used for the same action: to write = to sell = to short = to take a short position in, or to purchase = to buy = to take a long position in. Option is said to be *in the money* (ITM), if it would be profitable to exercise it at this moment (if allowed), *out of the money* (OTM), if it would not be profitable to exercise it at this moment, *at the money* (ATM), if it would produce neither profit nor loss. A trade is a *net credit trade* if the maximum award is received at the beginning (but could be lost afterwards) and a *net debit trade*, if the maximum risk (loss) is the amount paid at entering the spread.

For understanding the mechanism of option trading, it is useful to review the pro and contra of basic option trading strategies (spreads). We shall consider some examples of low risk vertical spreads (sometimes called 'income generating' by traders), where the maturity T is fixed, a stock is chosen with values S_0 (initial, known) and S_T (final, unknown), C_K and P_K denote the prices of European call and put with strike K:

- *Naked put*: short a put. This is a net credit trade with bullish outlook (you expect the price of a stock to grow). The maximum reward is P_K (you receive it up front), if $S_T > K$ and put becomes worthless at maturity. The maximum risk is $K - P_K$, if the stock becomes worthless at maturity, i.e. $S_T = 0$. Trading advise: use naked put spread OTM on a stock you'd like to have see Fig. 6.6.

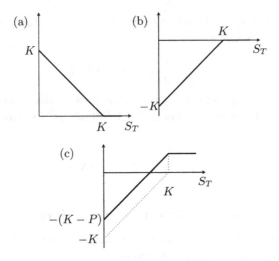

Figure 6.6 (a) Put $f = \max(K - S_T, 0)$, (b) short (naked) put, (c) naked put with income.

- *Covered call*: buy a share + sell OTM call (i.e. with $K > S_0$). This is a net debit transaction with bullish outlook. The maximum reward is $K - S_0 + C_K$, if $S_T > K$ and call is exercised. The maximum risk is $S_0 - C_K$ (that you paid up front), if the stock becomes worthless at maturity, i.e. $S_T = 0$.

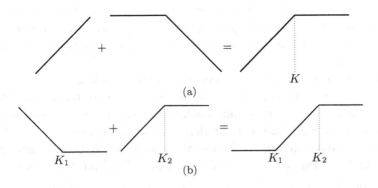

Figure 6.7 (a) Covered call, (b) bull put spread.

- *Bull put spread*: buy K_1-put and sell K_2-put with $K_1 < S_0 < K_2$. This is a net credit trade with bullish outlook, but with capped loss as compared with a naked put. The maximum reward is $P_{K_2} - P_{K_1}$

(net credit received), if $S_T > K_2$ and both puts become worthless at maturity. The maximum loss is $(K_2 - K_1) - (P_{K_2} - P_{K_1})$, if $S_T < K_1$ and both puts are exercised.

- *Long iron condor*: buy OTM K_1-put, sell OTM lower middle strike K_2-put, sell OTM higher middle strike K_3-call, buy OTM K_4-call with $K_1 < K_2 \leq S_0 \leq K_3 < K_4$. If $K_2 = K_3$ it is also called the *long iron butterfly*. This is a net credit trade, used when one expects S_T to stay near the interval $[K_2, K_3]$. The maximum reward is $R = (P_{K_2} - P_{K_1}) + (C_{K_3} - C_{K_4})$, if $S_T \in [K_2, K_3]$ and all four options become worthless at maturity. The maximum risk is $max(K_2 - K_1, K_4 - K_3) - R$.

The following are two examples of *volatility strategies*, used when you expect a move, but not sure in what direction, say after a tip on a big expectation, where disappointment would also imply a move:

- *Straddle*: buy ATM put + buy ATM call, i.e. both with $K = S_0$. This is a net debit trade with maximum risk $C_K + P_K$ and uncapped reward tending to ∞ as $S_T \to \infty$;
- *Strangle*: buy OTM K_1-put + buy OTM K_2-call, i.e. $K_1 < S_0 < K_2$ (turns to straddle if $K_1 - K_2 \to 0$). This is a net debit trade with maximum risk $P_{K_1} + C_{K_2}$ and uncapped reward tending to ∞ as $S_T \to \infty$.

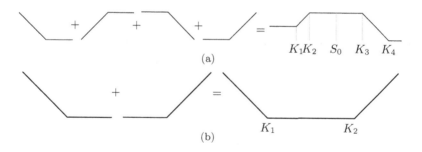

Figure 6.8 (a) Long iron condor, (b) strangle.

6.5.13 American options

We worked all the time with the so-called *European option*. This general term refers to options with payoffs only at the prescribed expiration date. In other words, the option holder has the right to exercise this option only at the expiration date. There exists a completely different class of options, referred generally to as *American options*, which gives to the holder the right to exercise it any time before the maturity. Thus, an American put confers to the holder the right to sell a share of a given stock at any time before the maturity N, for a fixed price K. The inductive approach for option prices developed in Theorem 6.5.3 has a natural extension to American options. Say, for the American put, the recursive pricing formula (6.63) for European options turns to the formula

$$V_n(S_n) = \max\left((K - S_n)^+, \frac{1}{1+r}[\tilde{p}V_{n+1}(S_n u) + \tilde{q}V_{n+1}(S_n d)] \right),$$
(6.63)

with $V_N = (K - S_N)^+$ for the expiration time N.

6.6 Introducing the central limit theorem and fat tails

6.6.1 Generating functions

Let us recall shortly the definitions and the main properties of generating functions.

For a random variable X its *generating function* (called also the *probability generating function*) $G_X(s)$ is defined for non-negative s as

$$G_X(s) = \mathbf{E}(s^X).$$

Generating functions represent a useful tool for working with random variable taking values in $\mathbf{Z}_+$ – the set of non-negative integers.

For continuous random variables X, more handy for analysis turns out to be other equivalent characteristics, namely the *moment generating function* M_X and the *Laplace transform* L_X, defined as

$$M_X(t) = \mathbf{E}e^{Xt}, \quad L_X(t) = \mathbf{E}e^{-Xt}.$$
(6.64)

Clearly

$$M_X(t) = L_X(-t) = G_X(e^t).$$

From the definition it follows straightforwardly that the moments $\mathbf{E}(X^k)$ are found from the derivatives of M_X as

$$M_X^{(k)}(0) = \mathbf{E}(X^k),$$

and expanding the exponential function in the Taylor series yields

$$M_X(t) = \sum_{k=0}^{\infty} \frac{t^k}{k!}\mathbf{E}(X^k)$$
(6.65)

(hence the name 'moment generating function').

If X is normal $N(\mu, \sigma^2)$, that is, its probability density function is given by (6.10), the well-known (and simple) calculations show that

$$M_X(t) = \mathbf{E}(e^{Xt}) = \exp\{t\mu + t^2\sigma^2/2\}. \tag{6.66}$$

The converse holds too (though this is not so obvious). Namely, if a random variable has the moment generating function of form (6.66), then it is normal $N(\mu, \sigma^2)$.

The key property of moment generating functions is given by the following result. If $X_1, X_2, \ldots, X_n$ is a finite sequence of independent random variables with the moment generating functions $M_{X_i}(s)$, the moment generating function of the sum $Y_n = X_1 + \cdots + X_n$ equals the product

$$M_{Y_n} = M_{X_1}(s) \cdots M_{X_n}(s). \tag{6.67}$$

6.6.2 Asymptotic normality

Let us sketch the deduction of the so-called *central limit theorem* (CLT) (representing one of the major strongholds of the probability theory), which states that the sum of independent identically distributed (i.i.d.) random variables becomes *asymptotically normal* as the number of terms increases. If they are properly normalized, then the limiting distribution is the standard normal one. Namely, assume $X_1, X_2, \ldots$ are i.i.d. random variables with $\mu = \mathbf{E}(X_i)$, $\sigma^2 = Var(X_i)$. Let

$$Y_n = \frac{X_1 + \cdots + X_n - n\mu}{\sigma\sqrt{n}} = \frac{X_1 - \mu}{\sigma\sqrt{n}} + \cdots + \frac{X_n - \mu}{\sigma\sqrt{n}}. \tag{6.68}$$

By linearity $\mathbf{E}(Y_n) = 0$ and $Var(Y_n) = 1$. The *central limit theorem* states that Y_n converges to the standard normal random variable in the sense that

$$\lim_{n\to\infty} \mathbf{P}(a \le Y_n \le b) = \frac{1}{\sqrt{2\pi}} \int_a^b \exp\{-\frac{x^2}{2}\} \, dx \tag{6.69}$$

for any $a < b$. To see why this result is plausible let us show that the moment generating function M_{Y_n} of Y_n converges to the moment generating function (6.66) (to transform this argument into a rigorous proof one would have of course to show that from this convergence one can deduce the convergence (6.69), which is beyond our scope). From (6.68) and using that the expectation of a product of i.i.d. random variables equals the product of their expectations it follows that

$$M_{Y_n}(t) = \left[M_{(X_1-\mu)/\sigma\sqrt{n}}(t) \right]^n.$$

By (6.65)

$$M_{(X_1-\mu)/\sigma\sqrt{n}}(t) = 1 + \frac{t^2}{2n} + \frac{\alpha(n)}{n}$$

with a certain function α that tends to zero as $n \to \infty$. Hence

$$M_{Y_n}(t) = \left[1 + \frac{t^2 + \alpha(n)}{2n} \right]^n \to \exp(t^2/2),$$

as required.

Since Y_n is asymptotically $N(0,1)$ normal, the sum $Z_n = X_1 + \cdots + X_n$ is asymptotically $N(n\mu, n\sigma^2)$ normal and the mean Z_n/n is asymptotically $N(\mu, \sigma^2/n)$ normal. Consequently, approximately, as $n \to \infty$, one gets the following equation allowing to calculate the distributions of arbitrary sums of i.i.d. random variables:

$$\mathbf{P}(Z_n \le x) = \Phi\left(\frac{x - n\mu}{\sigma\sqrt{n}} \right), \tag{6.70}$$

where Φ is the distribution function of a standard normal random variable.

Example: biased betting   Suppose you are betting on independent Bernoulli trials with success probability slightly less than $1/2$, say $p = 0.497$. Let Y_n be the total number of winning in n experiments (similar model would work in many situations, say when estimating the number of males or females born in a population in a given period). As p is close to $1/2$ one can expect that $P_n = \mathbf{P}(Y_n \ge n/2)$ should be slightly less than $1/2$. Can you guess an approximation, say for $n = 500\ 000$? By the asymptotic normality, Y_n is approximately $N(np, np(1-p)) = N(248\ 500, 354^2)$. Hence approximately

$$P_n = 1 - \mathbf{P}(Y_{500\ 000} \le 250\ 000) = 1 - \Phi\left(\frac{250\ 000 - 248\ 500}{354} \right)$$

$$= 1 - \Phi(4.24) = 0.000\ 01.$$

'Slightly' less than $1/2$ indeed!

6.6.3 Fat (or heavy) tails

The results of the previous section can lead to a misleading impression that essentially all distributions met in real life are normal. In fact, the practical applications of normal laws often go far beyond the situations, where this application can be justified by the above given central limit theorem. This often leads to superficial conclusions, based on, what one can call, the *prejudice of normality*.

Experiments with concrete statistical data give clear evidence that many real-life distributions (including in particular stock prices) have quite different qualitative features than Gaussian. In particular, they are often *heavy-tailed*, i.e. unlike the *light tails* of Gaussian laws, where $\mathbf{P}(X > x)$ decreases exponentially as $t \to \infty$, the random variable X with *heavy (or fat) tails* has a property that $\mathbf{P}(X > x)$ decreases as a power $x^{-\omega}$ as $x \to \infty$ for some $\omega > 0$. We are going to use the method of generating functions in order to show how heavy tailed distributions can appear as the limits of the sums of i.i.d. random variables when their second moment is not finite. We shall reduce our attention to positive random variable X (modeling, say, waiting times) such that

$$\mathbf{P}(X > x) \sim \frac{c}{x^\alpha}, \quad x \to \infty. \tag{6.71}$$

In other words, $\lim_{x \to \infty} \mathbf{P}(X > x)\, x^\alpha = c$ with some $c > 0$, $\alpha \in (0, 1)$. As one easily sees, even the expectation $\mathbf{E}(X)$ is not defined (it equals infinity) in this case, so that already the law of large numbers can not be applied. For positive random variables it is usually more convenient to work with the Laplace transform defined by the second equation in (6.64), rather than with the moment generating function.

Proposition 6.6.1 Let $X_1, X_2, \ldots$ be a sequence of positive i.i.d. random variable with the probability density function $p(x)$, $x \in (0, \infty)$, such that

$$\mathbf{P}(X_1 > x) = \int_x^\infty p(y)\, dy \sim \frac{c}{x^\alpha}, \quad x \to \infty, \tag{6.72}$$

with some $c > 0$, $\alpha \in (0, 1)$. Then the Laplace transforms $L_{Y_k}(t)$ of the normalized sums $Y_k = (X_1 + \cdots + X_k)/k^{1/\alpha}$ converge, as $k \to \infty$, to $\exp\{-c\beta_\alpha t^\alpha\}$ with a certain constant $\beta_\alpha > 0$.

Proof First one notes that, for a bounded differentiable function $g : [0, \infty) \to [0, \infty)$ such that $g'(0) = 0$ and $\int_1^\infty g(y) y^{-(1+\alpha)}\, dy < \infty$, it follows from (6.72) that

$$\lim_{k \to \infty} k \int_0^\infty g\left(\frac{x}{k^{1/\alpha}}\right) p(x)\, dx = \alpha c \int_0^\infty g(y) y^{-(1+\alpha)}\, dy.$$

In fact, by linearity and approximation, it is enough to show this for the indicators $\mathbf{1}_{[a,\infty)}$ with any $a > 0$. But in this case

$$k \int_0^\infty \mathbf{1}_{[a,\infty)} \left(\frac{x}{k^{1/\alpha}}\right) p(x)\, dx = k \int_{ak^{1/\alpha}}^\infty p(x)\, dx \to \frac{c}{a^\alpha}$$

$$= \alpha c \int_a^\infty y^{-(1+\alpha)}\, dy,$$

as required. Now one can follow the same arguments as for the central limit theorem above. Namely,

$$L_{Y_k}(t) = \left[L_{X_1}\left(\frac{t}{k^{1/\alpha}}\right)\right]^k = \left[\int_0^\infty \exp\{-\frac{xt}{k^{1/\alpha}}\}p(x)\, dx\right]^k$$

$$= \left[1 + \int_0^\infty (\exp\{-\frac{xt}{k^{1/\alpha}}\} - 1)p(x)\, dx\right]^k$$

$$= \left[1 + \alpha c \int_0^\infty (e^{-tx} - 1)\frac{dx}{x^{1+\alpha}}\frac{1 + \omega(k)}{k}\right]^k,$$

where $\omega(k) \to 0$, as $k \to \infty$. Consequently, since by the change of variables

$$\int_0^\infty (e^{-tx} - 1)\frac{dx}{x^{1+\alpha}} = t^\alpha \int_0^\infty (e^{-x} - 1)\frac{dx}{x^{1+\alpha}},$$

one concludes that

$$L_{Y_k}(t) \to \exp\{-c\beta_\alpha t^\alpha\},$$

as required with $\beta_\alpha = \alpha \int_0^\infty (1 - e^{-x})x^{-(1+\alpha)}\, dx$. $\qquad\square$

Remark 6.6.2 By the integration by parts

$$\beta_\alpha = \alpha \int_0^\infty (1 - e^{-x})\frac{dx}{x^{1+\alpha}} = \int_0^\infty x^{-\alpha}e^{-x}\, dx,$$

so that $\beta_\alpha = \Gamma(1 - \alpha)$, where Γ denotes the Euler Gamma function.

Positive random variables (and their distributions), whose Laplace transform equals $\exp\{-c't^\alpha\}$ with some constants $c' > 0$ and $\alpha \in (0, 1)$ are called α-*stable* with the *index of stability* α. One can show that such random variable X satisfies condition (6.71). In particular, it has fat tails. The term 'stable' stems from the observation that the sum of independent copies of such a random variable belongs to the same class, because the Laplace transform of this sum equals $\exp\{-c'nt^\alpha\}$, where n denotes the number of terms.

The above proposition shows that not only Gaussian random variables can describe the limits of the sums of i.i.d. random variables. Namely, one says that a random variable X belongs to the *domain of attraction* of an α-stable law, if (6.71) holds. By Proposition 6.6.1, the distributions of the normalized sums of i.i.d. copies of such random variables converge to the α-stable law. Unfortunately, the probability density functions of stable laws with $\alpha \in (0,1)$ can not be expressed in a closed form (in elementary function), which makes their analysis more involved than, say, of Gaussian or exponential distributions.

6.7 Black–Scholes option pricing

We shall sketch here the celebrated Nobel prize winning Black–Sholes option pricing theory that extends the CCR theory (binomial model) to continuous-time models. But first we prepare a couple of tools introducing log-normal laws and equivalent measures.

6.7.1 Log-normal random variables

A random variable X is called *log-normal* if its logarithm $Y = \ln X$ is normal $N(\mu, \sigma^2)$.

Suppose X is log-normal, so that $X = e^Y$ with Y being normal $N(\mu, \sigma^2)$. By (6.66),

$$\mathbf{E}(X) = \exp\{\mu + \sigma^2/2\}. \tag{6.73}$$

The following calculations form the basis for the deduction of the Black–Scholes option pricing formula.

Proposition 6.7.1 Suppose X is log-normal, so that $X = e^Y$ with a normal $N(\mu, \sigma^2)$ random variable Y, and let K be a positive constant. Then

$$\mathbf{E}\max(X - K, 0) = \tilde{\mu}\Phi\left(\frac{\ln(\tilde{\mu}/K) + \sigma^2/2}{\sigma}\right) - K\Phi\left(\frac{\ln(\tilde{\mu}/K) - \sigma^2/2}{\sigma}\right), \tag{6.74}$$

where

$$\tilde{\mu} = \mathbf{E}(X) = \exp\{\mu + \sigma^2/2\}.$$

Proof By (6.13),

$$\mathbf{E}\max(X - K, 0) = \mathbf{E}\max(\exp\{\mu + \sigma Z\} - K, 0)$$

with a standard $N(0,1)$ random variable Z. Consequently

$$\mathbf{E}\max(X - K, 0) = \int_{(\ln K - \mu)/\sigma}^{\infty} (e^{\mu + \sigma x} - K)\frac{1}{\sqrt{2\pi}} \exp\left\{-\frac{x^2}{2}\right\} dx$$

$$= \tilde{\mu} \int_{(\ln K - \mu)/\sigma}^{\infty} \frac{1}{\sqrt{2\pi}} \exp\left\{-\frac{(x-\sigma)^2}{2}\right\} dx$$

$$- K \int_{(\ln K - \mu)/\sigma}^{\infty} \frac{1}{\sqrt{2\pi}} \exp\left\{-\frac{x^2}{2}\right\} dx,$$

which rewrites as (6.74). $\hspace{3cm}$ □

6.7.2 Equivalent normal laws

Let Z be a strictly positive random variable on a probability space $(\Omega, \mathcal{F}, \mathbf{P})$ ($\mathcal{F}$ is the set of events and $\mathbf{P}$ the probability law on it) with the expectation 1, that is, $\mathbf{E}Z = 1$. We can then define another probability law on $\mathcal{F}$ via the equation

$$\tilde{\mathbf{P}}(A) = \int_A Z(\omega) d\mathbf{P}(\omega), \tag{6.75}$$

for all $A \in \mathcal{F}$, or, equivalently, denoting by $\tilde{E}$ the corresponding expectation, via the equation

$$\tilde{E}Y = \mathbf{E}(ZY) \tag{6.76}$$

for all random variables Y. Probability laws $\mathbf{P}$ and $\tilde{P}$ are called *equivalent* meaning that an event has zero probability in one of them if and only if it has zero probability in another. The random variable Z is referred to as the *density* (or Radon–Nykodime density) of $\mathbf{P}$ with respect to $\mathbf{P}$. In discrete setting this notion was already exploited when discussing deflators.

We shall need only equivalent normal laws, that is, the following fact.

Proposition 6.7.2 Let X be $N(\mu, \sigma^2)$ normal random variable defined on a probability space with the law $\mathbf{P}$. Let $\theta > 0$ and the equivalent measure $\tilde{P}$ be defined from the density

$$Z = \exp\{-\theta(X - \mu) - \frac{1}{2}\theta^2\sigma^2\}.$$

Then X is normal $N(\mu - \theta\sigma^2, \sigma^2)$ under $\tilde{\mathbf{P}}$.

Proof The proof will be based on the fact (already mentioned above) that a random variable has the moment generating function of form (6.66) if and only if it is normal $N(\mu, \sigma^2)$. Thus, to identify X as a normal random variable under $\tilde{\mathbf{P}}$, we will calculate its moment generating function.

Let us note also for completeness that $\mathbf{E}Z = 1$, as follows from (6.66) and as it should be to interpret Z as a density.

We have

$$\tilde{\mathbf{E}}(e^{Xt}) = \mathbf{E}(e^{Xt}Z) = \exp\{\theta\mu - \frac{1}{2}\theta^2\sigma^2\}\mathbf{E}\exp\{(t-\theta)X\}.$$

Using (6.66), this yields

$$\tilde{\mathbf{E}}(e^{Xt}) = \exp\{\theta\mu - \frac{1}{2}\theta^2\sigma^2\}\exp\{(t-\theta)\mu + \frac{1}{2}(t-\theta)^2\sigma^2\}$$

$$= \exp\{t(\mu - \theta\sigma^2) + \frac{1}{2}t^2\sigma^2\},$$

as required.   □

6.7.3 Continuous-time models for asset pricing

Thinking about stock prices as a result of a large number of influences of various sizes and sources, one naturally comes to the analogy with a small particle put in a liquid and moving under a constant bombardment of the immense number of molecules of the liquid. The erratic behavior of such a particle was observed and protocolled by R. Brown and is called Brownian motion. The analogy of share prices with Brownian motion was exploited by Bachelier at the dawn of the twentieth century. Referring to the central limit theorem (see Section 6.6), Bachelier assumed the prices S_T of a stock at time T to be normally distributed. However, the assumption of normality (distribution on all numbers, positive and negative) clearly contradicts the crucial positivity property of prices. Later on, the Bachelier model was improved by assuming that the rates of changes are normal, i.e. the logarithm of prices $\ln S_T$ is normal, in other words S_T themselves are log-normal. This was the starting point for the Black–Sholes theory.

Suppose now that a random variable X_t can be considered as the sum of independent inputs acting at shorter times, i.e. one can write

$$X_1 = X_t^1 + \cdots + X_t^n, \quad t = 1/n,$$

for any integer n with i.i.d random variable X_t^i, distributed like X_t. By the linearity of the expectation and variance for independent random variable it follows that

$$\mathbf{E}X_1 = n\mathbf{E}(X_t) = \frac{1}{t}\mathbf{E}(X_t), \quad Var(X_1) = n\,Var(X_t) = \frac{1}{t}\,Var(X_t),$$

so that

$$\mathbf{E}X_t = t\mathbf{E}(X_1), \quad Var(X_t) = t\,Var(X_1),$$

leading to the conclusion that the expectation and variance of such X_t should depend linearly on time. Applying this to $\ln S_T$ allows us to make our model for stock prices more precise by assuming that $\ln(S_T/S_0)$ is normal $N(\mu t, \sigma^2 t)$ and thus leaving only two parameters μ, σ to be specified (say, by observed data, see Section 6.2.2). The quadratic deviation σ in this model is usually referred to as the *volatility* of the stock.

6.7.4 Black–Scholes theory

European options for models with continuous time are defined in the same way as for discrete time models. Namely, a standard *European call option* confers to the holder the right (but unlike a futures contract, not the obligation) to buy a unit stock by certain time T in the future, called the *expiration date* or *maturity*, for a fixed price K, called the *strike price* or the *exercise price*. Equivalently, this means that at expiration the seller has to pay to the holder the premium $\max(S_T - K, 0)$.

We are interested in the fair price of an option at the initial time 0, assuming that the price S_T of the underlying asset at expiration is log-normal, more precisely that $\log(S_T/S_0)$ is normal $N(\mu T, \sigma^2 T)$. As in the discrete case, one assumes a fixed risk-free interest rate r, though now continuously compounded, i.e. borrowing the amount V, you have to return $e^{rt}V$ after time t, or equivalently, the discounted present cost of the amount V paid at the time t is $e^{-rt}V$).

We shall work by the analogy with the discrete case (the full story requires the heavy use of Ito's stochastic calculus). In discrete setting we found that the fair price of an option is given by the expectation of the discount final payoff with respect to the risk-neutral measure, for which the discount stock price is a martingale. In particular, the expectation of the stock return equals the return on the risk-free account, that is $\tilde{E}(S_N/S_0) = (1 + r)^N$. The risk-neutral probability and the real world probability were equivalent, in fact they were linked by the density, whose discounted value was called the state deflator. By the analogy, we are going to price European option in continuous time as the expectation $\tilde{E}$ of the discounted final payoff $\max(S_T - K, 0)$ under the normal law, which is equivalent to the original (real world) law $N(\mu T, \sigma^2 T)$, but for which the expectation of the stock return equals the return on the risk-free account, that is, $\tilde{E}(S_T/S_0) = e^{rT}$. According to Proposition 6.7.2 this can be easily achieved by using an appropriate exponential density. Moreover, under this change of probabilities, the variance $\sigma^2 T$ (and hence the volatility σ) remains unchanged. Finally we conclude that the fair price is given by the formula

$$e^{-rT} \tilde{E} \left(\max(S_T - K, 0) \right),$$

with respect to law $\tilde{E}$, under which $S_T = S_0 e^X$ with X being normal with variance $\sigma^2 T$ and such that $\tilde{E} e^X = e^{rT}$. Applying Proposition 6.7.1 yields

$$c = e^{-rT}[S_0 e^{rT} \Phi(d_1) - K\Phi(d_2)] = S_0 \Phi(d_1) - Ke^{-rT}\Phi(d_2), \quad (6.77)$$

where

$$d_1 = \frac{\ln \hat{E}(S_T)/K + \sigma^2 T/2}{\sigma\sqrt{T}} = \frac{\ln(S_0/K) + (r + \sigma^2/2)T}{\sigma\sqrt{T}},$$

$$d_2 = \frac{\ln \hat{E}(S_T)/K - \sigma^2 T/2}{\sigma\sqrt{T}} = \frac{\ln(S_0/K) + (r - \sigma^2/2)T}{\sigma\sqrt{T}}.$$

This is the celebrated *Black–Scholes formula*.

6.8 Credit risk and Gaussian copula

6.8.1 Probability of survival and hazard rate

Default time τ for a company are usually assessed (or stored) by the *cumulative probabilities of survival*:

$$V(t) = \mathbf{P}(\tau > t)$$

(no default till time t).

Suppose τ is a positive random variable with a differentiable distribution function. Then

$$\mathbf{P}(\tau > t | \tau > s) = \mathbf{P}(\tau > t)/\mathbf{P}(\tau > s).$$

Taking the logarithm and differentiating we obtain

$$\mathbf{P}(\tau > t | \tau > s) = e^{-\int_s^t \lambda(u)\, du}, \quad s \le t,$$

where $\lambda(u) = -\frac{d}{dt} \log \mathbf{P}(\tau > t)$ is a positive function. When τ denotes a default time, $\lambda(u)$ is called *default intensity* or *hazard rate*.

Thus hazard rates and probabilities of survival are linked by the differential equation

$$\frac{dV(t)}{dt} = -\lambda(t)V(t). \quad (6.78)$$

Approximately, for small s

$$\lambda(t) \sim \frac{V(t) - V(t+s)}{V(t)} = \mathbf{P}(\tau \in [t, t+s]|\tau > t),$$

so that the intensity specifies the conditional default probabilities on short intervals. From (6.78) one deduces that

$$V(t) = \exp\{-\int_0^t \lambda(s)\, ds\},$$

so that the distribution function of τ equals

$$F_\tau(t) = \mathbf{P}(\tau \le t) = 1 - \exp\{-\int_0^t \lambda(s)\, ds\}. \tag{6.79}$$

If a firm goes bankrupt, its creditors file claims against its assets to recover some of its dept. The *recovery rate* R for a bond is defined as the bond's market price immediately after default as a percent of its face value. It is reasonable to assume (at least for rough estimates) that the possibility of a default is the only reason for higher return rates of the bonds issued by a firm as compared to the corresponding risk-free bonds (say, issued by the government). Consequently, if s denotes the additional interest on a firm's bond as compared with an equivalent risk-free bond, and if p denotes the default probability per year so that $p(1-R)$ is the average rate of loss due to default, one should have

$$s = p(1-R),$$

yielding an important estimate for the default probability

$$p = s/(1-R) \tag{6.80}$$

via the prices of bonds the company has issued.

6.8.2 Credit default swaps (CDS)

Suppose you keep a portfolio of bonds belonging to 100 different corporations, and the probability of a default during the next three years for each of these corporations equals 0.02 (i.e. 2%). So-called *credit default swaps* (CDS) allows one to buy a protection against the first or, say, the tenth default inside your portfolio (default means a crucial devaluation of the bond and/or of the corresponding payments). If the firms are independent, the number of defaults are distributed like a binomial $(100, 0.02)$ random variable, implying that the probability of at least one (respectively at least 10) defaults equals 0.867 (respectively 0.003). Thus a first-to-default CDS should be valuable, but the ten-to-default CDS is worth almost nothing. On the other hand, if the defaults of your corporations are perfectly correlated, i.e. they occur simultaneously, then the probabilities of at least one or at least 10 (or 50) defaults coincide and equal 0.02.

This brief discussion shows the crucial importance of modeling and measuring the dependence of random variables.

6.8.3 Transformation of probability laws

The full knowledge of the structure of the dependence between the defaults of various firms requires specifying joint default probabilities, which are difficult to assess in practice. Only for normal or Gaussian random variables the dependence structure can be incorporate in a few key parameters (say, for two standard Gaussian random variables, only one parameter, their correlation coefficient, specifies the joint distribution uniquely).

However, the default times are principally different from Gaussian random variables (in particular, because they are positive), the idea is to transform them into Gaussians and then to specify the dependence structure on these Gaussian images.

Let us recall the corresponding results from elementary probability.

Proposition 6.8.1 Let $F(x)$ be a continuous and strictly increasing function on $\mathbf{R}$ (the latter meaning that $x < y$ always implies $F(x) < F(y)$) such that $\lim_{x \to \infty} F_X(x) = 1$, $\lim_{x \to -\infty} F_X(x) = 0$. Then $X(\omega) = F^{-1}(\omega)$ is a random variable on the standard probability space $[0, 1]$ having F as its distribution function.

Proof Under the given assumptions on F, the inverse function F^{-1} is clearly well defined and is a continuous strictly increasing function on $[0, 1]$ (recall that $x = F^{-1}(y)$ denotes the solution to the equation $F(x) = y$). And we have

$$F_X(x) = \mathbf{P}(\{\omega \in [0, 1] : X(\omega) \le x\})$$
$$= \mathbf{P}(\{\omega \in [0, 1] : \omega \le F(x)\}) = F(x),$$

as required, the latter equation is due to the fact that $\mathbf{P}$ is the uniform measure and the length of the interval $[0, F(x)]$ equals $F(x)$. $\qquad \square$

Remark 6.8.2 This statement can be extended to the case of general distribution functions, not necessarily continuous ones, which is crucial for modeling discrete random variables, see e.g. [9].

Proposition 6.8.3 Let X be a continuously distributed random variable with a distribution function F_X and let G be an increasing continuous function. Then the distribution function of the random variable $G(X)$ is the composition of F_X and G^{-1}, i.e. it equals $F_X \circ G^{-1}$, where G^{-1} is the inverse function to G.

Proof We have

$$\mathbf{P}(G(X) \leq y) = 1 - \mathbf{P}(G(X) > y) = 1 - \mathbf{P}(X > G^{-1}(y))$$

$$= \mathbf{P}(X \leq G^{-1}(y)) = F_X(G^{-1}(y)),$$

as required. □

Of importance are the following corollaries.

Proposition 6.8.4 Under the assumptions of the previous theorem assume additionally that G is itself a distribution function (i.e. it is right continuous and takes values 0 and 1 at $-\infty$ and ∞ respectively). Then (i) the random variable $F_X(X)$ is uniformly distributed on $[0,1]$ and (ii) the random variable $G^{-1}(F_X(X)) = (G^{-1} \circ F_X)(X)$ has the distribution function G.

Proof By Proposition 6.8.3 the random variable $F_X(X)$ has the distribution function $F_X \circ F_X^{-1}$. But since X is continuous random variable, its distribution function F_X is continuous, and hence $F_X \circ F_X^{-1}(x) = x$ for all $x \in [0,1]$, implying (i). Statement (ii) then follows from (i) and Proposition 6.8.1. □

Remark 6.8.5 It is worth noting that the assumption that X is continuous is essential for the validity of Proposition 6.8.4, because, say, if X takes only finite number of values, then so does the composition $F_X(X)$, which therefore can not be uniformly distributed.

6.8.4 Gaussian copulas

We are going to apply Proposition 6.8.4 with $G = \Phi$ being the distribution function (6.12) of a standard normal random variable implying that if τ is a random time to default for a firm and the distribution function of τ is F_τ (say, given by equation (6.79)), then

$$X = \Phi^{-1}(F_\tau(\tau))$$

is $N(0,1)$ normal random variable. For n firms with default times τ_i and their distribution functions F_i, the *Gaussian copula* assumption states that the random variables $X_i = \Phi^{-1}(F_i(\tau_i))$ are mutually Gaussian with the correlation coefficients between X_i and X_j being given numbers ρ_{ij}.

Avoiding here a discussion of multivariate distributions (Gaussian vectors), let us introduce only the simplest model of the Gaussian copula approach, namely the so-called *one factor model*, where one assumes that

$$X_i = a_i M + \sqrt{1 - a_i^2} Z_i, \tag{6.81}$$

where $M, Z_1, \ldots, Z_n$ are independent normal $N(0, 1)$ random variables and a_i are constants from $[-1, 1]$. Here M stands for the common factor affecting all firms and Z_i stands for their individual characteristics.

It is straightforward to check that each X_i is also $N(0, 1)$ and Cov $(X_i, X_j) = \rho_{X_i, X_j} = a_i a_j$ for $i \neq j$.

Consequently the event $(\tau_i \leq T)$ (the ith firm defaults before time T) can be described as

$$X_i = a_i M + \sqrt{1 - a_i^2} Z_i \leq \Phi^{-1}(F_i(T)),$$

or equivalently as

$$Z_i \leq \frac{\Phi^{-1}(F_i(T)) - a_i M}{\sqrt{1 - a_i^2}}.$$

Hence, as Z_i is $N(0, 1)$, for the probability $D_i(T|M)$ of the ith firm's default conditional on the value of the factor M one has

$$D_i(T|M) = \Phi\left(\frac{\Phi^{-1}(F_i(T)) - a_i M}{\sqrt{1 - a_i^2}}\right).$$

In particular, if all $F_i = F$ are the same (firms are chosen from a certain common class) and all correlations are the same, say $a_i = \sqrt{\rho}$, then all $D_i(T|M)$ coincide and equal

$$D(T|M) = \Phi\left(\frac{\Phi^{-1}(F(T)) - \sqrt{\rho}M}{\sqrt{1 - \rho}}\right). \tag{6.82}$$

In practice for a large portfolio of similar assets, this probability estimates the frequency of defaults, i.e. the ratio of the number of assets belonging to firms that defaulted to time T to the total number of assets of the portfolio. For risk assessment one is often interested in the worst scenario. Choosing a confidence level C (0.999, say, or 99.9%), one is interested in a bound for losses that can occur C almost certainly, i.e. with probability C (99.9% say). Since M is $N(0, 1)$ normal, $M \geq m$ with probability $1 - \Phi(m) = \Phi(-m)$. Hence $M \geq -\Phi^{-1}(C)$ with probability

C implying that with this probability the frequency of defaults over T years will be less than

$$V(C,T) = \Phi \left(\frac{\Phi^{-1}(F(T)) + \sqrt{\rho}\Phi^{-1}(C)}{\sqrt{1-\rho}} \right). \tag{6.83}$$

This is called the *Vasicek* formula.

Let us denote by $P(k,T|M)$ the probability of k defaults by time T conditioned on M. When M is fixed the default probabilities are independent. Hence, the binomial distribution can be applied to yield

$$P(k,T|M) = \frac{n!}{(n-k)!k!}[D(T|M)]^k[1-D(T|M)]^{n-k} \tag{6.84}$$

with $D(T|M)$ given by (6.82). To get unconditional probabilities it remains to integrate over the normally distributed factor M. These formulas become the standard market tool for valuing default ratios and hence pricing the corresponding nth-to-default CDS.

6.8.5 Markov model for default probabilities and defaultable bonds

This section is slightly more advanced than the rest of the exposition as it uses Markov chains and continuous-time martingales.

For a τ as above let us define a Markov chain X_t on the space of two points $\{D,N\}$ (default, no default) in such a way that D is an absorbing state and if $X_s = N$, then it stays there a random time τ_s s.t. $\mathbf{P}(\tau_s > t) = \mathbf{P}(\tau > t|\tau > s)$ and then jumps to D. Then this Markov chain has the following transition probabilities:

$$p_{s,t}(D,D) = 1, \quad p_{s,t}(D,N) = 0,$$

$$p_{s,t}(N,N) = e^{-\int_s^t \lambda(u)\,du}, \quad p_{s,t}(N,D) = (1 - e^{-\int_s^t \lambda(u)\,du}).$$

Consequently, for the conditional expectation we have

$$\frac{d}{dt}|_{t=s}\mathbf{E}(f(X_t)|X_s = D) = 0,$$

$$\frac{d}{dt}|_{t=s}\mathbf{E}(f(X_t)|X_s = N) = \lambda(s)(f(D) - f(N)),$$

for a function f on the state space $\{D,N\}$.

Hence this chain is described by the following (time-nonhomogeneous) Q-matrix (or Kolmogorov matrix):

$$Q(t) = \begin{pmatrix} -\lambda(t) & \lambda(t) \\ 0 & 0 \end{pmatrix}.$$

Remark 6.8.6 A natural extension of this model to the case of n states (credit ranking), where the last state, n, is absorbing (default) and other transitions are specified by certain rates $\lambda_{ij}(t)$ is known as the *Jarrow–Lando–Turnbull model*.

Let $B(t, T)$ denote the price at time t of a zero-coupon bond that matures at time T with payment π_T that equals 1 or δ in cases N (no default till time T) or D respectively. If $\lambda(t)$ above are given with respect to the *risk-neutral measure Q* (defined as measure with respect to which the discounted bond prices are martingales), and if at time t there is no default, then by analogy with option prices (as expectations of discounted final payoffs) one defines the bond price as

$$B(t, T) = e^{-r(T-t)} \mathbf{E}_Q(\pi_T).$$

Calculating the expectation yields

$$B(t, T) = e^{-r(T-t)} \left(\delta + (1-\delta)e^{-\int_t^T \lambda(u)\, du} \right),$$

implying the following important formula for hazard rates in terms of prices of zero-coupon bonds:

$$\lambda(T) = -\frac{\partial}{\partial T} \log[e^{r(T-t)} B(t, T) - \delta].$$

6.9 Revision exercises

[1]

(a) A utility is called CRRA if its RRA is a constant. Find all CRRA utility functions on $\mathbf{R}_+$. Which of them are increasing, and which of them are risk averse?

(b) A utility is called HARA (hyperbolic ARA), if its ARA is inverse to linear, that is of the form $1/(ax + b)$ with constants a, b. Find all HARA utilities.

[**2**] Let $X = \{a, b, c\}$ and $\leq$ be a preference order on simple lotteries on X such that

$$\left(\frac{1}{4}, \frac{1}{4}, \frac{1}{2}\right) \sim \left(\frac{1}{2}, \frac{1}{2}, 0\right),$$

$$\left(\frac{3}{4}, \frac{1}{4}, 0\right) < \left(0, \frac{1}{2}, \frac{1}{2}\right).$$

(a) Find all utility functions u such that the corresponding expected utility yields the preference $\leq$ and u is normalized by the condition $u(a) = 1$.

(b) For each of this u, find all lotteries on X that have the same utility as $(\frac{1}{4}, \frac{3}{4}, 0)$, but with the outcome a impossible.

[**3**] Suppose you are playing the following version of the St. Petersburg game. A coin is tossed until the first Head appears. If it happens at a time $k \leq K$ with a given natural number K, you win \$$2^k$. If it happens at a time $k > K$, you loose (you have to pay) \$$2^k$. Let R denote your random return in this game (positive or negative).

(a) What is the expectation of your return, that is, $\mathbf{E}(R)$?

(b) Find the shortfall probability with level zero, that is, the probability of losing money: $\mathbf{P}(R < 0)$.

(c) For $n > K$, find the probability of losing at least \$$2^n$ and the probability of losing more than \$$2^n$.

(d) The value at risk VaR for a random outcome of a game R at level p is defined via the equation $\mathbf{P}(R \geq -VaR) = p$. Find VaR for the game above at the level $1 - 2^{-20}$ assuming $K < 10$.

[**4**] *Second-order dominance and stop loss.* Show that if a random variable A dominates a random variable B (both with range $[a, b]$) in the second order, then (i) $\mathbf{E}(B - d)^- \geq \mathbf{E}(A - d)^-$ for all $d \in [a, b]$ and (ii) $\mathbf{E}\min(d, B) \leq \mathbf{E}\min(d, A)$ for all $d \in [a, b]$.

[**5**] The market M consists of two securities A, B with returns r_A, r_B and capitalization 10 and 30 million respectively. The standard deviations of return $\sigma_A = \sigma_B = 1$ and the correlation coefficient is $\rho_{AB} = 1/2$. Assume CAPM holds for M.

(a) Find the weights of the securities A and B in M and their β-coefficients β_A and β_B.

(b) Find the risk-free return r_f in terms of the expected returns $\bar{r}_A, \bar{r}_B$.

(c) Find the precise conditions on $\bar{r}_A$ and $\bar{r}_B$ ensuring that

$$0 \leq r_f \leq \min(\bar{r}_A, \bar{r}_B).$$

[**6**] A market consists of two risky assets A, B with capitalization 40 and 60 million respectively. It is known that $\sigma_A^2 = 0.2$, $\sigma_B^2 = 0.1$, $\sigma_{AB} = -0.1$, the expectation of the market rate of return is $\bar{r}_M = 28\%$ and the risk-free rate is $r_f = 8\%$. Assume that CAPM holds.

(a) Find $\sigma_M^2, \beta_A, \beta_B$.
(b) Calculate $\bar{r}_A, \bar{r}_B$ and the market price of risk.
(c) Assume an investor has a utility function $U_\alpha = \bar{r} - \alpha\sigma^2$, depending on its average return $\bar{r}$ and the variance σ^2, where α is a positive parameter. Supposing the investor is behaving optimally, express the average return r^* of such an investor in terms of α and then calculate it for $\alpha = 10$.

[**7**] Assume a risk-free interest rate of 4% p.a. (equivalent to 0.016% per trading day) continuously compounded, volatility $\sigma = 20\%$ p.a. Assume there are 250 trading days per year. A stock has an initial price of 100p. Set up a binomial model (with period one day) and consider an American put and an American call, both with exercise prices of 101p and maturity 2 days (they can be exercised either at the end of day 1 or at the end of day 2). At the end of day 1, before the investor is allowed to exercise, the company announces unexpectedly that a dividend of 3p per share will be paid at the end of day 2 immediately prior to the expiry of the options. Construct the binomial lattice of share prices allowing for the dividend payment. Explain the conditions under which the holders of the put or call options will exercise at the end of day 1 after the announcement of the dividend.

[**8**] Assume the current stock price is 62 (say, dollars) and the stock volatility is $\sigma = 0.20$ per anum. Interest rates are 1% per month (and compounded per month).

(a) Set up a two-period binomial model with expiration time 1 year (that is, with each period being 6 months) and compute the risk neutral probabilities $\tilde{p}$ and $\tilde{q}$. Suppose additionally that the expected return on the stock is 2% per month. Calculate the real world probabilities p, q and the state price deflator D_l (three values should be given corresponding to the number of higher jumps l being zero, one or two).

(b) On the basis of the binomial model set up above, determine the price of European put and American put, both with expiration 1 year and strike 60. Display the prices of stocks and both options for all periods in tables. On the table for American put stress the values (if any), where the earlier exercise is optimal.

(c) A lookback option in two periods is defined via the final payoff

$$V_2 = \max_{0 \le n \le 2} S_n - S_2.$$

Find the value V_0 of the option at the initial time. How many shares should an agent buy in order to hedge her short position in the option?

(d) Let $Y_n = \sum_{k=0}^{n} S_k$. On the basis of the two-period binomial model set up above, display all possible values of the pairs (S_k, Y_k), $k = 0, 1, 2$, in a table. Find the value A_0 of the Asian option at the initial time, if $K = 60$. How many shares should an agent buy in order to hedge her short position in the option?

(e) Consider the 'Chooser' or 'As you like it' option (based on the above two-period model) that confers the right to receive either call or put at the end of the first period, both having maturity time 2 and strike price 60. Calculate the price of this option for the model set up above.

6.10 Solutions to exercises (sketch)

[1]

(a) Condition CRRA states $u''/u' = c/x$ with a constant c, hence $\log(u') = \alpha \log x + c$ or $u' = wx^\alpha$ with constants w, α. Thus $u' > 0$ for $w > 0$ and $u'' < 0$ for $\alpha < 0$.

(b) Condition HARA yields $\log(u') = a^{-1} \log(x + b/a) + c$,

$$u'(x) = w(x + b/a)^{1/a}, \quad u(x) = \alpha(x + \beta)^\gamma + c,$$

i.e. a power function.

[2]

(a) Denote $\mu = u(b)$. Then

$$1 + \mu + 2u(c) = 2 + 2\mu \implies u(c) = \frac{1}{2}(\mu + 1).$$

$$\frac{3}{4} + \frac{1}{4}\mu < \frac{1}{2}\mu + \frac{1}{4}(\mu + 1) \implies \mu > 1.$$

(b) Lotteries with support on b, c have form $(0, p, 1 - p)$. Thus

$$\frac{1}{4} + \frac{3}{4}\mu = p\mu + (1 - p)\frac{1}{2}(\mu + 1) \implies p = 1/2.$$

[3]

(a)
$$\mathbf{E}(R) = \frac{1}{2} \times 2 + \frac{1}{2^2} \times 2^2 + \cdots + \frac{1}{2^K} \times 2^K - \frac{1}{2^{K+1}} \times 2^{K+1} - \cdots = -\infty.$$

(b)
$$\mathbf{P}(R < 0) = 1 - \left(\frac{1}{2} + \frac{1}{2^2} + \cdots + \frac{1}{2^K}\right) = 2^{-K}.$$

(c)
$$\mathbf{P}(R \le -2^n) = \sum_{j=n}^{\infty} 2^{-j} = 2^{-(n-1)}, \mathbf{P}(R < -2^n) = \sum_{j=n+1}^{\infty} 2^{-j} = 2^{-n}.$$

(d) It follows directly from (c) that $VaR = 2^{20}$.

[4] Use integration by parts.

[5] (a)–(b):

$$w_A = \frac{1}{4}, \quad w_B = \frac{3}{4};$$

$$\sigma_M^2 = \frac{1}{16} + \frac{9}{16} + \frac{3}{16} = \frac{13}{16}, \quad \sigma_{AM} = \frac{5}{8}, \quad \sigma_{BM} = \frac{7}{8},$$

$$\beta_A = \frac{\sigma_{AM}}{\sigma_M^2} = \frac{10}{13}, \quad \beta_B = \frac{14}{13},$$

$$r_f = \frac{\bar{r}_A - \beta_A \bar{r}_M}{1 - \beta_A} = \frac{1}{4}\frac{\bar{r}_A(4 - \beta_A) - 3\beta_A \bar{r}_B}{1 - \beta_A} = \frac{1}{2}(7\bar{r}_A - 5\bar{r}_B).$$

(c) Conditions are

$$7\bar{r}_A \ge 5\bar{r}_B, \quad 7\bar{r}_A - 5\bar{r}_B \le 2\bar{r}_A, \quad 7\bar{r}_A - 5\bar{r}_B \le 2\bar{r}_B.$$

Solving yields

$$\frac{5}{7}\bar{r}_B \le \bar{r}_A \le \bar{r}_B.$$

[6]

(a) Market capitalization is 100 million and $r_M = 0.4r_A + 0.6r_B$.

$$\sigma_M^2 = (0.4)^2 \times 0.2 - 2 \times 0.4 \times 0.6 \times 0.1 + (0.6)^2 \times 0.1 = 0.02,$$

$$\sigma_{AM} = 0.4{\times}0.2{-}0.6{\times}0.1 = 0.02, \quad \sigma_{BM} = 0.6{\times}0.1{-}0.4{\times}0.1 = 0.02,$$

$$\beta_A = \frac{\sigma_{AM}}{\sigma_M^2} = 1, \quad \beta_B = \frac{\sigma_{BM}}{\sigma_M^2} = 1$$

(b)

$$CAPM: \quad \bar{r}_{A,B} - r_f = \beta_{A,B}(\bar{r}_M - r_f).$$

$$\bar{r}_A = \bar{r}_B = 0.08 + 0.2 = 0.28.$$

The market price of risk is

$$(\bar{r}_M - r_f)/\sigma_M = 0.2/\sqrt{0.02} = \sqrt{2}.$$

(c) For

$$\bar{r} = (1 - x)r_f + x\bar{r}_M$$

we are looking for a maximum (over x) of

$$(1 - x)r_f + x\bar{r}_M - \alpha x^2\sigma_M^2,$$

which is attained at

$$x^* = \frac{\bar{r}_M - r_f}{2\alpha\sigma_M^2}$$

so that

$$r^* = r_f + \frac{(\bar{r}_M - r_f)^2}{2\alpha\sigma_M^2} = 0.08 + \frac{(0.2)^2}{2 \times 0.02\alpha} = 0.08 + \alpha^{-1} = 0.18.$$

[7] $u = e^{0.2/\sqrt{250}} = 1.01273$, $d = 1/u$, $1+r = e^{0.00016}$. The risk-neutral probability is

$$\tilde{p} = \frac{e^{0.00016} - d}{u - d} = 0.50316.$$

The values of stocks are (reduced by 3 after the payment)

$$
\begin{array}{ccc}
100 & 101.273 & 99.562 \\
 & 98.743 & 97 \\
 & & 94.502
\end{array}
$$

The call will be worthless at expiration. If the stock has risen in the first day, exercising the call would yield a profit.

If the share price is 101.273, exercising the put yields nothing, but holding it to the expiration yields a profit. If the share price is 98.743, exercising it yields $101 - 98.743 = 2.257$, and leaving it to expiration has the value

$$e^{-0.016}\left[\tilde{p}(101 - 97) + (1 - \tilde{p})(101 - 94.502)\right] > 2.257.$$

Thus it is better not to exercise the put earlier.

[8]

(a)

$$\Delta t = 1/2, \quad 1 + r = (1.01)^6 = 1.0615,$$

$$u = e^{0.2/\sqrt{2}} = 1.1519, \quad d = 1/u = 0.8681.$$

$$\tilde{p} = \frac{1 + r - d}{u - d} = \frac{1.0615 - 0.8681}{0.2838} = \frac{0.1934}{0.2838} = 0.6815, \quad \tilde{q} = 0.3185.$$

Since $\mathbf{E}(S_1/S_0) = (1.02)^6 = pu + (1-p)d$ (where p is the real world probability), it follows that

$$p = \frac{(1.02)^6 - d}{u - d} = \frac{1.1261 - 0.8681}{0.2838} = \frac{0.2580}{0.2838} = 0.9091, \quad q = 0.0909,$$

$$D_l = \frac{1}{(1.0615)^2} \left(\frac{\tilde{p}}{p}\right)^l \left(\frac{1-\tilde{p}}{1-p}\right)^{2-l}$$

$$= \frac{1}{(1.0615)^2} \left(\frac{0.6815}{0.9091}\right)^l \left(\frac{0.3185}{0.0909}\right)^{2-l},$$

$$D_0 = 10.937, \quad D_1 = 2.335, \quad D_2 = 0.498.$$

(b) Tables for stocks, and European and American puts are respectively

	62	71.4178	82.2662
		53.8222	62
			46.7231

1.1953	0	0
	3.9837	0
		13.2769

1.8536	0	0
	6.1778	0
		13.2769

(c) Table for the 'Lookback' option prices:

3.1896	2.8258	0
	9.4178	
	4.5838	0
		15.2769

$$V_1(u) = (1+r)^{-1}[\tilde{p}\,V_2(uu) + \tilde{q}\,V_2(ud)] = \frac{0.3185 \times 9.4178}{1.0615} = 2.8258,$$

$$V_1(d) = \frac{1}{1.0615}[\tilde{p}\,V_2(du) + \tilde{q}\,V_2(dd)] = \frac{0.3185 \times 15.2769}{1.0615} = 4.5838,$$

$$V_0 = \frac{1}{1.0615} [0.6815 \times 2.8258 + 0.3185 \times 4.5838] = 3.1896,$$

$$\Delta_0 = \frac{V_1(u) - V_1(d)}{S_0(u - d)} = -\frac{4.5838 - 2.8258}{17.5956} = -0.0999.$$

(d)

(62, 62)	(71.4178, 133.4178)	(82.2662, 215.6840)
		(62, 195.4178)
	(53.8222, 115.8222)	(62, 177.8222)
		(46.7231, 162.5453)

Using the recurrent equations for Asian option pricing, we get

$$A_2(S(du), Y(du)) = A_2(S(dd), Y(dd)) = 0,$$

$$A_2(S(ud), Y(ud)) = 5.1393,$$

$$A_2(S(uu), Y(uu)) = 11.8949, \quad A_1(S(d), Y(d)) = 0,$$

$$A_1(S(u), Y(u)) = (1 + r)^{-1} [\tilde{p} A_2(uu) + \tilde{q} A_2(ud)]$$

$$= \frac{1}{1.0615} [0.6815 \times 11.8949 + 0.3185 \times 5.1393] = 9.1787,$$

$$A_0 = \frac{1}{1.0615} \times 0.6815 \times 9.1787 = 5.8928,$$

$$\Delta_0 = \frac{A_1(u) - A_1(d)}{S_0(u - d)} = \frac{9.1787}{17.5956} = 0.5216.$$

(e) By the put-call parity, the time-zero price of the Chooser is the sum of the time-zero put with expiration 2 and strike 60 and a call with expiration 1 and strike $60/(1 + r)$. Hence we find the price of the Chooser to be

$$1.1953 + \frac{0.6815}{1.0615} \left(71.4178 - \frac{60}{1.0615}\right)^+ = 10.7572.$$

6.11 Concluding remarks

We have only touched upon continuous-time models. Their accessible expositions can be found in [3], [18], [19]. Recent trends in financial economics are strongly linked with power laws (as introduced in Section 6.6.3) and related application of Lévy processes, see e.g. [14], and the methods of econophysics, see e.g. [20] and [13]. Together with probability theory, a natural tool for analyzing finance is game theory, see e.g. [2], [10], [21]. Closely related to financial economics is also the theory of risk (touched upon in Section 6.2.5) and insurance mathematics, for which we refer to books [4] and [11].

References

[1] S. Barbera et al (eds.) 1998–2004. *Handbook of Utility Theory.* Kluwer Academic.

[2] P. Bernhard, J.-P. Aubin, P. Saint-Pierre, J. Engwerda, V. N. Kolokoltsov. *The Interval Market Model in Mathematical Finance: Game-Theoretic Methods.* Birkhäuser, 2012.

[3] N. H. Bingham and R. Kiesel 2004. *Risk-neutral Valuation. Pricing and Hedging of Financial Derivatives,* 2nd edition. Springer, 2004.

[4] P. J. Boland. *Statistical and Probabilistic Methods in Actuarial Sciences.* Chapman and Hall/CRC, 2007.

[5] K. Cuthbertson, D. Nitzsche. *Quantitative Financial Economics,* 2nd edition. John Wiley, 2005.

[6] E. Elton et al. *Modern Portfolio Theory and Investment Analysis,* 6th edition. John Wiley, 2003.

[7] D. Foster, S. Hart. An operational measure of risk. *Journal of Political Economy,* 2009.

[8] J. Hull. *Options, Futures and other Derivatives,* 6th edition. Prentice-Hall, 2006.

[9] V. N. Kolokoltsov. *Mathematical Analysis of Financial Markets: Probability Theory and Financial Mathematics – A brief introduction.* (In English and Russian). Moscow Economic Institute, 2010.

[10] V. N. Kolokoltsov, O. A. Malafeyev. *Understanding Game Theory: Introduction to the Analysis of Many Agent Systems with Competition and Cooperation.* World Scientific, 2010.

[11] V. Yu. Korolev, V. E. Bening, S. Ya. Shorgin. *Mathematical Foundations of the Risk Theory.* FIZMATLIT (in Russian), 2007.

[12] L. C. MacLean, E. O. Thorp and W. T. Ziemba. *The Kelly Capital Growth Investment Criterion: Theory and Practice* (World Scientific Handbook in Financial Economics Series). World Scientific, 2012.

[13] V. P. Maslov. *Quantum Economics.* Nauka (in Russian), 2006.

[14] A. Lipton and A. Rennie (eds.) *Credit Correlation: Life After Copulas.* World Scientific, 2008.

[15] D. G. Luenberger. *Investment Science.* Oxford University Press, 1998.

[16] J. A. Paulos. *Mathematician Plays the Stock Market.* Basic Books, 2003.

[17] W. F. Sharpe and G. J. Alexander. *Investments,* 4th edition. Prentice Hall, 1990.

[18] A. N. Shiryaev. *Essentials of Stochastic Finance: Facts, Models, Theory.* World Scientific, 1999.

[19] S. E. Shreve. *Stochastic Calculus for Finance I. The Binomial Asset Pricing Model. Stochastic Calculus for Finance II. Continuous-Time Models.* Springer, 2005.

[20] L. T. Wille (ed.) *New Directions in Statistical Physics. Econophysics, Bioinformatics, and Pattern Recognition.* Springer, 2004.

[21] A. Ziegler. *A Game Theory Analysis of Options.* Springer, 2004.

7

Space-time phases

Robert S. MacKay

Abstract

Complexity science is the study of systems with many interdependent components. One of the main concepts is "emergence": the whole may be greater than the sum of the parts. The objective of this chapter is to put emergence on a firm mathematical foundation in the context of dynamics of large networks. Both stochastic and deterministic dynamics are treated. To minimise technicalities, attention is restricted to dynamics in discrete time, in particular to probabilistic cellular automata and coupled map lattices. The key notion is *space-time phases*: probability distributions for state as a function of space and time that can arise in systems that have been running for a long time. What emerges from a complex dynamic system is one or more space-time phases. The *amount of emergence* in a space-time phase is its distance from the set of product distributions over space, using an appropriate metric. A system exhibits *strong emergence* if it has more than one space-time phase. Strong emergence is the really interesting case.

This chapter is based on MSc or PhD courses given at Warwick in 2006/7, Paris in April 2007, Warwick in Spring 2009 and Autumn 2009, and Brussels in Autumn 2010. It was written up during study leave in 2010/11 at the Université Libre de Bruxelles, to whom I am grateful for hospitality, and finalised in 2012.

The chapter provides an introduction to the theory of space-time phases, via some key examples of complex dynamic system.

I am most grateful to Dayal Strub for transcribing the notes into LaTeX and for preparing the figures. Also to Piotr Slowinski and Manuela

Bujorianu for critical reading and to the Sloan Foundation for their support in the final stages.

7.1 Stochastic dynamics — probabilistic cellular automata (PCA)

Although there is a wide range of types of stochastic dynamic system, we will restrict attention here to probabilistic cellular automata (PCA). These are discrete-time Markov processes on large products of finite-state units, for which the transitions at different sites given the whole current configuration are independent. It is usual, but not essential, to assume that the transition probability for the state of site s in network S is independent of state on $S \setminus N(s)$ for some finite subset $N(s)$ called its neighbourhood.

In this chapter we will study two main examples: Stavskaya's PCA and the NEC majority voter PCA, but we will derive more general results in the process of treating them.

Some suggested background reading is [7] for a short introduction and [26] for an extended treatment. There is an overlap in phenomenology with the continuous-time systems presented in Chapter 3.

7.1.1 Stavskaya's PCA

Stavskaya's PCA [24] is also known as directed percolation, oriented percolation, asymmetric contact process, discrete Regge field, Domany-Kinzel model, and possibly more names. It is a PCA in 1D, in which each unit at each time is in one of two states: 1 representing healthy and 0 representing infected. Each healthy unit can catch infection from its left-hand neighbour only.

The state is $x = (x_s)_{s \in \mathbb{Z}} \in \{0,1\}^{\mathbb{Z}}$. The parameter $\lambda \in (0,1)$ is both the probability of recovery in one step and the probability of avoiding infection from one's left-hand neighbour (the probabilities can be taken different if desired, and also site-dependent, and the recovery rate could be taken to depend on whether the neighbour is infected, but to reduce notational complexity we restrict to this case). At each time $t \in \mathbb{Z}_+$, for each site $s \in \mathbb{Z}$,

$$\text{if } (x_{s-1}^t, x_s^t) = (1,1) \text{ then } x_s^{t+1} = 1,$$

$$\text{if } (x_{s-1}^t, x_s^t) \neq (1,1) \text{ then } x_s^{t+1} = \begin{cases} 1, \text{ with probability } & \lambda, \\ 0 & 1-\lambda. \end{cases}$$

independently of all other $x_{s'}^{t'}$. Let $\underline{1}$ denote the state of all 1, which is an absorbing state, and $\delta_{\underline{1}}$ the probability distribution supported on $\underline{1}$.

For the finite version of this system on $\mathbb{Z}_N = \mathbb{Z}$ mod N, we have the probability

$$P\{x^{t+1} = \underline{1}|x^t\} \geq \lambda^N,$$

for all x^t (x^t means the state of the whole line at time t; we sometimes denote it by $\underline{x}^t$ to emphasise that it is the state of the whole system). Therefore $P\{\text{not absorbed by time } t\} \leq (1 - \lambda^N)^t \to 0$ as $t \to \infty$, and so $P\{\text{eventually } \underline{1}\} = 1$.

This discussion of the finite case misses an important feature, however, namely that there exists a critical parameter value $\lambda_c \in (0,1)$, such that for $\lambda > \lambda_c$, the time to absorption is of order $\log_\gamma N$ for some $\gamma(\lambda) > 0$, and for $\lambda < \lambda_c$, the time to absorption is instead of order $e^{\gamma N}$. This is best understood by considering the infinite system, for which the following hold:

- for $\lambda \geq \lambda_c$, the stationary probability measure $\delta_{\underline{1}}$ attracts all initial probability measures;
- for $\lambda < \lambda_c$, there exists an additional stationary probability ν_λ with $\nu_\lambda (x_s = 0) = c(\lambda) > 0$ ("endemic infection" or "eternal transient"), such that for all initial probabilities ν, we have

$$\nu P^t \to \sigma \delta_{\underline{1}} + (1 - \sigma) \nu_\lambda,$$

where $\sigma = \nu\{\text{eventual absorption}\}$, P is the transition operator for the process (definition to be given in Section 7.1.1) and P^t is its tth power.

We will first treat λ near 1 and then λ near 0.

λ **near 1**

This is the regime of low infectivity and high recovery rate, so we expect any initial infection to die out. We will introduce a nice metric on a space of multivariate probabilities to prove the (exponential) attraction of νP^t to $\delta_{\underline{1}}$ for any initial probability ν, based on ideas of Dobrushin [8] (see [17]).

We introduce the metric in the following general setting. Let S be a countable set, representing the sites of a "network" (on which we can have a metric, but it is not required at this stage). An element $s \in S$ is called a "site", and for all $s \in S$, (X_s, d_s) is a Polish (i.e. complete separable metric) space of diameter at most some Ω. For example, let $S = \mathbb{Z}$, $X_s = \{0,1\}$ and $d_s(0,1) = 1$. Let $X = \prod_{s \in S} X_s$ with product topology.[1] Let $\mathcal{P}$ be the set of Borel probabilities[2] on X, and Z be the set of zero-charge measures[3] on X. The transition probability for updating the state of site s is denoted $p_s(x'_s|\underline{x})$ (or $p_s(\mathrm{d}x'_s|\underline{x})$ if X_s is not discrete), and their product is denoted $p(\mathrm{d}\underline{x}'|\underline{x})$.

Let BC denote the set of bounded continuous functions $f : X \to \mathbb{R}$. The transition operator P on BC is defined by

$$(Pf)(x) = \int f(x')p(\mathrm{d}x'|x).$$

Note that $P1 = 1$, where $1(x) = 1, \forall x \in X$.

The transition operator P can be extended to act on measures ρ on X by

$$(\rho P)(f) = \rho(Pf) = \int (Pf)(x)\, d\rho(x)$$

(in general $\rho(f)$ denotes the integral of f with respect to ρ). We write P as acting to the left on measures to fit with standard matrix notation, considering measures as row-vectors.

A probability measure $\rho \in \mathcal{P}$ is *stationary* for P if $\rho P = \rho$.

For $f \in BC$, $s \in S$, let

$$\Delta_s(f) = \sup \frac{f(x) - f(x')}{d_s(x_s, x'_s)},$$

over states x, x' such that $x_s \neq x'_s$ and $x_r = x'_r \; \forall r \neq s$, i.e. the best Lipschitz constant of f with respect to the state x_s on s (if finite). Let $F = \{f \in BC : \|f\|_F = \sum_{s \in S} \Delta_s(f) < \infty\}$ modulo addition of constants, which is a normed linear space. For $\mu \in Z$, let

$$\|\mu\|_Z = \sup_{f \in F \backslash C} \frac{\mu(f)}{\|f\|_F},$$

where C is the set of constant functions. Note that $(Z, \|\cdot\|_Z)$ is a Banach space (complete normed linear space), because it is the dual of a normed linear space.

Finally we define

$$D(\rho, \rho') = \|\rho - \rho'\|_Z,$$

---

[1] That is, the open sets are those generated from the "cylinder sets" $\prod_{s \in S'} C_s \times \prod_{s \in S \backslash S'} X_s$ for $S' \subset S$ finite, C_s an open subset of X_s, by arbitrary union and finite intersection

[2] Probabilities which are defined for all subsets of X generated from the open subsets by countable unions and intersections and complementation

[3] "Signed" Borel measures μ for which $\mu(X) = 0$

for $\rho, \rho' \in \mathcal{P}$, which makes $\mathcal{P}$ a complete metric space. We call D the *Dobrushin metric*.

For $L : Z \to Z$ linear, e.g. the restriction of P to Z, let

$$\|L\|_Z = \sup_{\mu \in Z \backslash 0} \frac{\|\mu L\|_Z}{\|\mu\|_Z}.$$

The point of $\|\cdot\|_Z$ is to make a small parameter change in the Stavskaya PCA (for example) produce a small change in the transition operator P, uniformly in the size of S.

The metric $D(\rho, \rho')$ is useful because, e.g. for ρ_λ equal to the product of N independent probabilities $1 - \lambda$ on 0, λ on 1, then the speed of change of ρ_λ with respect to λ is

$$v = \left\| \frac{\mathrm{d}\rho}{\mathrm{d}\lambda} \right\|_Z = 1.$$

In contrast, most of the standard metrics on probability spaces give speeds that go to infinity as $N \to \infty$: for the total variation (TV) metric, Kullback–Leibler (KL) divergence, Hellinger metric, and Fisher information metric, one obtains $v \sim \sqrt{N}$, and for projective metric and l_∞-transportation metric, one obtains $v \sim N$ [17]. One could divide these metrics by $\sqrt{N}$ or N respectively, but then many localised changes would have small distances, and even all changes would have small distances for some of the above (e.g. the TV metric has diameter 1, so $1/\sqrt{N}$ after division by $\sqrt{N}$)!

To bound a transition operator L (or their differences) we can use that $L : Z \to Z$ bounded and linear induces $L : F \to F$ by

$$\mu(Lf) = (\mu L)(f) \ \forall \mu \in Z,$$

and so

Proposition 7.1.1

$$\|L\|_Z \leq \|L\|_F.$$

Proof

$$\frac{|\mu L f|}{\|f\|_F} \leq \frac{\|L\|_F}{\|Lf\|_F} |\mu L f| \leq \|L\|_F \|\mu\|_Z,$$

so

$$\|\mu L\| \leq \|L\|_F \|\mu\|.$$

$\square$

Actually $\|L\|_Z = \|L\|_F$, using the Hahn–Banach theorem, but we will not need this observation.

In particular, let us define the *dependency matrix* k, indexed by sites $r, s \in S$, by

$$k_{rs} = \sup_{x, \tilde{x}} \frac{D_T \left(p_r^x, p_r^{\tilde{x}}\right)}{d_s \left(x_s, \tilde{x}_s\right)},$$

with $x, \tilde{x}$ agreeing off s, and differing on s, where

$$p_r^x \left(x_r'\right) = p_r(x_r'|\underset{\sim}{x}),$$

and the *transportation metric* D_T on probabilities on a metric space X is

$$D_T(\rho, \tilde{\rho}) = \sup_{f \in L \backslash C} \frac{\rho(f) - \tilde{\rho}(f)}{\|f\|_L},$$

with L the Lipschitz functions $f : X \to \mathbb{R}$, and $\|f\|_L$ the best Lipschitz constant of f. The matrix element k_{rs} bounds the influence of the state at site s now on the state at site r next step, e.g. $k_{rs} = 0$ if $s \notin N(r)$.

D_T is called transportation metric because of its origins in Monge's earth movement questions [11]. For example, see Figure 7.1, where we want to minimise the integral of the mass moved times the distance moved. In 1D this is simple to find (the area between the cumulative distribution functions, Kolmogorov's formula); however, in higher dimensions (even 2D) it is not easy, and the literature is frustrating as it strays from the original problem to other cost functions.

Figure 7.1 Example of Monge's earth movement problem in 1D.

We obtain the useful bound:

Proposition 7.1.2

$$\|P\|_Z \le \delta = \sup_{r \in S} \sum_{s \in S} k_{rs} = \|k\|_\infty,$$

Proof Use an alternative way of writing the transportation distance

$$D_T \left(\rho, \tilde{\rho}\right) = \inf_{\tau} \int d(x, \tilde{x}) \tau(dx, d\tilde{x}),$$

over those $\tau \in \mathcal{P}(X \times X)$ such that the marginals of τ on the two copies of X are $\rho, \tilde{\rho}$ respectively. See Figure 7.2. Such τ are usually called "couplings", but "joinings" is a better name because for me "couplings" implies some non-trivial effect of at least one on the other. Kantorovich and Rubinstein [11] proved equality of the two definitions of the transportation distance.

Figure 7.2 Illustration of the marginals property of a joining of two probabilities.

Take $x, \tilde{x}$ agreeing off s and take optimal joinings $\tau_r^{x,\tilde{x}}$ of p_r^x to $p_r^{\tilde{x}}$ $\forall r$ (if there is no optimal joining, take τ close to optimal and restrict the functions f to be independent of the state on all but finitely many sites). By the marginals property of $\tau_r^{x,\tilde{x}}$, for any $f \in F$

$$(Pf)(x) - (Pf)(\tilde{x}) = \int (f(x') - f(\tilde{x}')) \prod_{r \in S} \tau_r^{x,\tilde{x}}(dx'_r, d\tilde{x}'_r).$$

But

$$f(x') - f(\tilde{x}') \leq \sum_r \Delta_r(f) d_r(x'_r, \tilde{x}'_r),$$

and

$$\int d_r(x'_r, \tilde{x}'_r) \tau_r^{x,\tilde{x}}(dx'_r, d\tilde{x}'_r) \leq k_{rs} d_s(x_s, \tilde{x}_s).$$

therefore

$$(Pf)(x) - (Pf)(\tilde{x}) \leq \sum_r \Delta_r(f) k_{rs} d_s(x_s, \tilde{x}_s),$$

and

$$\Delta_s(Pf) \leq \sum_r \Delta_r(f) k_{rs}.$$

Now sum this result over $s \in S$:

$$\|Pf\|_F = \sum_s \Delta_s(Pf) \leq \|k\|_\infty \sum_r \Delta_r(f),$$

i.e. $\|Pf\|_F \leq \|k\|_\infty \|f\|_F$, cf. Maes [18]. $\qquad\qquad \square$

Example 7.1.3 (An optimal joining)   Let $\rho, \tilde{\rho}$ be probabilities on $\{0, 1\}$ with $\rho(0) = 1 - \lambda$, $\rho(1) = \lambda$, $\tilde{\rho}(0) = 1 - \mu$, $\tilde{\rho}(1) = \mu$, $\lambda > \mu$, and use the metric d with $d(0, 1) = 1$. Then we have a unique optimal joining τ and $D_T(\rho, \tilde{\rho}) = \lambda - \mu$. See Figure 7.3.

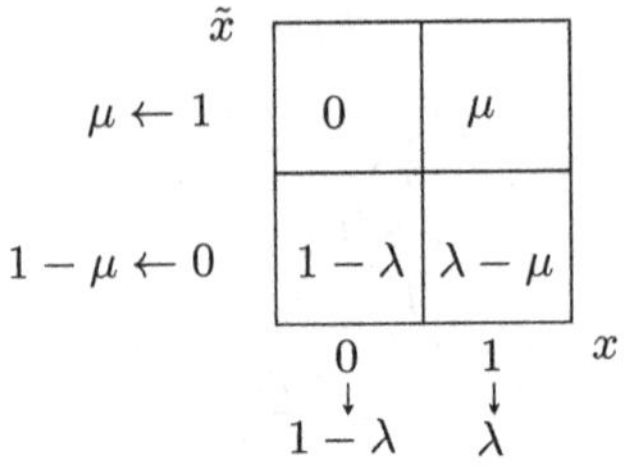

Figure 7.3 Joining τ for Example 7.1.3.

Now we will prove existence of exponentially attracting, unique stationary probabilities for "weakly dependent" PCAs , i.e. those for which $\|P\| < 1$. This is simply because P is then a contraction on $\mathcal{P}$, so it has a unique stationary probability ρ ($\mathcal{P}$ is complete with respect to our norm) and ρ attracts all $\nu \in \mathcal{P}$ exponentially:

$$D\left(\nu P^t, \rho\right) \leq \|P^t\| D\left(\nu, \rho\right) \leq \|P\|^t D\left(\nu, \rho\right).$$

Example 7.1.4   Show that $\|P\| \leq 2(1 - \lambda)$ for Stavskaya's PCA. This can be seen by using the dependency matrix k_{rs}. Recall from Example 7.1.3 that $D_T\left((1 - \lambda, \lambda), (1 - \mu, \mu)\right) = |\lambda - \mu|$. $\|P\| \leq \sup_r \sum_s k_{rs}$, where the dependency matrix is

$$k_{rs} = \sup_{x, \tilde{x}} \frac{D_T\left(p_r^x, p_r^{\tilde{x}}\right)}{d_s\left(x_s, \tilde{x}_s\right)}, \quad \text{where } x, \tilde{x} \text{ agree off } s, \text{ and differ on } s.$$

Now, for:

- $k_{r,r}$:
  if $x_{r-1} = 0$ then $p_r^x = (1 - \lambda, \lambda)$ independently of x_r, so $D_T = 0$,
  else if $x_{r-1} = 1$ then $p_r^x = (1 - \lambda, \lambda)$ if $x_r = 0$, or $p_r^x = (0, 1)$ if $x_r = 1$,
  so $D_T = 1 - \lambda$,
  so $k_{r,r} = 1 - \lambda$.
- $k_{r,r-1}$:
  if $x_r = 0$ then $D_T = 0$,
  else if $x_r = 1$ then $D_T = 1 - \lambda$,
  so $k_{r,r-1} = 1 - \lambda$.

- $k_{r,s} = 0$ for $s \notin \{r, r-1\}$.

So the dependency matrix has diagonal and sub-diagonal entries $1 - \lambda$ and all the rest are zero. So $\|k\|_\infty = 2(1 - \lambda)$, and $\|P\| \leq 2(1 - \lambda)$. Hence for all $\lambda > \frac{1}{2}$ there exists a unique stationary probability and it attracts everything exponentially. However, we already know one stationary probability, $\delta_{\underline{1}}$, so $\delta_{\underline{1}}$ attracts all of $\mathcal{P}$ exponentially. See Figure 7.4.

Figure 7.4 The part of parameter space for the Stavskaya PCA for which absorption is established from weak dependence.

Rapid absorption for finite system when $\lambda > \frac{1}{2}$ On $\mathbb{Z}_N$, we still have $\|P\| \leq \delta = 2(1 - \lambda) < 1$. Take $f(\underline{x})$ to be the number of zeros in $\underline{x}$. So $\|f\|_F = N$, from the definition, and $\|P^t f\|_F \leq \delta^t N$. Thus $P\{\text{not absorbed at time } t\} \leq \delta^t N$. See Figure 7.5. So, in any meaningful sense, the time to absorption is of order $\frac{\log N}{\log 1/\delta}$, which is relatively short (compared with N).

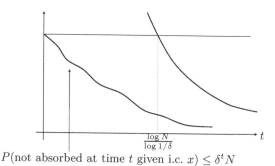

$P(\text{not absorbed at time } t \text{ given i.c. } x) \leq \delta^t N$

Figure 7.5 Absorption time is logarithmic in N for $\lambda > \frac{1}{2}$.

Robustness of exponentially attracting stationary probability
So far we have shown existence of unique and exponentially attracting stationary probability for all weakly dependent PCA, but the same result

holds for some PCA that are not weakly dependent. In particular, it is an open property, with the following explicit estimates.

Say a (stationary) probability $\rho_0 \in \mathcal{P}$ *attracts exponentially* for a transition operator P_0 if there exist C and $r < 1$ such that

$$D(\sigma P_0^t, \rho_0) \leq Cr^t D(\sigma, \rho_0).$$

Theorem 7.1.5 *If P_0 has stationary probability ρ_0 which attracts exponentially, then all P near P_0, i.e. $\|P - P_0\|_Z = \delta < \frac{1-r}{C}$, also have exponentially attracting stationary probability ρ.*

Proof Introduce the adapted norm on Z,

$$\|\mu\|_r = \sup_{n \geq 0} \|\mu P_0^n\| r^{-n}.$$

Then

$$\|\mu\| \leq \|\mu\|_r \leq C\|\mu\|,$$

so the norms are equivalent, and

$$\|\mu P_0\|_r = \sup_{n \geq 0} \|\mu P_0^{n+1}\| r^{-n} \leq r\|\mu\|_r,$$

so $\|P_0\|_r \leq r$. Next

$$\|P - P_0\| = \delta \Rightarrow \|P - P_0\|_r \leq C\delta,$$

so

$$\|P\|_r \leq \|P_0\|_r + \|P - P_0\|_r$$
$$\leq r + C\delta < 1,$$

since $\delta < \frac{1-r}{C}$. So $\|P^n\|_r \leq (r + C\delta)^n$ and hence $\|P^n\| \leq C(r + C\delta)^n$. So P has a unique stationary probability ρ and it attracts all exponentially. $\square$

Also, we show that

$$\|\rho - \rho_0\| \leq \frac{C\|\rho_0(P - P_0)\|}{1 - r - C\delta}.$$

Proof $\rho_0 P_0 = \rho_0$ and $\rho P = \rho$, so

$$(\rho - \rho_0)(Id - P) = \rho_0(P - P_0).$$

Hence, using the r-norm,

$$\|\rho - \rho_0\|_r \leq \frac{1}{1 - \|P\|_r}\|\rho_0(P - P_0)\|_r,$$

where $\|P\|_r \leq r + C\delta < 1$. Use results in the previous proof to obtain the result. $\square$

It would be nice to give an explicit example of use of these results, e.g. to extend the parameter range for unique stationary probability of the Stavskaya PCA to some larger interval than $(\frac{1}{2}, 1]$, but there isn't an easy way to achieve this (though it can be done by the methods of Maes and Shlosman, see [18], which consist roughly speaking in showing that $\|P^n\| < 1$ for some $n > 1$). Nevertheless, the results of this subsection have conceptual importance.

λ near 0

This is the low recovery rate, high infectivity case for the Stavskaya PCA. We will show, following Toom et al. [26], that

$$\delta_{\underset{\sim}{0}} P^t \nrightarrow \delta_{\underset{\sim}{1}} \text{ as } t \to \infty \text{ for } \lambda < \frac{1}{54},$$

where $\delta_{\underset{\sim}{0}}$ is the probability distribution supported on all infected and $\delta_{\underset{\sim}{1}}$ is that on all healthy. Note that $1/54$ is not an intrinsically important number, it is just to show that one can do concrete calculations.

Start from $\underset{\sim}{0}$ (all infected), the "worst case", and we ask for $P\{x_s^t = 1\}$ for $t > 0$. Without loss of generality, $s = 0$. The trick is to consider an equivalent way of generating the probability distribution on space-time configurations $\underset{\sim}{x} \in \{0, 1\}^{\mathbb{Z} \times \mathbb{Z}^+}$, where $\mathbb{Z} \times \mathbb{Z}^+$ represents space $\times$ time, starting from $\underset{\sim}{0}$. We generate probability distributions over space-time configurations from $\underset{\sim}{0}$ at $t = 0$ by putting "stoppers" in $\mathbb{R}^2$ (embedding the discrete space-time $\mathbb{Z} \times \mathbb{Z}^+$ in $\mathbb{R}^2$) from $\left(s - \frac{1}{2}, t - \frac{1}{4}\right)$ to $\left(s + \frac{1}{2}, t - \frac{1}{4}\right)$ with independent probabilities λ for each (s, t) and using the deterministic rules given in Figure 7.6 (the numbers $1/2, 1/4$ are chosen such that the stoppers "block" straight lines joining the space-time points). The no stopper rule is the case $\lambda = 0$ of the Stavskaya PCA; the stopper rule generates "healthy" in all circumstances. For example, see Figure 7.7.

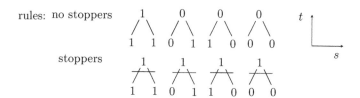

Figure 7.6 Stopper generating rules for construction of space-time configurations.

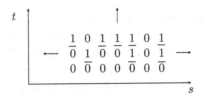

Figure 7.7 Example of space-time configuration constructed using stoppers.

This generates the same probability distribution $P_t^{\underset{\sim}{0}}$ as the PCA with initial condition $\underset{\sim}{0}$.

Therefore $x_0^t = 1$ (recall that without loss of generality $s = 0$) if and only if every path along ups and up-rights from $t = 0$ encounters a stopper, that is if and only if there exists a "fence" around $(0, T)$ starting at $\left(\frac{1}{2}, T + \frac{3}{4}\right)$ formed from stoppers (going from left to right) and down-left and ups, e.g. see Figure 7.8.

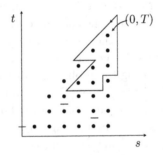

Figure 7.8 Example of a "fence" around $(0, T)$.

If a fence has k stoppers, then it has precisely k ups and k down-lefts. Let N_k be the number of fences with k stoppers, then

$$N_k \le \binom{3k}{k\ k\ k} \le 27^k$$

(a large over-estimate because we could exclude those which self-intersect and those which don't surround $(0, T)$, and without loss of generality one could always start with a down-left and end with an up). So

$$P^{\underset{\sim}{0}}\{x_0^T = 1\} \le \sum_{k \ge 1} N_k \lambda^k \le \frac{27\lambda}{1 - 27\lambda}, \text{ for } \lambda < \frac{1}{27},$$

$$< 1, \qquad \text{for } \lambda < \frac{1}{54},$$

independently of T.

Thus $P^0\{x_0^T = 1\} \nrightarrow 1$ as $T \to \infty$, and $P^0\{x_0^T = 0\} \geq c(\lambda) = \frac{1-54\lambda}{1-27\lambda} > 0 \ \forall T$. So $\delta_0 P^t \nrightarrow \delta_1$ as $t \to \infty$.

Where does $\delta_0 P^t$ go as $t \to \infty$? More generally what about νP^t for arbitrary initial ν?

To analyse this question we will use "monotonicity" of Stavskaya's PCA (also called "attractivity"). We define a partial order $\underline{x} \leq \underline{x}' \in \{0, 1\}^{\mathbb{Z}}$ if for all $s \in \mathbb{Z}$ we have that $x_s \leq x'_s$ (with $0 < 1$). Also, say $f : \{0, 1\}^{\mathbb{Z}} \to \mathbb{R}$ is *non-decreasing* if $\underline{x} \leq \underline{x}'$ implies that $f(\underline{x}) \leq f(\underline{x}')$. Say that probabilities $\rho \leq \rho'$ on $\{0, 1\}^{\mathbb{Z}}$ if for all non-decreasing f, we have that $\rho(f) \leq \rho'(f)$. Say that transition operator P is *monotone* if $\rho \leq \rho'$ implies that $\rho P \leq \rho' P$, equivalently if f non-decreasing implies Pf non-decreasing.

We prove Stavskaya's PCA is monotone by joining ("coupling") processes from two initial conditions $\underline{x} \leq \underline{x}'$ in such a way that for all $t > 0$ we have that $\underline{x}^t \leq \underline{x}'^t$, so it follows that

$$\left(P^t f\right)(\underline{x}) \leq \left(P^t f\right)(\underline{x}') \ \forall \text{ non-decreasing } f.$$

We denote the state $x_s^t x_s'^t$ of the joint process by $00, 01, 11$. The state 10 does not occur because initially $\underline{x} \leq \underline{x}'$ and our choice of joining never generates it. We choose the transition rates for the joint process to be:

$$00 \to \begin{cases} 11 \ P\{0 \to 1|\underline{x}\} \\ 01 \ P\{0 \to 1|\underline{x}'\} - P\{0 \to 1|\underline{x}\} \\ 00 \ P\{0 \to 0|\underline{x}'\}, \end{cases}$$

$$01 \to \begin{cases} 11 \ P\{0 \to 1|\underline{x}\} \\ 01 \ P\{0 \to 0|\underline{x}\} - P\{1 \to 0|\underline{x}'\} \\ 00 \ P\{1 \to 0|\underline{x}'\}, \end{cases}$$

$$11 \to \begin{cases} 11 \ P\{1 \to 0|\underline{x}\} \\ 01 \ P\{1 \to 0|\underline{x}\} - P\{1 \to 0|\underline{x}'\} \\ 00 \ P\{1 \to 0|\underline{x}'\}. \end{cases}$$

It can be checked that the marginals on the first and second components are just copies of the Stavskaya process. This proves monotonicity of the Stavskaya process.

Then $\delta_0 P \geq \delta_0$, so $\delta_0 P^t$ is a non-decreasing sequence. It is bounded above by $\delta_{\underline{1}}$, and so converges to its supremum in the partial order $\leq$, see Figure 7.9 (consider $\delta_{\underline{0}} P^t$ acting on arbitrary non-decreasing f). Call the limit ν_λ, so $\delta_{\underline{0}} P^t \to \nu_\lambda$ as $t \to \infty$.

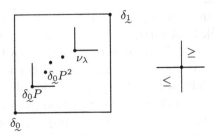

Figure 7.9 Monotonicity of the dynamics of P on $\mathcal{P}$ (represented schematically as a square).

In addition:

- ν_λ is non-decreasing with respect to λ (prove this by joining processes for $\lambda < \lambda'$ to make $\underline{x}^t(\lambda) \leq \underline{x}^t(\lambda')$).
- There exists a value $\lambda_c \in \left(\frac{1}{54}, \frac{1}{2}\right)$ such that $\nu_\lambda = \delta_{\underline{1}}$ for all $\lambda \geq \lambda_c$, and $\nu_\lambda < \delta_{\underline{1}}$ for all $\lambda < \lambda_c$ (this follows from ν_λ non-decreasing with λ).
- If $\lambda < \lambda_c$ then $\nu_\lambda(\underline{1}) = 0$.

Proof of the latter Else if $\nu_\lambda(\underline{1}) = \rho > 0$, then the conditional probability $\mu = \nu_\lambda(\cdot \mid \text{not } \underline{1})$ is stationary and $\nu_\lambda = p\delta_{\underline{1}} + (1+p)\mu$. So $\mu < \nu_\lambda$. However, $\mu \geq \delta_{\underline{0}}$, so

$$\mu = \mu P^t \geq \delta_{\underline{0}} P^t \to \nu_\lambda \text{ as } t \to \infty,$$

and finally, we reach the contradiction $\mu \geq \nu_\lambda$. $\qquad\square$

- For $\lambda < \lambda_c$ and any subset $A \subset \mathbb{Z}$, let

$$\alpha_A = P\{\text{infection never dies} \mid \text{initial infected set is } A\},$$

then $\alpha_A = 1$ if A is infinite (if A is finite $\alpha_A \leq 1 - \lambda^{2|A|} < 1$).
- For all $\nu \in \mathcal{P}$,

$$\nu P^t \to \gamma\delta_{\underline{1}} + (1-\gamma)\nu_\lambda \text{ as } t \to \infty,$$

with $\gamma = P\{\text{eventual absorption}\}$ $(= 1 - \alpha_A$ if $\nu = \delta_A)$, so in particular there are no other stationary probabilities than the convex combinations of ν_λ and $\delta_{\underline{1}}$. See Figure 7.10.

Figure 7.10 Schematic of $\mathcal{P}$ indicating the line segment of stationary probabilities from ν_λ to δ_1.

For finite versions (on $\mathbb{Z}_N$) the estimate of $P^0\{x_0^T = 1\}$ does not apply for $T > N$ because fences can wrap round the cylinder $\mathbb{Z}_N \times \mathbb{Z}$ and the counting has to be changed, but one can obtain estimates depending on T, N and prove the exponentially long time to absorption (with respect to N) for λ small.

Variations on Stavskaya's PCA

The following are some possible variations on Stavskaya's PCA:

- One can make the recovery probability λ_1 and avoiding infection probability λ_2 differ (and make recovery depend on the state of the left-hand neighbour) and obtain similar results.
- One can add infectivity from the other side: by monotonicity one obtains similar results.
- I don't know about the effects of small changes breaking monotonicity (one would need to go back to drawing fences), but the large λ regime is still easy.
- One can go to higher dimensions and inhomogeneous networks: see [13] for continuous time. To go to higher dimensions, a useful method is to compare the "radial process" of the extent of infection with Stavskaya's PCA on the half-line $\mathbb{Z}_+$, using monotonicity.
- There is an alternative treatment of Stavskaya's PCA (under the name "oriented percolation") by Durrett [9], which involves looking at the velocity of propagation of a front between infected and healthy zones.

7.1.2 The majority voter PCA

Stavskaya's PCA has the feature of an absorbing state, so is not communicating (a Markov process is *communicating* if it is possible to get from any state to any other and back). The majority voter PCA has a non-unique phase even though communicating. It is important to distinguish it from the many other voter models (see e.g. [13]) with absorbing states, which I prefer to call "opinion-copying models", the randomness coming solely from the choice of which neighbour's opinion to copy.

The majority voter PCA was proposed and studied numerically by [27] and analysed by Toom. We concentrate on the NEC (north-east-centre) model: $S = \mathbb{Z}^2$, $x_s \in \{0,1\} \approx \{-,+\}$ and

$$x_s^{t+1} = \begin{cases} \text{majority of } x_s^t, x_{s+E}^t, x_{s+N}^t, \text{ with probability } 1 - \lambda, \\ \text{the opposite with probability } \lambda, \end{cases}$$

where $E = (1,0)$ and $N = (0,1)$.

Our treatment will be much more sketchy than for Stavskaya's PCA. Key properties of the model are as follows:

- There is an exponentially attracting, unique stationary probability for λ near $1/2$ (because it is weakly dependent).
  Exercise: prove that $|\lambda - \frac{1}{2}| < \frac{1}{6}$ implies weak dependence.

- P is monotone for $\lambda \leq \frac{1}{2}$, so $\delta_0 P^t$ approaches some $\underline{\mu}$, and $\delta_1 P^t$ approaches some $\bar{\mu}$, where $\underline{\mu} \leq \bar{\mu}$ (reflections of each other via $+ \leftrightarrow -$). See Figure 7.11.

- For small enough λ, we have $\underline{\mu} < \bar{\mu}$.

Figure 7.11 The space $\mathcal{P}$ of probabilities for the NEC majority voter PCA.

Idea of proof of the latter For $\lambda = 0$, no islands of 1 in a sea of 0 will survive. Toom calls the deterministic cellular automaton an "eroder". Given an island of 1 in a sea of 0, draw its south-west envelope: the island can't grow beyond it. However, every north-east corner is eroded, so after finitely many steps the island will disappear. See Figure 7.12.

Now turn on $\lambda > 0$ and show that

$$\underline{\mu}\left(x_{00}^t = 1\right) \leq \text{ some } \psi(\lambda) \to 0 \text{ as } \lambda \to 0,$$

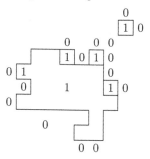

Figure 7.12 Island of 1s in a sea of 0s for the NEC majority voter model.

where x_{00}^t is the state at site $(0,0) \in \mathbb{Z}^2$ at time t, by showing that

$$P^0\left(x_{00}^t = 1\right) = \sum_{m=0}^{\infty} \sum_{\Gamma} P\left(G(\underline{x}) = \Gamma\right), \quad \text{where } \Gamma \text{ are certain graphs on}$$
$$\text{S} \times \text{T with } m \text{ edges}$$

$$\leq \sum_m N_m \, P(\text{error at each vertex})$$

$$\leq \sum_m 48^{2m} \lambda^{m/4+1}$$

$$= \frac{\lambda}{1 - 48^2 \lambda^{1/4}} \quad \text{for } \lambda < \frac{1}{48^8}$$

$$< \frac{1}{2} \text{ for } \lambda \text{ small enough,}$$

meaning that $1 - 48^2 \lambda^{1/4} > 2\lambda$, e.g. $\lambda < \frac{1}{2} 48^{-8}$. □

For details, see Toom's original proof [25], or lecture notes [26] or [12], from which the above estimates are taken.

Thus, we have at least two stationary probabilities (plus their convex combinations). I call these "ferromagnetic" phases (cf. statistical mechanics). There may be yet more stationary probabilities.

Possible variations

- For λ near 1 the model can be called an "antimajority" voter. The model is equivalent to parameter value $1 - \lambda$ by recoding x^t for odd t as $-x^t$. So if $\delta_0 P_\lambda^t \to \underline{\mu}$, then $\delta_0 P_{1-\lambda}^t \to$ a 2-cycle, with $\underline{\mu}$ at even t and $\bar{\mu}$ at odd t. So on $\{0,1\}^{S \times T}$ we have (at least) two probabilities, which we call "period-2" phases. This is an example of "nontrivial collective behaviour", or "asymptotic periodicity".

- We can also make PCA which have "antiferromagnetic" phases, by considering x_s^{t+1} = majority of $(x_s^t, -x_{s+E}^t, -x_{s+N}^t)$. One can prove this by recoding the usual NEC PCA by multiplication by $(-)^{\text{parity of } s}$ where the parity of s is $s_1 + s_2 \mod 2$.
- By taking λ near 1 in the antiferromagnetic model we obtain "antiferromagnetic period-2" phases. See Figure 7.13.
- One can break the $+ \leftrightarrow -$ symmetry and still prove non-unique stationary probabilities for λ small.
- One can make the interaction anisotropic (see [4]) and still obtain non-unique stationary probability.

$$
\begin{array}{ccc}
+ & - & + \\
- & + & - \\
+ & - & +
\end{array}
\qquad\qquad
\begin{array}{ccc}
- & + & - \\
+ & - & + \\
- & + & -
\end{array}
$$
$$
\text{even } t \qquad\qquad\qquad \text{odd } t
$$

Figure 7.13 Antiferromagnetic NEC voter PCA taking λ near 1 to obtain "antiferromagnetic period-2" phases.

For the last two points, contrast the 2D Ising model where the non-unique phase for low temperature is lost as soon as one adds a magnetic field.

Many other variants of the NEC majority voter PCA are possible, e.g. [7].

Open questions

- What about adding south and west neighbour influence? One can allow the error rate to depend weakly on the state of the S and W neighbours and still keep non-unique stationary probability, but numerically one obtains the same with the strong perturbation of enlarging the majority neighbourhood to NECSW, yet we are not aware of any proof.
- Majority voter models in continuous time? Note that there exist continuous-time voter models (see e.g. [13]), but they are very different because they choose to duplicate a random neighbour's state precisely, so $\widetilde{+}, \widetilde{-}$ are absorbing.

7.1.3 General properties for phases of PCA

Phases and emergence

Define a *phase* of a PCA as a limit point of probability on space-time configurations started from an initial space probability in the distant past. Any convex combination of phases is a phase, so it suffices to consider "extremal" phases, i.e. those which are not combinations of any others. Thus, Stavskaya's PCA has two extremal phases for λ small enough: those generated by the time evolution of $\delta_{\underset{\sim}{1}}$ and ν_λ. Phases do not have to be time-translation invariant, e.g. the period-2 phases of the NEC voter PCA for $\lambda = 1 - \varepsilon$, ε small.

Emergence maps the dynamical model to the set of phases. The amount of emergence expressed by a phase μ is

$$D(\mu, \{\text{products of independent dynamics}\}),$$

using the Dobrushin metric D (now on space-time phases, rather than on probabilities on configuration space). This can be considered to be the distance from all possible mean field phases. Weak emergence means we get $D > 0$.

For strong emergence the set of phases is not a singleton, i.e. a non-unique phase. The amount of strong emergence is given by the diameter of the set of phases, e.g. for Stavskaya's PCA,

$$\text{diam}\left(\{\text{phases}\}\right) = D(\nu_\lambda, \delta_{\underset{\sim}{1}}),$$

which by analogy to a computation in [7] is $c(\lambda)$ (the probability of a given site being infected in ν_λ). See Figure 7.14.

Figure 7.14 Amount of strong emergence for Stavskaya's PCA.

One word of caution: a system may exhibit strong emergence and yet none of its space-time phases exhibit weak emergence. For example, consider a weakly coupled network of bistable units. The good conjecture would be that for indecomposable systems exhibiting strong emergence, at least one of the phases exhibits weak emergence.

See my paper in *Nonlinearity* for some more open problems [16].

Gibbsian view of space-time phases

The space-time phases for PCA can be re-defined as the probabilities μ on the set of space-time configurations $\underset{\sim}{x}$ such that for all bounded

$\Lambda \subset S \times T$ and configurations $\underset{\sim}{x}_{\Lambda^c}$ on the complement $\Lambda^c = (S \times T)\backslash\Lambda$, the conditional probability of μ given $\underset{\sim}{x}_{\Lambda^c}$ is proportional to

$$\prod_{R:R\cap\Lambda\neq\emptyset} P_R(\underset{\sim}{x}),$$

where R ranges over subsets of $S \times T$ of the form $\{(s, t+1)\}\cup(N(s)\times\{t\})$ and $P_R(\underset{\sim}{x}) = p_s(x_s^{t+1}|\underset{\sim}{x}^t)$. See Figure 7.15.

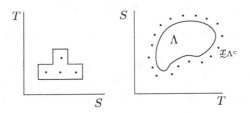

Figure 7.15 Gibbsian view of space-time phases.

One can rewrite the product as

$$\exp\left(-\sum_{R:R\cup\Lambda\neq\emptyset}\phi_R(\underset{\sim}{x})\right), \text{ with } \phi_R(\underset{\sim}{x}) = -\log P_R(\underset{\sim}{x}).$$

This is the defining (Dobrushin–Lanford–Ruelle) condition for Gibbs phases in equilibrium statistical mechanics (where we would have configurations on only S and not $S \times T$), with "energy" function $\sum_R \phi_R(\underset{\sim}{x})$. Recall from equilibrium statistical mechanics that for energy function $H = \sum_R H_R$ and "coolness" $\beta \in \mathbb{R}^+$ (inverse temperature), the conditional probability for configuration $\underset{\sim}{x}$ with respect to counting measure is given by

$$P(\underset{\sim}{x}_\Lambda|\underset{\sim}{x}_{\Lambda^c}) = \frac{1}{Z_\Lambda(x_{\Lambda^c})}e^{-\beta\sum_{R:R\cap\Lambda\neq\emptyset}H_R(\underset{\sim}{x})}$$
$$= e^{-\beta(\sum_{R:R\cap\Lambda\neq\emptyset}H_R(\underset{\sim}{x})-F_\Lambda(\underset{\sim}{x}_{\Lambda^c})|\Lambda|)},$$

where Z_Λ is a normalisation constant called the partition function and $F = \lim_{|\Lambda|\to\infty} F_\Lambda$ is called the free energy per site, see e.g. Lebowitz, Maes and Speer [12]. One can absorb β into H and F, so in our interpretations $\beta = 1$.

Note, however, that the "energy functions" for PCA have the special feature that $F = 0$, because

$$\sum_{\underset{\sim}{x}\text{ on }\Lambda}\prod_{R:R\cap\Lambda\neq\emptyset} P_R(\underset{\sim}{x}) = P\text{ (upper boundary values | lower ones)},$$

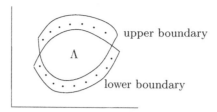

Figure 7.16 Surface to volume effect.

which scales like the surface area of Λ, not the volume of Λ. See Figure 7.16.

So generic statements for equilibrium statistical mechanics do not apply, e.g. Gibbs' phase rule in equilibrium statistical mechanics says that generically the co-existence of N extremal phases is of codimension $(N-1)$, such as in the $2D$ Ising model.

One can ask how the phase or set of phases varies with parameters. In the exponentially attracting regime the (unique) phase varies smoothly with parameters (interpreted as in [17]). In general, the set of phases depends upper semi-continuously on parameters, but it need not be lower semi-continuous. For example, consider discrete-time Glauber dynamics of the 2D Ising model below the critical temperature: as the magnetic field h crosses zero the set of phases jumps. The non-unique phase for the NEC majority voter PCA is robust to small changes in the model, because the perturbations can be compared with the NEC majority voter with slightly larger error rate. This gives an explicit example where Gibbs' phase rule fails for PCA, but the argument depends on monotonicity, and so the question of the generic robustness of the non-unique phase for PCA remains open.

Indecomposability

If the state-space decomposes into two or more components which do not communicate then we obtain non-unique phases trivially, because each communicating component supports at least one phase, but don't count this as strong emergence, because it suffices to check in which component you started to understand which phase you will see.

Say a PCA is "indecomposable" [15] if there exists a $D_0 \in \mathbb{R}_+$ such that for all finite subsets $A, B \subset S \times T$ with separation greater than or equal to D_0 (with respect to a metric on $S \times T$, e.g. the sum of the displacements in S and T), and two realisations x, x' then there exists a realisation z agreeing with x on A and x' on B, cf. "specification

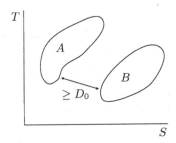

Figure 7.17 Indecomposability of PCA.

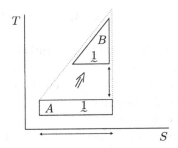

Figure 7.18 Failure of Stavskaya's PCA to be indecomposable.

property" in dynamical systems theory, e.g. for the NEC voter, one can take $D_0 = 1$. See Figure 7.17.

Stavskaya's PCA does not satisfy this indecomposability property because of the absorbing state $\underline{1}$, but deserves to be called strongly emergent (for λ small enough) because the phase corresponding to ν_λ is not associated with another communicating component but with the transient set (Figure 7.18). The finite versions of Stavskaya's PCA have a single communicating component (the absorbing state) even though it is not the whole state space. Therefore, we say a PCA is "pre-indecomposable" if it has a unique indecomposable subset and it can be attained in bounded time and we count the non-unique phase as strong emergence for pre-indecomposable systems.

7.2 Deterministic case – coupled map lattices (CML)

Now we turn to deterministic dynamical systems. We restrict attention to the simplest case of coupled map lattices: discrete-time dynamics on a product of manifolds. First we will show how to translate PCA examples into coupled map lattices. Second we will develop a general theory of space-time phases for uniformly hyperbolic coupled map lattices; this requires a rapid summary of finite-dimensional theory.

7.2.1 Examples with non-unique space-time phase

A CML (coupled map lattice) is a map $f : M \to M$, where $M = \prod_{s \in S} M_s$, M_s is a finite-dimensional manifold, e.g. $[-1, +1]$, and S is a countable metric space. We want to turn our PCA examples into CML examples.

Given a PCA on $\Sigma = \prod_{s \in S} \Sigma_s$, with transition probabilities p_s^σ for $s \in S, \sigma \in \Sigma_s$, partition $[-1, +1]$ into $|\Sigma_s|$ equal length intervals labelled by the possible values of $\sigma_s \in \Sigma_s$ (Figure 7.19), define function $\sigma : [-1, +1] \to \Sigma_s$ correspondingly, and construct a piecewise affine map $f : M \to M$ with

$$\frac{\partial f_s}{\partial x_s}(x) = \frac{1}{p_s^{\sigma(x)}(\sigma_s')}, \quad \frac{\partial f_s}{\partial x_r} = 0 \text{ for } r \neq s, \text{ for } x_s \in \sigma_s, x_r \in \sigma_r,$$

where σ_s' is the symbolic state of $f_s(x)$, e.g. for the NEC voter, see Figure 7.20.

Figure 7.19 Partition of $[-1, +1]$ into two intervals for $\Sigma_s = \{-, +\}$.

Then for any initial probability on M with absolutely continuous marginals on finite subsets Λ of S and density h_Λ satisfying a Hölder condition

$$|\log h_\Lambda(x) - \log h_\Lambda(y)| \leq C \sum_{s \in \Lambda} |x_s - y_s|^\delta$$

in the distant past, $\sigma(x)$ is distributed according to a phase of the PCA [10]. So, for example, one can make CML with ferromagnetic phases.

The idea is that the initial probability is repeatedly stretched and cut, and so becomes more uniform in each cylinder set (set with prescribed symbols), and the action on probabilities that are uniform on cylinder sets is precisely the PCA (see Figure 7.21) (cf. random number generators).

The Hölder condition on the densities is certainly stronger than necessary, but we did not determine a better condition.

This construction provided a response to a challenge by Bunimovich

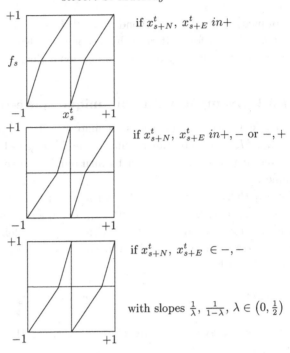

Figure 7.20 CML for the NEC voter model.

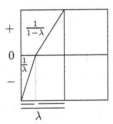

Figure 7.21 Action of f on probabilities.

and Sinaĭ [6] to construct a CML with non-unique phase. A similar construction was made by [22] to simulate the discrete-time Glauber dynamics of the 2D Ising model, with variable partition according to the symbolic state of the neighbours and constant slope within the partition element, but it is not clear that paper contained a proof.

Similarly, we can make CML with period-2 phase (simulate the NEC voter with $\lambda \in (1/2, 1)$), anti-ferromagnetic, persistent infection phases, etc.

One can make invertible examples by replacing $[-1, +1]$ by a solid torus $[-1, +1] \times D^2$ modulo identification of the ends, and using distorted versions of the solenoid map, depending on the symbolic state of the neighbours. Recall the standard (Smale–Williams) solenoid map [20]: $(x, w) \in S^1 \times D^2$, where $S^1 = \mathbb{R}/2\mathbb{Z}$ (note we're taking S^1 of length 2), $D^2 = \{w \in \mathbb{C} : |w| \leq 1\}$, and

$$\begin{cases} x' = 2x, \\ w' = \lambda w + \mu e^{i\pi x}, \end{cases}$$

with $\lambda < \mu$, $\mu + \lambda \leq 1$, to make it map $S^1 \times D^2$ into itself and one-to-one. The intersection of the forward images of the solid torus under the solenoid map is a strange attractor – a Cantor set of lines locally (see Figure 7.22).

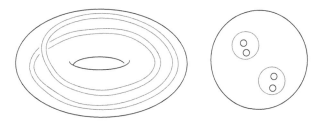

Figure 7.22 Solenoid map.

Our distorted solenoid maps are maps on $[-1, +1] \times D^2$ depending on the symbolic state of neighbours, where we set $\sigma_s = \text{sign}(x_s)$, e.g. for $++$ neighbours:

$$\begin{cases} x' = g_{++}(x), \\ w' = \lambda w + \mu e^{i\pi x}, \end{cases}$$

where g_{++} denotes the first map of Figure 7.20. The addition of the D^2 dynamics makes the map one-to-one. To make it surjective, we need to extend the solenoid maps outside $S^1 \times D^2$ to say $S^1 \times \sqrt{2}D^2$ to make backward orbits exist for all initial points, e.g. by

$$w' = (2 - |w|^2)(\lambda w + \mu e^{i\pi x}) + (|w|^2 - 1)w \text{ for } 1 \leq |w| \leq \sqrt{2};$$

see Figure 7.23. Alternatively, regard D^2 as the southern hemisphere of S^2 and extend the map to the northern hemisphere with a repellor at the north pole.

Our CML is not continuous, however, because of the jumps at changes

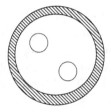

Figure 7.23 Extending solenoid map outside $S^1 \times D^2$.

of symbolic state of neighbours. Bardet and Keller [3] overcame this defect, they even made C^∞ examples. Their main idea was to replace symbolic coupling by discrete diffusion

$$x_s^{t+1} = (1 - \varepsilon)e(x_s^t) + \frac{\varepsilon}{2}(e(x_{s+N}^t) + e(x_{s+E}^t)),$$

with e a carefully chosen map, but it is not clear how to make it invertible.

7.2.2 Statistical phases for uniformly hyperbolic attractors of finite-dimensional deterministic dynamical systems

We shall now review the statistical theory of "chaos". Thus for this section we drop the study of spatially extended systems and consider finite-dimensional dynamical systems only. Instead of space-time phases, we will obtain just time-phases. This is only "complexity science" if we think of the states at successive times as the interdependent components. It is a warm-up for the theory of space-time phases for spatially extended deterministic differentiable dynamical systems. The key idea is to introduce a symbolic description of the orbits, so the approach is limited to systems for which this is possible. We shall see, however, that this is a large class, in particular C^1-open.

The idea (due to Sinaĭ [23]) is that, under suitable conditions, the deterministic dynamics on an attractor is equivalent to a stochastic dynamics on a finite symbol set, of a form called "Gibbsian".

The setting is an iteration of a C^1-diffeomorphism $x^{t+1} = f(x^t)$, $x^t \in M$, a C^1 manifold with norm $|\cdot|$ on tangent vectors and associated Finsler metric. A key trick is to note that its orbits $\underline{x} = (x^t)_{t \in \mathbb{Z}}$ are the fixed points of a "super-map" $F : M^{\mathbb{Z}} \to M^{\mathbb{Z}}$ defined by $F(\underline{x})^t = f(x^{t-1})$. F is a C^1 map when $M^{\mathbb{Z}}$ is endowed with the supremum metric over its components. The tangent space $T(M^{\mathbb{Z}})$ at a sequence x is the set of sequences ξ of tangent vectors ξ^t to M at x^t with

$$\|\underline{\xi}\|_{\infty} = \sup_{t\in\mathbb{Z}} |\xi^t| < \infty.$$

We sometimes denote it $TM_{\infty}^{\mathbb{Z}}$.

Definition 7.2.1 (UH)   An orbit $\underline{x}$ is *uniformly hyperbolic* if it is a non-degenerate fixed point of F, i.e. $F(\underline{x}) = \underline{x}$ and

$$I - DF_{\underline{x}} = \begin{pmatrix} \ddots & & & & & \\ & \ddots & I & & & \\ & & -f'_{-1} & I & & \\ & & & -f'_0 & I & \\ & & & & -f'_1 & I \\ & & & & & \ddots & \ddots \end{pmatrix}$$

has a bounded inverse, where f'_t denotes $f'(x^t)$. A set Λ of orbits is uniformly hyperbolic if $\|(I - DF_{\underline{x}})^{-1}\|$ is bounded for $\underline{x} \in \Lambda$.

Example 7.2.2 (Arnold's cat map)   Consider

$$f = \begin{pmatrix} 2 & 1 \\ 1 & 1 \end{pmatrix} \text{ on } M = \mathbb{T}^2 = \mathbb{R}^2/\mathbb{Z}^2.$$

The whole of $\mathbb{T}^2$ is uniformly hyperbolic, which is seen by solving

$$(I - DF)\underline{\xi} = \underline{\eta} \tag{7.1}$$

for $\underline{\xi} \in TM_{\infty}^{\mathbb{Z}}$, given $\underline{\eta} \in TM_{\infty}^{\mathbb{Z}}$. This can be done by splitting ξ^t, η^t into components $\xi^t_{\pm}, \eta^t_{\pm}$ along the eigenvectors of f. See Figure 7.24.

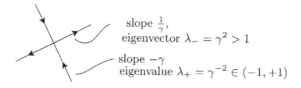

slope $\frac{1}{\gamma}$,
eigenvector $\lambda_- = \gamma^2 > 1$

slope $-\gamma$
eigenvalue $\lambda_+ = \gamma^{-2} \in (-1, +1)$

Figure 7.24 Eigenvectors of f.

The notation is: $+$ means forward contracting and $-$ means backward contracting. Many books use $+ = s$ and $- = u$, standing for stable and unstable. This is poor terminology, however, as can be seen by considering points near the eigen-directions, and noticing that actually the "stable" direction is unstable and vice-versa.

The splitting disconnects equations (7.1) into separate equations for the $+$ components and the $-$ components:

$+$ component: $\xi_+^t - \lambda_+ \xi_+^{t-1} = \eta_+^t$. Rewrite it as $\xi_+^t = \eta_+^t + \lambda_+ \xi_+^{t-1}$ to see it has a unique bounded solution $\xi_+^t = \sum_{n \geq 0} \lambda_+^n \eta_+^{t-n}$.

$-$ component: $\xi_-^t - \lambda_- \xi_-^{t-1} = \eta_-^t$. Rewrite it as $\xi_-^{t-1} = \frac{1}{\lambda_-}(\xi_-^t \eta_-^t)$ which has a unique bounded solution $\xi_-^t = -\sum_{n \geq 1} \lambda_-^{-n} \eta_-^{t+n}$.

Thus we have a unique bounded solution $\underline{\xi} = \underline{\xi}_+ + \underline{\xi}_-$ and with respect to Euclidean norm on the tangent space $T\mathbb{T}^2$, for example, we have that

$$\|\underline{\xi}_+\| \leq \frac{\|\eta_+\|}{1 - \lambda_+}, \quad \|\underline{\xi}_-\| \leq \frac{\|\eta_-\|}{\lambda_- - 1}, \quad \text{and so} \quad \|(I - DF)^{-1}\| \leq \frac{1}{1 - \lambda_+}.$$

The usual definition of uniform hyperbolicity is:

Definition 7.2.3 (UH$'$) Diffeomorphism $f : M \to M$ is *uniformly hyperbolic* if there exists an invariant splitting of

$$TM_{x^t} = E_{x^t}^+ \oplus E_{x^t}^-$$

along orbit $\underline{x}$, and $C \in \mathbb{R}^+, \lambda \in (0, 1)$ such that $\xi^t \in E_{x^t}^{\pm}$ implies

$$|\xi^{t+s}| \leq C\lambda^{|s|}|\xi^t|,$$

for all $s > 0$ for E^+, all $s < 0$ for E^-. See Figure 7.25.

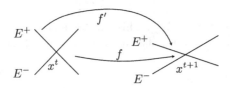

Figure 7.25 Invariant splitting of uniform hyperbolicity.

This definition of uniform hyperbolicity is more or less equivalent to definition UH (Definition 7.2.1). In particular, it is implied by UH if $f_t' = f'(x^t), t \in \mathbb{Z}$ is bounded (which is automatic if M is compact).

Proof Suppose $(I - DF_{\underline{x}})\underline{\xi} = \underline{\eta}$ has unique bounded solution $\underline{\xi}$ for any bounded $\underline{\eta}$. In particular, take $\eta^t = 0$ for all $t \neq 0$. Let $\tilde{\xi}^t = z^{|t|}\xi^t$, for some $z > 1$ and try to prove $\underline{\tilde{\xi}}$ is bounded.

Now $\xi^t - f'_{t-1}\xi^{t-1} = \eta^t$, therefore

$$\begin{cases} \tilde{\xi}^t - zf'_{t-1}\tilde{\xi}^{t-1} = 0, \ t > 0, \\ \tilde{\xi}^t - \frac{1}{z}f'_{t-1}\tilde{\xi}^{t-1} = \eta^t, \ t \leq 0 \ (= 0 \text{ for } t < 0). \end{cases}$$

So $(I - DF - E)\tilde{\xi} = \eta$ with

$$E = \begin{pmatrix} \ddots & & & & & \\ & \ddots & & 0 & & \\ & & \left(\frac{1}{z} - 1\right)f'_{-1} & 0 & & \\ & & & (z-1)f'_0 & 0 & \\ & & & & \ddots & \ddots \end{pmatrix}.$$

Then $\|E\| \leq |z - 1|l$ with $l = \sup_t|f'_t|$ and so $(I - DF - E)$ has bounded inverse if

$$\|E\| < K := \|(I - DF)^{-1}\|^{-1},$$

and then

$$\|(I - DF - E)^{-1}\| \leq \frac{1}{K - \|E\|}.$$

So $\|\tilde{\xi}\| \leq \frac{|\eta^0|}{K - |z-1|l}$, i.e. $|\xi^t| \leq \frac{z^{-|t|}|\eta^0|}{K - |z-1|l}$ decays exponentially both ways in t. One obtains the same for any initial time s.

Then split $\eta^0 = \eta^0_+ + \eta^0_- = \xi^0 - f'_{-1}\xi^{-1}$ and check that this defines a complementary pair of projections $P^\pm$ with $P^+\eta^0 = \eta^0_+$ and $P^-\eta^0 = \eta^0_-$, $P^2 = P$, $P^+ + P^- = I$ and invariance of splitting and uniform exponential decay estimates. $\qquad\square$

Similarly, definition UH′ (Definition 7.2.3) plus some extra hypotheses (notably that the angle of splitting is bounded away from 0) if M is non-compact implies UH.

A nice feature of uniformly hyperbolic orbits is "robustness": for all $\tilde{f}$ C^1-close to f, there exists a unique orbit $\tilde{x}$ of $\tilde{f}$ uniformly near to $\underline{x}$, i.e. $d(\tilde{x}^t, \underline{x}^t) \leq \delta$ (prove by applying implicit function theorem to $F(\underline{x}) = \underline{x}$). Furthermore, $\tilde{x}^0(x^0)$ is Hölder continuous and one-to-one, so given a uniformly hyperbolic set $\Lambda \subset M$ (which is the union of points of a uniformly hyperbolic set of orbits), for all $\tilde{f}$ C^1-close to f, there exists $h : \Lambda \to M$, a near-identity homeomorphism onto $h(\Lambda)$ such that

$$\tilde{f}h = hf \text{ on } \Lambda.$$

Similarly for t-dependent perturbations (but with h time-dependent).

Now we will show that the dynamics on a uniformly hyperbolic attractor is equivalent to a Gibbsian stochastic process on a symbol space (a generalisation of a Markov chain). We follow Adler's choice of definition of Markov partition [1].

Definition 7.2.4 A *Markov partition* of an invariant set Λ is a finite set of disjoint open subsets A_i whose closures cover Λ, such that, letting Γ be the graph with nodes A_i and edges $i \to j$ if

$$f(A_i) \cap A_j \neq \emptyset,$$

every doubly infinite path in Γ occurs as the symbolic trajectory of one and only one orbit in Λ.

Example 7.2.5 (Solenoid attractor)   Consider the attractor of $f :$ $S^1 \times D^2 \to S^1 \times D^2$ $(S^1 = \mathbb{R}/2\mathbb{Z})$ given by

$$\begin{cases} x' = 2x \in S^1, \\ w' = \lambda w + \mu e^{i\pi x} \in D^2 \subset \mathbb{C}. \end{cases}$$

This has a Markov partition: $0 = - = \{(x,w) : x \in (-1,0)\}$, $1 = + = \{x \in (0,1)\}$. See Figures 7.26 and 7.27.

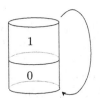

Figure 7.26 Markov partition for solenoid map.

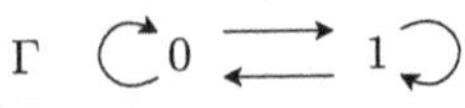

$$\Gamma \quad \overset{\curvearrowright}{0} \underset{\longleftarrow}{\overset{\longrightarrow}{}} 1 \overset{\curvearrowright}{}$$

Figure 7.27 The graph of allowed transitions for the solenoid attractor.

Example 7.2.6 (Arnold's cat map)   Consider

$$f = \begin{pmatrix} 2 & 1 \\ 1 & 1 \end{pmatrix} \text{ on } \mathbb{T}^2 = \mathbb{R}^2/\mathbb{Z}^2.$$

The partition $\{1, 2\}$ of Figure 7.28 is almost a Markov partition, but to achieve a unique orbit for each symbol sequence one has to subdivide region 1, e.g. into the three strips $\{1a, 1b, 1c\}$ (not shown) which form the parts of 1 which make the transitions to the first copy of 1, region 2, and the second copy of 1. Rather than draw the graph of allowed transitions on $\{1a, 1b, 1c, 2\}$ we prefer to use multiple edges in the graph on $\{1, 2\}$ to indicate the two transitions $1a \to 1, 1c \to 1$ which have to be distinguished to achieve a unique orbit, see Figure 7.29.

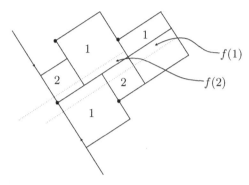

Figure 7.28 An almost Markov partition for Arnold's cat map.

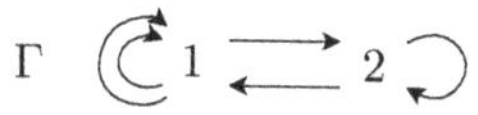

Figure 7.29 The graph Γ of allowed transitions.

Then f is topologically conjugate[4] to the shift σ on the set of doubly infinite paths in Γ

$$\sigma : \quad \cdots e_{-1}.e_0 e_1 \cdots \quad \mapsto \quad \cdots e_{-1} e_0.e_1 \cdots ,$$

where e_i label edges in Γ, with product topology (two paths are close if they agree on a long finite portion $e_M, \ldots, e_N$), modulo identifications of paths which correspond to the same orbit of f.

A general way to obtain symbolic dynamics is to show that for any locally maximal, compact uniformly hyperbolic set, there exists a Markov partition. This can be proved by using the following theorem:

---

[4] $f : M \to M$ is topologically conjugate to $g : N \to N$ if there exists a homeomorphism $h : M \to N$ such that $hf = gh$.

Theorem 7.2.7 ("Shadowing theorem")   *Given a uniformly hyperbolic set Λ there exist $K, \varepsilon_0 > 0$ such that for all $\varepsilon \leq \varepsilon_0$ every ε-pseudo-orbit is $K\varepsilon$-shadowed by a true orbit of f.*

Here, $\underset{\sim}{y}$ is an ε-pseudo-orbit for f if $d(fy_n, y_{n+1}) \leq \varepsilon$ for all n, i.e. $d(F\underset{\sim}{y}, \underset{\sim}{y}) \leq \varepsilon$ in supremum metric, and $\underset{\sim}{x}$ δ-shadows $\underset{\sim}{y}$ if $d(x_n, y_n) \leq \delta$ for all n, i.e. $d(\underset{\sim}{x}, \underset{\sim}{y}) \leq \delta$ in supremum metric. See Figure 7.30.

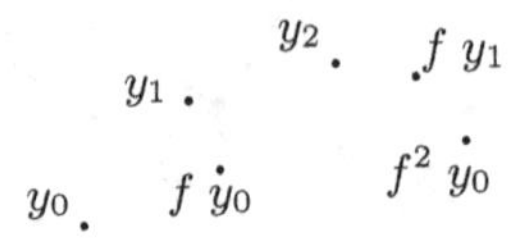

y_2

y_1 . $f\,y_1$

y_0 . $f\,\dot{y}_0$ $f^2\,\dot{y}_0$

Figure 7.30 An ε-pseudo-orbit.

One can prove the shadowing theorem by constructing a good approximation J to $(I - DF_{\underset{\sim}{y}})^{-1}$ and then showing $\underset{\sim}{x} \mapsto \underset{\sim}{x} - JF(\underset{\sim}{x})$ is a contraction in a neighbourhood of $\underset{\sim}{y}$ (in some local coordinate system about $\underset{\sim}{y}$).

To make a Markov partition, choose "enough" periodic orbits of f on Λ as the "symbols" σ and use the shadowing theorem to make the whole of Λ from true orbits shadowing a sequence of segments of the chosen periodic orbits.

One can also show that $x^t(\underset{\sim}{\sigma})$ depends exponentially weakly on σ^s, for $|s - t| > N$ large.

We are now ready to sketch why there is a *natural measure* on a uniformly hyperbolic attractor and how it corresponds to a Gibbsian process. Suppose $f \in C^{1+\text{Hölder}}$ and Λ is a uniformly hyperbolic, bounded topologically mixing attractor (*topologically mixing* means there exists an N such that for all $A, B \neq \emptyset$ open, $f^n A \cap B \neq \emptyset$ for all $n \geq N$). Then there exists a probability ρ on Λ (called the *SRB measure*, after Sinai, Ruelle and Bowen) such that for ν any absolutely continuous[5] probability on the basin $B(\Lambda)$ of attraction of Λ, $\nu f_*^t \to \rho$ as $t \to +\infty$ in weak-* topology.

The simplest way to understand weak-* topology on $\mathcal{P}(C)$, a compact neighbourhood of attraction for Λ, is to metrise it by

$$d(\mu, \nu) = \sup_{g \in \text{Lip}\backslash C} \frac{\mu(g) - \nu(g)}{\text{Lip}(g)},$$

---

[5] With respect to Lebesgue class, i.e. if λ denotes a Lebesgue class measure then $\lambda(A) = 0 \Rightarrow \nu(A) = 0$. Alternatively ν has a density h in $L^1(\lambda)$.

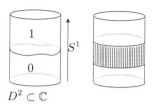

Figure 7.31 The set which will make transition $0 \to 1$.

C denoting the constant functions, and Lip the best Lipschitz constant.

The probability ρ can be obtained as a Gibbs phase for symbol sequences with "energy" contribution from time t

$$\phi^t_-(\underline{\sigma}) = \log|\det Df_{E^-}(x^t(\underline{\sigma}))|,$$

where Df_{E^-} is the restriction of the derivative to the unstable bundle, i.e.

$$\rho\{\sigma^{-T} \cdots \sigma^R| \left(\sigma^t\right)_{t<-T, t>+R}\} \propto \exp\{-\sum_{t\in\mathbb{Z}} \left(\phi^t_-(\underline{\sigma}) - \phi^t_-(\underline{\bar{\sigma}})\right)\},$$

for any allowed reference sequence $\bar{\sigma}$ satisfying the given past and future.

The Gibbs phase is unique because Λ is topologically mixing, the dependence of ϕ^t_- on σ^s decays exponentially with $|s - t|$ and $\mathbb{Z}$ is 1D so one can apply a 1D statistical mechanics result to deduce the unique phase [21].

Idea of proof We sketch a proof that $\nu f^t_* \to \rho$ as $t \to \infty$, given by the Gibbs potential $\log|\det Df_{E^-}(x^t(\underline{\sigma}))|$.

Given an absolutely continuous initial probability ν on $B(\Lambda)$ in the distant past $t = -M$, what are the relative probabilities of seeing allowed paths $\sigma^0 \cdots \sigma^N$? (We adopt the convention that σ^t labels the edge in Γ taken from time t to $t + 1$.)

We illustrate the relevant subsets for a nonlinear distortion of the solenoid map, see Figure 7.31. So $\sigma_0 = 0 \to 1$ means that at $t = 0$ we are in the shaded region in Figure 7.31. Then $\sigma^0 \cdots \sigma^N$ equal to a specified sequence corresponds to a thin slice at $t = 0$, see Figure 7.32. Go back to $t = -M$, but without specifying σ^t, $t < 0$. $P(\sigma^0 \cdots \sigma^N) = \nu$(set at $t = -M$ giving rise to this sequence) $= \nu$(union of 2^M disks), as shown in Figure 7.33.

The dependence of $P(\sigma^0 \cdots \sigma^N)$ on sequence $\sigma^0 \cdots \sigma^N$ is via the thicknesses of these disks. So

$$P(\sigma^0 \cdots \sigma^N) \propto \prod |\det Df_{E^-}(x^t)|^{-1}$$

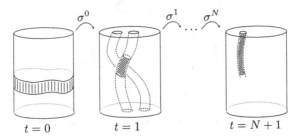

Figure 7.32 The sets of points at times $0, 1 \cdots N + 1$ corresponding to symbol sequence $\sigma^0 \cdots \sigma^N$.

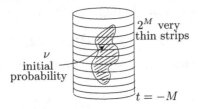

Figure 7.33 The 2^M thin disks corresponding to initial conditions at $t = -M$ which will perform sequence $\sigma^0 \cdots \sigma^N$.

along associated backwards orbits. More precisely, letting $\mathcal{C}$ be any choice of $\sigma^{-M/2} \cdots \sigma^{-1}, \sigma^{N+1} \cdots \sigma^{N+M/2}$,

$$\lim_{M \to \infty} P(\sigma^0 \cdots \sigma^N | \mathcal{C}, \nu \text{ at } t = -M) = \frac{1}{Z(\mathcal{C}, \bar{\sigma})} \prod_{t=-\infty}^{+\infty} \frac{|\det Df_-(x^t(\bar{\sigma}))|}{|\det Df_-(x^t(\underline{\sigma}))|},$$

where $\bar{\sigma}$ is some particular choice of $\sigma^0 \cdots \sigma^N$ completed by the chosen conditions $\mathcal{C}$ and Z is a normalisation constant. We choose to write this as

$$\frac{1}{Z} \exp\{-\sum_t (\phi^t(\underline{\sigma}) - \phi^t(\bar{\underline{\sigma}}))\},$$

with $\phi^t(\underline{\sigma}) = \log|\det Df_-(x^t(\bar{\sigma}))|$.

The function ϕ^t depends exponentially weakly on σ^s with respect to $|s - t|$, so the infinite sum makes sense and standard statistical mechanics (e.g. [21]) implies that there exists a unique probability with these conditionals. $\qquad\square$

7.2.3 Uniformly hyperbolic dynamics on networks

Now we move to the spatially extended context. A *network* is a countable metric space (S, d). For all $s \in S$, we suppose as local state space a finite-dimensional manifold M_s with a norm $|\cdot|_s$ on tangent vectors and the induced Finsler metric. We let $M = \prod M_s$ with supremum metric, and hence tangent vectors ξ satisfying $\|\xi\| = \sup_{s \in S} |\xi_s| < \infty$. The state $x^t \in M$ evolves by $x^{t+1} = f(x^t)$ (f is a CML). Suppose $f \in C^1$, in particular

$$\sup_{r \in S} \sum_{s \in S} |\frac{\partial f_r}{\partial x_s}| < \infty.$$

If all $M_s = V$ a vector space, then a standard class of examples is

$$x'_s = f_s^\varepsilon(x_s) = f_s^0(x_s) + \varepsilon \sum_{r \in S} C_{sr}(x_r - x_s),$$

with the matrix C satisfying $\sup_s \sum_r |C_{sr}| < \infty$. But one can also couple maps of nontrivial manifolds, like cat maps and solenoid maps.

The orbits of f are precisely the fixed points of the super-map $F : M^{\mathbb{Z}} \to M^{\mathbb{Z}}$ defined by $\underline{x} = (x_s^t) \mapsto f_s(\underline{x}^{t-1})$ for all $(s,t) \in S \times \mathbb{Z}$. We endow $M^{\mathbb{Z}}$ with supremum metric and then tangent vectors are those $\underline{\xi}$ with finite supremum norm. F is C^1 for $\underline{x}$ in bounded subsets.

Definition 7.2.8 An orbit $\underline{x}$ of the CML f is *uniformly hyperbolic* if it is a non-degenerate fixed point of F (using supremum norm on $M^{\mathbb{Z}}$), i.e. $\|(I - DF)^{-1}\| < \infty$.

One could apply the same results as in the previous section (the existence of a splitting, robustness, shadowing, Markov partition) except for deducing the unique natural probability if $|S| = \infty$ because $M = \prod_{s \in S} M_s$ is infinite dimensional and so $\det Df_-$ requires interpretation. Also one would not deduce anything useful about dependence on $s \in S$, e.g. spatial correlations of ρ.

So we do a space-time version of the uniform hyperbolicity theory (cf. [14], [19]). Assume that Df is *exponentially local* (which includes the case of finite-range interaction), i.e.

$$\exists \zeta > 1, \ \phi : [0, \zeta) \to \mathbb{R} \ \text{s.t.} \ \sup \sum_r \left| \frac{\partial f_r}{\partial x_s} \right| z^{d(r,s)} \leq \phi(z) \ \forall z \in [0, \zeta).$$

Then the uniformly hyperbolic splitting $TM_{\underline{x}} = E_{\underline{x}}^+ \oplus E_{\underline{x}}^-$ constructed as in the previous section is given by exponentially local projections $P_{\underline{x}}^\pm$.

One can prove this via the following theorem (the application will be to $S = S \times \mathbb{Z}$).

Theorem 7.2.9 *For each $s \in S$, let X_s, Y_s be Banach spaces and let* $X = \{x \in \prod_{s \in S} X_s : \|x\|_\infty < \infty\}$, $Y = \{y \in \prod_{s \in S} Y_s : \|y\|_\infty < \infty\}$.

Let $L : X \to Y$ be a linear map, ϕ-exponentially local, with bounded inverse, and let $y \in Y$ be (C, λ)-exponentially localised around $o \in S$, i.e. $|y_s| \le C\lambda^{d(s,o)}$. Then $x = L^{-1}y$ is (WC, μ)-exponentially localised around o for some W, μ, functions of $\|L^{-1}\|, \lambda, \phi$.

Proof (following Baesens and MacKay [2]).

Let $z \in [1, \zeta)$, with $z\lambda < 1$, $\tilde{y}_r = y_r z^{d(r,o)}$, $\tilde{x}_r = x_r z^{d(r,o)}$. Then $\|\tilde{y}\| < \infty$, and we want to bound $\|\tilde{x}\|$. Now $Lx = y$ implies that

$$\sum_s L_{rs} z^{d(r,o) - d(s,o)} \tilde{x}_s = \tilde{y}_r.$$

So $(L + L^o)\tilde{x} = \tilde{y}$ with $L^o_{rs} = L_{rs} \left(z^{d(r,o) - d(s,o)} - 1\right)$.

Let $\tilde{L}_{rs} = L_{rs} z^{d(r,s)}$, so $\|\tilde{L}\| < \phi(z) < \infty$. Now $d(r,o) - d(s,o) \le d(r,s)$, by the triangle inequality, so $|L^o_{rs}| \le |\tilde{L}_{rs} - L_{rs}|$, and so $\|L^o\| \le \|\tilde{L} - L\|$. For $z = 1$, $\tilde{L} = L$. $\tilde{L}(z)$ depends continuously on z because given $1 \le z_1 \le z_2 < \zeta$

$$(\tilde{L}(z_2) - \tilde{L}(z_1))_{rs} = L_{rs}(z_2^{d(r,s)} - z_1^{d(r,s)}).$$

Therefore

$$|(\tilde{L}(z_2) - \tilde{L}(z_1))_{rs}| \le |L_{rs}| \, d(r,s) z_2^{d(r,s)} \log \frac{z_2}{z_1}$$

$$\le \frac{1}{e} \frac{\log z_2 / z_1}{\log z_3 / z_1} z_3^{d(r,s)} |L_{rs}|, \text{ for any } z_3 > z_2,$$

and so

$$\|\tilde{L}(z_2) - \tilde{L}(z_1)\| \le \frac{1}{e} \frac{\log z_2 / z_1}{\log z_3 / z_1} \|\tilde{L}(z_3)\|$$

$$\le \frac{1}{e} \frac{\log z_2 / z_1}{\log z_3 / z_1} \phi(z_3).$$

In particular, $\|\tilde{L}(z) - L\| \le \beta(z) = \min_{z_3} \frac{1}{e} \frac{\log z}{\log z_3 / z} \phi(z_3) \to 0$ as $z \searrow 1$, and so is less than $\|L^{-1}\|^{-1}$ for z near enough to 1. Hence $\|(L + L^0)^{-1}\| \le (\|\tilde{L}\|^{-1} - \|\tilde{L} - L\|)^{-1}$ for z near 1. Finally

$$\|\tilde{x}\| \le \frac{\|\tilde{y}\|}{\|L^{-1}\|^{-1} - \|\tilde{L} - L\|}.$$

□

Apply this to the uniformly hyperbolic splitting for CML by taking $S = S \times \mathbb{Z}$ and considering DF exponentially local with respect to

$$\bar{d}((s_1, t_1), (s_2, t_2)) = d(s_1, s_2) + |t_2 - t_1|,$$

and repeat the construction of the splitting and see that the resulting projections $P^{\pm} : TM_x \to TM_x$ are exponentially local with respect to d.

As for the case of time only, we obtain the robustness of uniformly hyperbolic orbits in $S \times T$ with respect to exponentially local perturbations in space. Also one can prove $S \times T$ shadowing. Similarly one can obtain Markov partitions for uniformly hyperbolic sets of CML (cf. [19]). This means a coding $\underline{x} = \underline{x}(\underline{\sigma})$ of the orbits $\underline{x} = (x_s^t)_{s \in S}^{t \in T = \mathbb{Z}}$ of the uniformly hyperbolic set by $S \times T$ symbol tables $\underline{\sigma}$ (see Figure 7.34) from some allowed set Σ such that every allowed table occurs and there exists a unique $\underline{x}$ for each $\underline{\sigma}$.

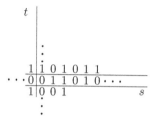

Figure 7.34 A space-time symbol table for a CML of solenoid maps.

The simplest case is $\Sigma = (\prod_{s \in S} \Sigma_s)^{\mathbb{Z}}$ with each Σ_s finite, e.g. an uncoupled lattice of solenoid maps $\Sigma_s = \{0, 1\}$ and

$$x_s^t = X_s \left(\cdots \sigma_s^{t-1} . \sigma_s^t \sigma_s^{t+1} \cdots \right),$$

the unique point whose orbit visits partition elements $0, 1$ of $\mathbb{S}^1 \times D^2$ in sequence $(\sigma_s^t)_{t \in \mathbb{Z}}$. Now make a C^1-small exponentially local coupling. By the robustness of uniformly hyperbolic sets, we can deform the coding $\underline{x} = \underline{x}(\underline{\sigma})$ but now x_s^t depends in general on all $\sigma_{s'}^{t'}$ (not just $\sigma_s^{t'}$), though exponentially weakly on distant (s', t').

The same holds if the uncoupled dynamics has a general finite graph of allowed transitions, e.g. Example 7.2.6. Also one can use the $S \times T$ shadowing theorem to construct a Markov partition for locally maximal uniformly hyperbolic sets without assuming proximity to an uncoupled case.

7.2.4 Natural measures on uniformly hyperbolic attractors for coupled map lattices

Finally we sketch how the phases of uniformly hyperbolic attractors of CML can be analysed. The phases are the probability measures on space-time orbits which arise by starting in the distant past with a probability measure ν whose marginals on all finite subsets of S are absolutely continuous and have Hölder density (plus perhaps a little more).

They are the Gibbs phases for an "energy" with contribution

$$\phi_s^t(\underline{x}) = \text{tr}[\log Df_-(\underline{x}^t(\underline{x}))]_{ss},$$

from space-time site (s,t). By uniform hyperbolicity theory, this ϕ_s^t depends exponentially weakly on $\sigma_{s'}^{t'}$ with respect to $\bar{d}((s',t'),(s,t))$.

The connection with the theory for a single dynamical system (SRB measures) is that

$$\log|\det A| = \text{tr}[\log A] = \sum_{s \in S} \text{tr}[\log A]_{ss},$$

plus a constant from the choice of branch of log, see [5].

Thus phases for uniformly hyperbolic attractors of CML correspond to equilibrium statistical mechanics for a special class of spin systems on $S \times T$. When the dimension of $(S \times T)$ is greater than 2 (or perhaps even equal to 2), one can expect to make differentiable examples with a non-unique Gibbs phase if sufficiently coupled (just as we did for piecewise affine CML).

Challenge: make a CML of solenoid maps with a uniformly hyperbolic attractor exhibiting a non-unique space-time phase, cf. [3] who came close.

7.2.5 Conclusion

In conclusion, we have seen that the space-time phases of a large class of spatially extended differentiable dynamical systems can be reduced to Gibbs phases for an associated spin system on space-time. This opens up the possibility of understanding the set of space-time phases for such coupled map lattices.

References

[1] Adler, R. 1998. Symbolic dynamics and Markov partitions. *Bull. Amer. Math. Soc.*, **35**, 1–56.

[2] Baesens, C., and MacKay, R. S. 1997. Exponential localization of linear response in networks with exponentially decaying coupling, *Nonlinearity* **10**, 931–940.

[3] Bardet, J. B., and Keller, G. 2006. Phase transitions in a piecewise expanding coupled map lattice with linear nearest neighbour coupling. *Nonlinearity*, **19**(9), 2193.

[4] Bennett, C. H., and Grinstein, G. 1985. Role of irreversibility in stabilizing complex and nonergodic behavior in locally interacting discrete systems. *Phys. Rev. Lett.*, **55**, 657–660.

[5] Bricmont, J., and Kupiainen, A. 1996. High temperature expansions and dynamical systems. *Comm. Math. Phys.*, **178**, 703–732.

[6] Bunimovich, L. A., and Sinaĭ, Ya. G. 1988. Spacetime chaos in coupled map lattices. *Nonlinearity*, **1**(4), 491–516.

[7] Diakonova, M., and MacKay, R. S. 2011. Mathematical examples of space-time phase. *Int J. Bifurcat. Chaos*, **21**(08), 2297–2304.

[8] Dobrushin, R. L. 1971. Markov processes with a large number of locally interacting components: existence of a limit process and its ergodicity. *Probl. Inf. Transm.*, **7**, 149–164.

[9] Durrett, R. 1984. Oriented percolation in two dimensions. *Ann. Probab.*, **12**(4), 999–1040.

[10] Gielis, G., and MacKay, R. S. 2000. Coupled map lattices with phase transition. *Nonlinearity*, **13**(3), 867–888.

[11] Kantorovich, L. V., and Rubinstein, G. Sh. 1958. On a space of completely additive functions (in Russian). *Vestnik Leningrad Univ.*, **13**(7), 52–59.

[12] Lebowitz, J. L., Maes, C., and Speer, E. R. 1990. Statistical mechanics of probabilistic cellular automata. *J. Stat. Phys.*, **59**, 117–170.

[13] Liggett, T. M. 2005. *Interacting Particle Systems*. Classics in Mathematics. Berlin: Springer-Verlag. Reprint of the 1985 original.

[14] MacKay, R. S. 1996. Dynamics of networks: features which persist from the uncoupled limit. Pages 81–104 of: Strien, S. J. Van, and Verduyn Lunel, S. M. (eds.), *Stochastic and Spatial Structures of Dynamical Systems*. North Holland.

[15] MacKay, R. S. 2005. Indecomposable coupled map lattices with non-unique phase. Pages 65–94 of: Chazottes, J. R., and Fernandez, B. (eds.), *Dynamics of Coupled Map Lattices and of Related Spatially Extended Systems*. Lecture Notes in Physics 671. Springer.

[16] MacKay, R. S. 2008. Nonlinearity in complexity science. *Nonlinearity*, **21**(12), T273–T281.

[17] MacKay, R. S. 2011. Robustness of Markov processes on large networks. *J. Difference Equ. Appl.*, **17**(8), 1155–1167.

[18] Maes, C. 1993. Coupling interacting particle systems. *Rev. Math. Phys.*, **5**(3), 457–475.

[19] Pesin, Ya. B., and Sinaĭ, Ya. G. 1991. Space-time chaos in chains of weakly interacting hyperbolic mappings. *Adv. Sov. Math.*, **3**, 165–198.

[20] Robinson, R. C. 1999. *Dynamical Systems*, 2nd edn. Studies in Advanced Mathematics. Boca Raton, FL: CRC Press.

[21] Ruelle, D. 1978. *Thermodynamic Formalism*. Addison-Wesley.

[22] Sakaguchi, H. 1988. Phase transitions in coupled Bernoulli maps. *Prog. Theor. Phys.*, **80**, 7–12.

[23] Sinaĭ, Ya. G. 1972. Gibbs measures in ergodic theory. *Russ. Math. Surv.*, **28**, 21–64.

[24] Stavskaya, O. N. 1973. Gibbs invariant measures for Markov chains on finite lattices with local interaction (in Russian). *Matematicheskii Sbornik*, **134**(3), 402–419.

[25] Toom, A. L. 1980. Stable and attractive trajectories in multicomponent systems. Pages 549–576 of: Dobrushin, R. L., and Sinai, Ya. G. (eds), *Multicomponent Random Systems, Adv. Probab. Related Topics 6.*

[26] Toom, A. L., Vasilyev, N. B., Stavskaya, O. N., Mityushin, L. G., Kurdyumov, G. L., and Pirogov, S. A. 1990. Discrete local Markov systems. Pages 1–182 of: Dobrushin, R. L., Kryukov, V. L., and Toom, A. L. (eds), *Stochastic Cellular Systems: Ergodicity, memory, morphogenesis.* Manchester: Manchester University Press.

[27] Vasilyev, N. B., Petrovskaya, M. B., and Piatetski-Shapiro, I. I. 1969. Simulation of voting with random errors (in Russian). *Automatica i Telemekh*, **10**, 103–7.

8
Selfish routing
Robert S. MacKay

Abstract

Traffic flow is an important example of a complex system. This chapter studies an idealised case where time-dependence and granularity are ignored, and interactions are solely with the flux of traffic on the same edges. The theory has been presented very well by Roughgarden [1], and I merely summarise his exposition here. The theory can be viewed in the wider context of the question whether the free market leads to a desirable situation. The answer is no, but depending on the circumstances its deficiencies might not be too large to merit central control.

I am most grateful to Dayal Strub for typing up the notes and for preparing the figures.

8.1 Introduction

Question: Does the free market lead to a desirable situation? We consider this question in the context of traffic flow where each driver chooses their route to minimise some notion of cost (e.g. travel time) subject to the congestion effects of other drivers. The answer is no, as illustrated by two fundamental examples.

Example 8.1.1 (Pigou's example)   See Figure 8.1. The selfish solution $x_1 = 1$ has cost 1 for everybody, whereas $x_1 = 1/2$ has average cost $3/4$ and cost ≤ 1 for everybody.

Example 8.1.2 (Braess' example)   See Figure 8.2. Selfish solution $x_2 = x_4 = 1$ has cost 2 for everyone, whereas $x_2 = x_4 = 1/2 = x_3 = x_4$ has cost $3/2$ for everyone.

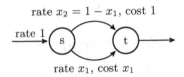

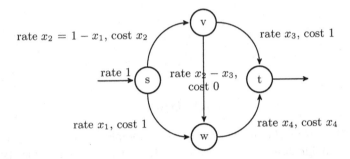

Figure 8.1 Pigou's example.

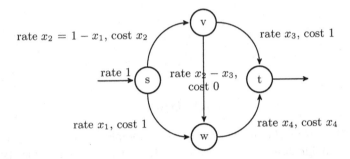

Figure 8.2 Braess' example.

8.2 Basic theory of Nash flows

Consider a finite directed graph G with vertices V and edges E. A *commodity* i is a pair s_i, t_i (source and target) of nodes and a traffic rate r_i. A *path* is a route in G with no cycles. Let $\mathcal{P}_i$ be the set of paths from s_i to t_i, and $\mathcal{P} = \cup_i \mathcal{P}_i$. A *flow* is a function $f : \mathcal{P} \to \mathbb{R}_+ :$ $p \mapsto f_p$. This induces a *flow on edges*, that is a function again denoted $f : E \to \mathbb{R}_+ : e \mapsto f_e$. Now $f_e = \sum_{p \in \mathcal{P}\text{st. } e \in p} f_p$.

Say f is *feasible* if $\sum_{p \in \mathcal{P}_i} f_p = r_i$.

The *cost* c_e for using an edge e is a non-negative, continuous, non-decreasing function of f_e. Also we suppose no cost from junctions, so the cost of a path is $c_p(f) = \sum_{e \in p} c_e(f_e)$, and the total cost is

$$C(f) = \sum_{p \in \mathcal{P}} c_p(f) f_p = \sum_{e \in E} c_e(f_e) f_e.$$

We say that f^* is *optimal* if it minimises $C(f)$ over feasible flows f. There exists an optimal flow because C is a continuous function on a compact set.

Say f is a *Nash flow* (Wardrop equilibrium) if it is feasible and for all i and $p, p' \in \mathcal{P}_i$ with $f_p > 0$, and $\delta \in (0, f_p]$ then $c_p(f) \le c_{p'}(\hat{f})$, where $\hat{f}$ is obtained from f by switching δ flow from p to p', i.e. no unilateral improvement is possible. It follows that a feasible f is Nash if and only if for all i, $p, p' \in \mathcal{P}_i$ with $f_p > 0$, we have that $c_p(f) \le c_{p'}(f)$ (all flows travel on minimum cost paths). In particular, f Nash implies that all paths used in $\mathcal{P}_i$ have the same cost $c_i(f)$ and $C(f) = \sum_{p \in \mathcal{P}} c_i(f) r_i$.

Theorem 8.2.1 *Every (G, r, c) admits a Nash flow and is "essentially unique", i.e. if f and $\hat{f}$ are Nash, then $c_e(f_e) = c_e(\hat{f}_e)$ for all $e \in E$.*

To prove this, we will first characterise optimal flows under stronger conditions on C, then prove Nash flows are optimal for a modified cost function satisfying the stronger conditions. So temporarily suppose c_e is *semi-convex*, i.e. $h_e(x) = x c_e(x)$ is convex, that is

$$h_e(\lambda x + (1 - \lambda) y) \le \lambda h_e(x) + (1 - \lambda) h_e(y) \quad \forall \lambda \in (0, 1),$$

and suppose h_e is C^1. Then C is convex. For a path P, let $h'_p(f) = \sum_{e \in p} h'_e(f_e)$, that is, the derivative of $h_p = \sum_e h_e$ with respect to f_p.

Theorem 8.2.2 *Flow f^* is optimal if and only if $h'_p(f^*) \le h'_{p'}(f^*)$ for all i and $p, p' \in \mathcal{P}_i$ with $f^*_p > 0$.*

Proof ($\Rightarrow$): Suppose that f^* is optimal, $p, p' \in \mathcal{P}_i$ and $f^*_p > 0$.
Transferring $\lambda > 0$ flow from p to p' gives cost

$$\sum_e h_e(f^*_e) + \lambda \left(\sum_{e \in p'} h'_e(f^*_e) - \sum_{e \in p} h'_e(f^*_e) \right) + o(\lambda) \ge C(f^*) \text{ for } \lambda > 0,$$

so $h'_p(f^*) \le h'_{p'}(f^*)$.
($\Leftarrow$): Suppose $h'_p(f^*) \le h'_{p'}(f^*)$ for all $p, p' \in \mathcal{P}_i$ and $f^*_p > 0$.
By convexity of h_e:

$$C(f) = \sum_e h_e(f_e) \ge \sum h_e(f^*_e) + h'_e(f^*_e)(f_e - f^*_e)$$

$$= C(f^*) + H(f) - H(f^*),$$

where

$$H(f) = \sum_e h'_e(f^*_e) f_e = \sum_p h'_p(f^*) f_p.$$

A feasible flow f minimises H if and only if for all $p \in \mathcal{P}$ with $f_p > 0$, it minimises $h'_p(f^*)$ over $p' \in \mathcal{P}_i$, so f^* minimises H. Thus $C(f) \ge C(f^*)$. □

Corollary 8.2.3 *If $h_e(x) = x c_e(x)$ is convex and C^1 then flow f is optimal for C if and only if f is Nash for $c^*_e = h'_e$, i.e. $c^*_e(x) = c(x) + x c'(x)$ "marginal cost function".*

In particular, Nash flows optimise $\sum_e f_e c_e^*(f)$, but not necessarily C. Also, we can make optimal flow for c arise as Nash flow if we add "marginal cost taxes" $f_e c_e'(f_e)$.

Remark Note:

- we can also subtract a constant, c_i say, for each commodity i;
- Corollary 8.2.3 holds even if h is not convex.

Proof of Theorem 8.2.1 (i) Existence of Nash flow: let $k_e(x) = \frac{1}{x}\int_0^x c_e(t)\,dt$, which has $xk_e(x)$ convex and C^1. Then there exists an optimal flow f for $K = \sum_e x_e k_e(x_e)$ (because K is a continuous function on a compact set) and thus f is Nash for $[xk_e(x)]' = c_e(x)$.
(ii) Two Nash flows have equal edge costs: take K convex. $f, \tilde{f}$ being optimal flows for K implies that K is constant on $\lambda f + (1-\lambda)\tilde{f}$, for $\lambda \in [0,1]$. Each $xk_e(x)$ being convex implies that each is affine on this line, which in turn implies that $c = (xk)'$, constant between f_e and $\tilde{f}_e$. □

Corollary 8.2.4 *Flow f is Nash if and only if*

$$\sum_e c_e(f_e)\,f_e \le \sum_e c_e(f_e)\,\tilde{f}_e$$

for all feasible $\tilde{f}$.

Proof Theorem 8.2.2 applied to $h_e(x) = \int_0^x c_e(t)\,dt$. □

8.3 Bounding the price of anarchy

Define the *price of anarchy* as

$$\rho = \frac{C(f_{Nash})}{C(f^*)}.$$

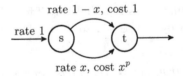

Figure 8.3 Modified Pigou's example.

Example 8.3.1 (Modified Pigou example)   See Figure 8.3. Nash: $x = 1$, optimal: $x = (p+1)^{-1/p}$ (equality of h'_p for both paths, where $h(x) = xc(x)$).

So we find that the price of anarchy is

$$\rho = \frac{C(f_{Nash})}{C(f^*)} = \frac{1}{1 - p(p+1)^{-(p+1)/p}} \to \infty \text{ as } p \to \infty,$$

and there is no bound to the price of anarchy!

8.3.1 Affine cost functions

An affine cost function satisfies $c_e(x) = a_e x + b_e$.

We will prove the following theorem:

Theorem 8.3.2   *For affine cost functions, $\rho \le \frac{4}{3}$.*

Remark   Equality was attained for Pigou's ($p = 1$) and Braess' example.

To prove the theorem, apply two previous results:

- f is Nash if and only if for all i and $p, p' \in \mathcal{P}_i$ with $f_p > 0$ then

$$\sum_{e \in p} a_e f_e + b_e \le \sum_{e \in p'} a_e f_e + b_e.$$

- f^* is optimal if and only if for all i and $p, p' \in \mathcal{P}_i$ with $f_p > 0$ then

$$\sum_{e \in p} 2a_e f_e^* + b_e \le \sum_{e \in p'} 2a_e f_e^* + b_e.$$

So f Nash flow for r implies that $\frac{f}{2}$ is optimal for $\frac{r}{2}$, and $C\left(\frac{f}{2}\right) \ge \frac{1}{4}C(f)$ because

$$\sum_{e \in p} \left(\frac{1}{2}a_e f_e + b_e\right) \frac{f_e}{2} \ge \sum (af + b)f.$$

Now augment $\frac{f}{2}$ to a flow for r. Recall the marginal cost functions $c_e^*(x) = 2a_e x + b_e$. To complete the proof we need the following lemma:

Lemma 8.3.3   *If f^* is optimal for r then for all $\delta > 0$ and f feasible for $(1 + \delta)r$, we have that*

$$C(f) \ge C(f^*) + \delta \sum_e c^*(f_e^*) f_e^*.$$

Proof $xc_e(x)$ is convex, so

$$h_e(f_e) = c_e(f_e)f_e \geq c_e(f_e^*)f_e^* + (f_e - f_e^*)c_e^*(f_e^*). \qquad (8.1)$$

See Figure 8.4.

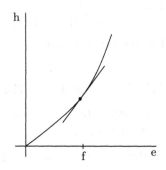

Figure 8.4 Illustration of (8.1).

Therefore $C(f) \geq C(f^*) + \sum_e (f - f^*)c^*(f^*).$

In the proof of Theorem 8.2.2, we showed that

$$\sum h'(f^*)f^* \leq \sum h'(f^*)f$$

for all feasible f for rates r. Now $h' = c^*$, so for f feasible for $(1+\delta)r$, we have that

$$\sum c^*(f^*)f^* \leq \sum c^*(f^*)\frac{f}{1+\delta},$$

hence the result.   □

Proof of Theorem 8.3.2 We want to show that $\rho \leq 4/3$ for affine cost functions.

Flow f is Nash implies that $f/2$ is optimal for $r/2$ and has $c^*(f/2) = c(f)$. Take $\delta = 1$ in the lemma. Then f^* is feasible for $(1+\delta)\frac{r}{2}$, so

$$C(f^*) \geq C\left(\frac{f}{2}\right) + \sum c^*\left(\frac{f}{2}\right)\frac{f}{2}$$

$$\geq \frac{1}{4}C(f) + \frac{1}{2}\sum c(f)f$$

$$= \frac{3}{4}C(f).$$

□

8.3.2 General cost functions

For a cost function c, let the *anarchy value* α be

$$\alpha(c) = \sup_{x,r \geq 0} \frac{rc(r)}{xc(x) + (r-x)c(r)},$$

that is, the worst case (over k and r) for ρ for Pigou's example with an edge with constant cost $k = c(r)$, and the other edge with cost function c, see Figure 8.5.

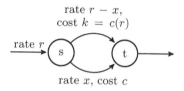

rate $r - x$,
cost $k = c(r)$

rate r s t

rate x, cost c

Figure 8.5 Pigou's example for the definition of anarchy value.

Proposition 8.3.4 *If cost c is C^1 and semi-convex, then the anarchy value*

$$\alpha(c) = \sup_{r \geq 0} \frac{1}{\lambda \mu + 1 - \lambda},$$

where λ solves $c^*(\lambda r) = c(r)$ and $\mu = \frac{c(\lambda r)}{c(r)}$.

For a set $\mathcal{C}$ of cost functions, let the anarchy value be

$$\alpha(\mathcal{C}) = \sup_{c \in \mathcal{C}} \alpha(c).$$

Theorem 8.3.5 *The price of anarchy and the anarchy value satisfy the relation $\rho \leq \alpha(\mathcal{C})$.*

Proof Note that $xc(x) \leq \frac{r(r)}{\alpha(\mathcal{C})} + (x-r)c(r)$. Let f^* be optimal, and f be Nash. Then

$$C(f^*) = \sum_e c(f^*)f^* \geq \frac{1}{\alpha}\sum c(f)f + \sum(f^* - f)c(f) \geq \frac{C(f)}{\alpha},$$

where we used Corollary 8.2.4. $\qquad\square$

By the definition of the anarchy value α, the price of anarchy ρ arbitrarily close to $\alpha(\mathcal{C})$ occurs for the set-up in Figure 8.6.

Definition 8.3.6 Set $\mathcal{C}$ is *diverse* if for all $\gamma > 0$ there exists $c \in \mathcal{C}$ with $c(0) = \gamma$. Then for all $\varepsilon > 0$, $\rho > \alpha(\mathcal{C}) - \varepsilon$ occurs for the set-up in Figure 8.7, close to realising $\alpha(\mathcal{C})$.

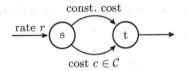

Figure 8.6 Set-up for ρ arbitrarily close to $\alpha(\mathcal{C})$.

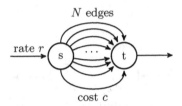

Figure 8.7 Set-up close to realising $\alpha(\mathcal{C})$ for $\mathcal{C}$ diverse.

We can weaken the requirement on $\mathcal{C}$ further and still keep $\rho \leq \alpha(\mathcal{C})$ as the best possible bound. Say $\mathcal{C}$ *inhomogeneous* if $\tilde{c}(0) \neq 0$ for some $\tilde{c} \in \mathcal{C}$. Then for all $\varepsilon > 0$, $\rho \geq \alpha(\mathcal{C}) - \varepsilon$ occurs for a "union of paths" (Figure 8.8).

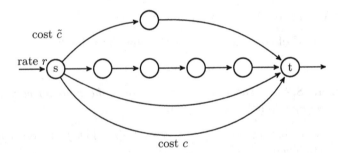

Figure 8.8 Set-up close to realising $\alpha(\mathcal{C})$ for $\mathcal{C}$ inhomogeneous.

Next we address a different type of bound: by what factor could rates be increased by an optimal controller to realise no more cost than Nash?

Theorem 8.3.7   *If f is Nash for r, and f^* is optimal for $2r$, then*

$$C(f) \leq C(f^*).$$

Proof $C(f) = \sum_i c_i(f) r_i.$

Let

$$\bar{c}_e(x) = \begin{cases} c_e(f_e) & \text{if } x \le f_e, \\ c_e(f_e) & \text{if } x \ge f_e. \end{cases}$$

See Figure 8.9.

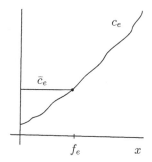

Figure 8.9 Modification of cost function c_e to $\tilde{c}_e$.

Then $\bar{C}(f) = C(f)$ and for all feasible f^* for $2r$, we have that

$$\sum \bar{c}_e(f_e^*) f_e^* - C(f_e^*) = \sum f_e^* (\bar{c}(f_e^*) - c_e(f_e^*))$$

$$\le \sum c_e(f_e) f_e$$

$$= C(f),$$

and

$$\sum_p \bar{c}_p(f^*) f_p^* \ge \sum_i \sum_{p \in \mathcal{P}_i} c_i(f) f_p^*$$

$$= \sum 2 c_i(f) r_i$$

$$= 2 C(f),$$

because f^* is feasible for $2r$.

Therefore

$$C(f^*) \ge \sum_p \bar{c}_p(f^*) f_p^* - C(f)$$

$$\ge 2 C(f) - C(f)$$

$$= C(f).$$

$\square$

Corollary 8.3.8 *Let $\tilde{f}$ be Nash for $\tilde{c}$ and f^* be optimal for c where $\tilde{c}_e(x) = \frac{1}{2} c_e\left(\frac{x}{2}\right)$. Then $\tilde{C}(\tilde{f}) \le C(f^*)$.*

Proof Let f be Nash for $\left(\frac{r}{2}, c\right)$, and f^* be feasible for (r, c). By the theorem: $\sum c_e(f_e) f_e \leq \sum c_e(f_e^*) f_e^*$.

Let $\hat{f} = 2f$, which is feasible for $r, \tilde{c}$. It is Nash for $r, \tilde{c}$ and $\sum \tilde{c}(\hat{f})\hat{f} = \sum c(f) f$. $\qquad\qquad\qquad\qquad\qquad\qquad\qquad\qquad\qquad\qquad\qquad\qquad\square$

8.4 Avoiding Braess' paradox

Recall Braess' example (with the numbers of Roughgarden [1]), see Figure 8.2. The selfish solution with $x_2 = x_4 = 1$ has cost 2 for everyone (all use the short-cut), whereas the optimal solution with $x_2 = x_4 = 1/2$ (and no traffic on the short-cut) has cost $3/2$ for everyone. The paradox is that adding the short-cut makes the effect of selfish routing worse for everyone.

Let us restrict our attention to a single commodity. One solution is to look for a sub-graph H of G which minimises $d(H, r, c)$, the cost per unit of Nash flow on H (or at least an H which is close to minimising) and forbid the remaining edges (restrict to H that connect s to t).

First, we study the severity of the problem.

Theorem 8.4.1 *If (G, r, c) is single commodity and H is a sub-graph of G then $d(G, r, c) \leq \alpha(\mathcal{C}) d(H, r, c)$.*

Proof Let $f, \tilde{f}$ be Nash for G and H. $C(f) = rd(G, r, c)$, $C\left(\tilde{f}\right) = rd(H, r, c)$. $\tilde{f}$ is feasible for G, so $C(f) \leq \alpha(\mathcal{C}) C\left(\tilde{f}\right)$. $\qquad\square$

Note that we can sometimes do better if $\alpha(\mathcal{C})$ is large.

Definition 8.4.2 $S \subset E$ is *sparse* if no two edges in S share an endpoint and no edge in S has s or t as an endpoint. See Figure 8.10.

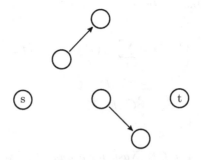

Figure 8.10 A sparse graph.

Theorem 8.4.3 *Let (G, r, c) be a single commodity, H be a sub-graph, and S be the set of edges in G but not in H. If every sparse subset of S has k or less edges then $d(G, r, c) \leq (kH) d(H, r, c)$.*

In particular, every sparse set has at most $\lfloor \frac{n-2}{2} \rfloor$ edges (n is the number of vertices), hence $k + 1 \leq \lfloor \frac{n}{2} \rfloor$.

Reference for proof. See Roughgarden [1, pp.125–128]. A lot of discrete mathematics is involved. Roughgarden also shows that this bound is optimal. □

Remark Braess' paradox has been observed in Stuttgart and Winnepeg (the authorities closed some roads and the traffic flow improved!).

8.5 Possible research directions

We finish with some possible research directions:

Remark (Taxes and incentives) We saw that adding taxes $f_e^* c_e'(f_e^*)$ to c_e, where f^* is optimal for r, makes f^* into a Nash flow. We can then subtract any set of constants k_i for each commodity i if desired (e.g. to make the total revenue equal to zero). Alternatively, we can implement this via incentives λ_e on edges such that

$$\sum_{e \in p} \lambda_e = k_i \ \forall p \in \mathcal{P}_i.$$

However, there is a "fairness" question about k_i. Also the optimal taxes depend on $(r_i, s_i, t_i)_i$ via f^*, so how do we devise taxes that take this into account?

Figure 8.11 Examples of flows at a junction.

Remark (Junction costs and interaction effects) See Figure 8.11. One could include junction costs by letting the cost of passing through a junction depend on all flows along the paths using the vertex. Also consider two-way roads, see Figure 8.12.

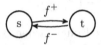

Figure 8.12 Example of two-way road.

One trick – think of edges of G as nodes of a graph G' and vertices of G as providing edges of G', see Figure 8.13. That is, the edge plus junction cost on G gives the costs for flows on u_{12}, u_{13}, etc. However, they are functions of all the flows using the same vertex u, so they can be reduced to the case of interacting flows only. Analyse this extension.

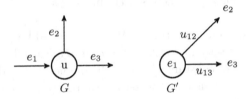

Figure 8.13 Possible trick to include interaction costs.

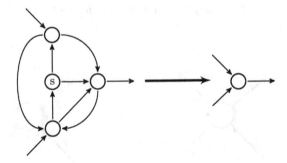

Figure 8.14 Example of aggregation of components of a sub-graph in a traffic flow.

Remark (Aggregation procedures) It could be computationally efficient (and/or insightful) to compute selfish flow hierarchically, e.g. aggregate all of some sub-graph into one "super-node" with effective junction costs, see Figure 8.14.

Questions:

- What is a good choice of parts to aggregate?
- What are good computational procedures?

References

[1] Roughgarden, T. 2005. *Selfish Routing and the Price of Anarchy*. The MIT Press.

Index

Printed in the United States
by Baker & Taylor Publisher Services